Communications in Computer and Information Science 2600

Series Editors

Rationale

The CCIS series is devoted to the publication of proceedings of computer science conferences. Its aim is to efficiently disseminate original research results in informatics in printed and electronic form. While the focus is on publication of peer-reviewed full papers presenting mature work, inclusion of reviewed short papers reporting on work in progress is welcome, too. Besides globally relevant meetings with internationally representative program committees guaranteeing a strict peer-reviewing and paper selection process, conferences run by societies or of high regional or national relevance are also considered for publication.

Topics

The topical scope of CCIS spans the entire spectrum of informatics ranging from foundational topics in the theory of computing to information and communications science and technology and a broad variety of interdisciplinary application fields.

Information for Volume Editors and Authors

Publication in CCIS is free of charge. No royalties are paid, however, we offer registered conference participants temporary free access to the online version of the conference proceedings on SpringerLink (http://link.springer.com) by means of an http referrer from the conference website and/or a number of complimentary printed copies, as specified in the official acceptance email of the event.

CCIS proceedings can be published in time for distribution at conferences or as post-proceedings, and delivered in the form of printed books and/or electronically as USBs and/or e-content licenses for accessing proceedings at SpringerLink. Furthermore, CCIS proceedings are included in the CCIS electronic book series hosted in the SpringerLink digital library at http://link.springer.com/bookseries/7899. Conferences publishing in CCIS are allowed to use Online Conference Service (OCS) for managing the whole proceedings lifecycle (from submission and reviewing to preparing for publication) free of charge.

Publication process

The language of publication is exclusively English. Authors publishing in CCIS have to sign the Springer CCIS copyright transfer form, however, they are free to use their material published in CCIS for substantially changed, more elaborate subsequent publications elsewhere. For the preparation of the camera-ready papers/files, authors have to strictly adhere to the Springer CCIS Authors' Instructions and are strongly encouraged to use the CCIS LaTeX style files or templates.

Abstracting/Indexing

CCIS is abstracted/indexed in DBLP, Google Scholar, EI-Compendex, Mathematical Reviews, SCImago, Scopus. CCIS volumes are also submitted for the inclusion in ISI Proceedings.

How to start

To start the evaluation of your proposal for inclusion in the CCIS series, please send an e-mail to ccis@springer.com.

Qingtang Liu · Jing Lei · Liming Zhang ·
Yantao Wei
Editors

Educational Innovation Through Technology

13th International Conference, EITT 2024
Macau, China, November 8–10, 2024
Proceedings

Springer

Editors
Qingtang Liu
Central China Normal University
Wuhan, China

Jing Lei
Syracuse University
Syracuse, NY, USA

Liming Zhang
University of Macau
Macau, China

Yantao Wei
Central China Normal University
Wuhan, China

ISSN 1865-0929 ISSN 1865-0937 (electronic)
Communications in Computer and Information Science
ISBN 978-981-95-1524-0 ISBN 978-981-95-1525-7 (eBook)
https://doi.org/10.1007/978-981-95-1525-7

This Springer imprint is published by the registered company Springer Nature Singapore Pte Ltd.
The registered company address is: 152 Beach Road, #21-01/04 Gateway East, Singapore 189721, Singapore

Preface

The proceedings of the 13th International Conference on Educational Innovation through Technology (EITT 2024) record the research presented in Macao, P.R. China, on 8–10 November 2024. EITT 2024 was organized by the Society of International Chinese in Educational Technology (SICET) and hosted by the University of Macao and Central China Normal University. The conference is organized annually by SICET. Founded in 2003, SICET is a nonpolitical, non-profit academic organization, and it is affiliated with the Association for Educational Technology and Communications (AECT). The goal of SICET is to provide face-to-face and online forums for international scholars, especially Chinese scholars in the field of educational technology, to meet, support, collaborate, and exchange ideas. The mission of SICET is to establish and strengthen international academic connections, exchanges, research, and studies in educational technology for teaching and learning, as well as to promote the application of educational technology in education. The first twelve EITT conferences have been successfully held since 2012, with EITT 2012 in Beijing; EITT 2013 in Williamsburg, Virginia, USA; EITT 2014 in Brisbane, Australia; EITT 2015 in Wuhan, China; EITT 2016 in Tainan, China; EITT 2017 in Osaka, Japan; EITT 2018 in Auckland, New Zealand; EITT 2019 in Biloxi, MS, USA; EITT 2020, EITT 2021, and EITT 2022 virtually due to the COVID-19 global pandemic; and EITT 2023 in Fujian, China.

The meeting continued the EITT tradition of bringing together scholars and practitioners to discuss "Educational Innovation and Development in the Age of AI". All papers published here were selected through a rigorous single-blind review. Every submission received at least three independent reports; final decisions were taken in an online TPC meeting. In total, 78 manuscripts were submitted and 24 were accepted (17 full, 7 short), giving an overall acceptance rate of $\approx 31\%$.

The invited contributions were subjected to the same scholarly scrutiny as regular submissions. Manuscripts co-authored by Program Committee members were reassigned to senior reviewers with no conflicts of interest, and the relevant chairs recused themselves from discussions, thus ensuring objectivity throughout the selection process.

We thank every author for choosing EITT 2024, every reviewer for timely and constructive feedback, and all keynote speakers, session chairs, student volunteers, and local organizers for their indispensable support. We are also grateful to Springer's editorial team for guiding the preparation of this volume.

November 2024

Qingtang Liu
Jing Lei
Liming Zhang
Yantao Wei

Organization

Conference Chairs

Qingtang Liu	Central China Normal University, China
Xiaojun Chen	St. John's University, USA

Technology Support Committee Chairs

Jui-Long (Andy) Hung	Boise State University, USA
Hong Zhan	Embry-Riddle Aeronautical University, USA
Jining Han	Southwest University, China

Conference Financial Chairs

Hong Zhan	Embry-Riddle Aeronautical University, USA
Yingxiao Qian	University of South Carolina, USA

Program Committee Chairs

Jing Lei	Syracuse University, USA
Liming Zhang	University of Macau, China
Yantao Wei	Central China Normal University, China

Committee Members

Lin Zhong	Southern Illinois University, USA
Yinxiao Qian	Arizona State University, USA
Hong Zhan	Embry-Riddle Aeronautical University, USA
Dongqing Wang	South China Normal University, China
Jui-Long (Andy) Hung	Boise State University, USA
Shen Ba	University of Hong Kong, China
Hsiu-Jen Cheng	Kaohsiung Normal University, China
Jining Han	Southwest University, China
Jing Ma	Zhengzhou University, China

Grace Yue Qi	Massey University, New Zealand
Yi Sun	Kobe Institute of Computing, Japan
Ping Wang	Great Basin College, USA
Fati Wu	Beijing Normal University, China
Linjing Wu	Central China Normal University, China
Qingzhi Zhang	Falmouth University, UK
Li Cheng	Worcester Polytechnic Institute, USA
Tian Luo	Old Dominion University, USA
Ye Chen	University of West Georgia, USA
Tianxiao Yang	Texas Tech University, USA
Yuqin Yang	Central China Normal University, China
Si Zhang	Central China Normal University, China
Xiuhan Li	Central China Normal University, China

Contents

AI-Powered Educational Innovation

Empowering Elementary School English Review Lessons with Generative AI: Constructing the RCPA Human-Machine Collaborative Teaching Model

Jiawei Li[1](✉), Youru Xie[2], Yi Lu[2], Wei Chen[2], Ziru Ouyang[3], and Lihong Zeng[4]

[1] Department of Counseling, Educational Psychology and Special Education, Michigan State University, East Lansing, US
lijiawei@msu.edu

[2] School of Educational Information Technology, South China Normal University, Guangzhou, China

[3] School of Arts and Media, The University of New South Wales, Sydney, Australia

[4] Guangzhou Tiyudong Education Group, Guangzhou 510000, China

Abstract. The implementation of new curriculum reform and the "double reduction" educational policies in China have called for reducing classroom burdens while enhancing teaching quality and efficiency. The innovative application of Generative AI (GenAI) in classroom teaching marks a significant milestone in advancing educational digitalization. This study examined the impact of GenAI on elementary school English review lessons, addressing critical issues such as the absence of precise data-driven instruction, ineffective student collaboration, and inadequate personalized feedback. Guided by Large Unit Teaching (LUT) instructional design focused on developing core English competencies and the innovative "Teacher-Student-Machine" triadic structure empowered by GenAI, this study used literature research, theoretical deduction, action research, and evaluation research methods. The study followed a structured process of status investigation, theoretical analysis, model construction, and effect analysis. Initially, it clarified the teaching process of elementary school large unit English review lessons, establishing the "Review-Check-Practice-Advance (RCPA)" teaching sequence. A preliminary human-machine collaborative RCPA model was constructed and refined through two rounds of action research. Results evaluated by artificial intelligence analysis systems indicated that the RCPA teaching model effectively stimulated students' interest in learning English, enhanced core English competencies and transformed classroom activities.

Keywords: Generative AI · elementary school English · review lessons · human-machine collaboration · the RCPA teaching model

1 Introduction

In the context of new curriculum reforms and the "double reduction" educational policies in China, reducing excessive homework and off-campus tutoring while enhancing teaching quality and efficiency has become imperative. This urgency stems from the modern

Q. Liu et al. (Eds.): EITT 2024, CCIS 2600, pp. 3–15, 2026.
https://doi.org/10.1007/978-981-95-1525-7_1

era's demand for cultivating innovative talent. Consequently, the innovation of English classroom teaching has become a critical task. The advent of Generative AI (GenAI) has significantly impacted human-machine collaborative teaching, transforming the traditional "Teacher-Student" dynamic into a "Teacher-Student-Machine" triadic structure. This shift is essential for improving teaching quality and preparing students for future challenges. The review lesson, a fundamental lesson type in elementary school English education, is crucial for reinforcing knowledge and facilitating its application to new contexts. However, in today's elementary and secondary schools in China, these lessons often lack robust data-driven methods, meaningful student interaction, and personalized feedback. These gaps hinder the development of core English competencies. By fully utilizing the innovative capabilities of GenAI, review lessons can be transformed to more effectively stimulate students' interest in English learning, enhance knowledge transfer, and develop core competencies. In response to the pressing need for reform in contemporary elementary English teaching, this study aims to refine the teaching process of large unit review sessions. It seeks to clarify the empowering role of GenAI and develop a collaborative teaching model that integrates both human and machine efforts. The goal is to provide both theoretical and practical insights to improve classroom teaching quality and nurture innovative talents suited for the intelligence era.

2 Literature Review

2.1 Elementary School English Review Lessons

The review lesson, a fundamental lesson type in elementary school English education, helps students consolidate learned materials, resolve misunderstandings, address learning gaps, strengthen skills, and extend understanding. Recent research on elementary English review lessons has explored various teaching strategies and the use of mind maps. For example, Shen proposed review strategies based on unit context maps [1], while Wang examined different applications of mind maps in elementary English review lessons [2]. Despite these efforts, several issues persist in elementary school English review lessons, including a lack of integration with overall unit instructional design, insufficient attention to students' limited attention spans, inadequate class time, and monotonous teaching models that primarily rely on written exercises [3, 4]. The new curriculum reforms and the "double reduction" educational policies necessitate innovative review lesson models that emphasize core competencies and leverage digital technologies to reduce burdens and enhance both quality and efficiency.

2.2 GenAI Empowering Elementary English Teaching

GenAI has garnered significant interest in education, promising transformative changes in teaching practices. Advanced technologies have alleviated some of these burdens by reducing students' reliance on teachers and providing access to a wealth of language resources as well as to intelligent learning platforms that better align with their interests and needs. For instance, Lee, et al. found that AI-based content generators could boost students' enjoyment and interest in reading English [5]. Dong, et al. found that the integration of IoT and GenAI can provide students with a personalized English learning experience [6]. Integrating GenAI into elementary English teaching marks a paradigm shift, supporting teachers and fostering student autonomy, thus becoming a vital tool in modern classrooms.

2.3 Human-Machine Collaborative Teaching Models and Applications

Human-machine collaboration seeks to extend human intelligence by integrating machine intelligence. Maximizing the complementary strengths of both will create a synergistic effect greater than the sum of their parts. For example, Zhang outlined the general process of human-machine collaboration and constructed a "self-content-society" human-machine learning interactions model empowered by GenAI. The model can enhance learner interactions and MOOC learning quality [7]. Similarly, Xie, et al. introduced the "MOOC Plus Theme Lecture-Artificial Intelligence Tutoring-Project Practice (MOOC+TAIP)" model to meet specific AI literacy needs of preschool teachers, demonstrating significant improvements in AI literacy and high participant satisfaction with the program [8]. In summary, optimizing the interaction between teachers and students and machines in teaching practice, giving full play to the complementary advantages of human intelligence and machine intelligence, is crucial for improving the effectiveness of teaching. Therefore, this study will explore the innovative application of GenAI in elementary school English classroom teaching, clarify the role of teachers, students, and machines in elementary school English review lessons, and improve the quality and efficiency of elementary school English review lessons.

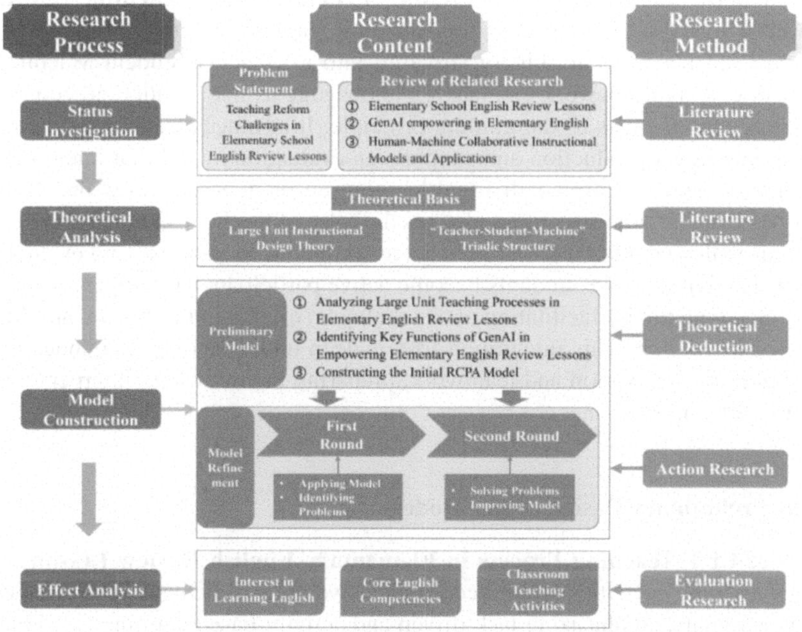

Fig.1. Research process and methods.

3 Research Process and Methods

This study employed a multi-method approach structured across four phases: status investigation, theoretical analysis, model construction, and effect analysis (see Fig. 1).

4 Model Construction

4.1 Key Principles

LUT Instructional Design Focused on Core English Competencies. The English Curriculum Standards for Compulsory Education (2022 Edition) emphasizes cultivating core subject competencies and proposes the concept of LUT to improve knowledge integration and structure. LUT encourages starting with a guiding theme at the macro level and then organizing content around this core to ensure thematic coherence within each unit. This instructional design approach unifies teaching, learning, and assessment by setting clear unit and lesson objectives, defining internal unit logic, and restructuring content. By addressing real-life problems, LUT connects lessons to students' everyday lives, enhancing their ability to apply the "large unit-task clusters-problem chains" strategy and develop "great ideas". Using basic questions to guide student thinking, this approach sequences them into interconnected problems. Ultimately, this holistic instructional design framework allows for comprehensive teaching activities and assessments and finally, effectively measures how well students achieve core competencies by teaching evaluation.

The "Teacher-Student-Machine" Triadic Structure with Teaching Innovation Empowered by GenAI. GenAI is transforming teaching practices by evolving the traditional "Teacher-Student" binary structure into a "Teacher-Student-Machine" triadic structure. This framework emphasizes knowledge as the foundation, competence as the priority, and values as the guiding principle. It advances teaching content from manual to intelligent production and supports an evaluation model that integrates both knowledge and skills to meet individual learning needs. Consequently, the "Teacher-Student-Machine" triadic structure redefines the roles of teachers and the nature of their interactions with students. Teachers now act as designers, facilitators, and evaluators of educational activities while students become active participants. Intelligent technology supports this dynamic by facilitating active dialogue, effective interaction, and human-machine collaboration. This integration fosters deep understanding and enhances students' abilities in innovation and teamwork, preparing them for both interpersonal and human-machine interactions.

4.2 The Preliminary Design of the Model

Analysis of LUT Teaching Process in Elementary English Review Lessons. LUT organizes teaching activities around comprehensive units to achieve core competency development goals, emphasizing task-driven and activity-based learning [9]. This approach not only provides a more logical and coherent structure than traditional review lessons but also helps implement theme-guided teaching principles. By aligning review lessons with unit themes, teachers can effectively highlight key knowledge points and their logical relationships. Elementary English large unit review lessons build on newly taught unit lessons by focusing on core unit concepts. This approach uses progressive tasks and increasingly difficult questions to guide students through the review process [10], helping them solidify foundational knowledge, deepen their understanding, and facilitate knowledge transfer and expansion. Ultimately, it enhances core English competencies. The teaching process, depicted in Fig. 2, can be divided into four stages:

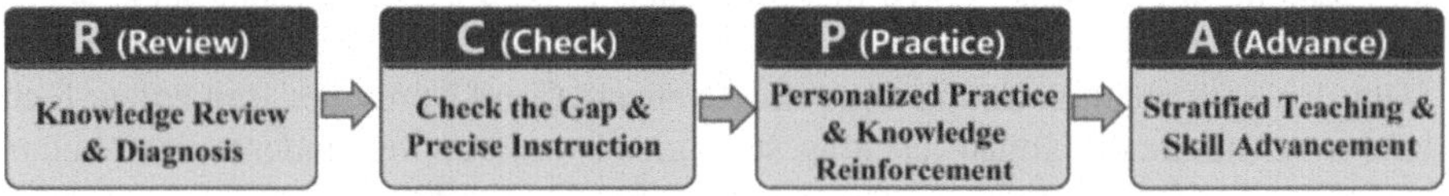

Fig. 2. The RCPA teaching process in elementary English.

a. Review (Knowledge Review and Diagnosis): In this stage, teachers guide students through a systematic review of the unit's core concepts using questions, interactive discussions, and quizzes. This process helps students build a comprehensive knowledge framework while teachers assess their mastery and depth of understanding.
b. Check (Check the Gap and Precise Instruction): Based on the diagnostic results, teachers identify specific areas where students struggle and provide precise interventions. Different review strategies are employed to address knowledge gaps and misunderstandings, ensuring each student receives the necessary support.
c. Practice (Personalized Practice and Knowledge Reinforcement): In this stage, students engage in practice tasks tailored to their individual learning needs. These tasks are designed to deepen their understanding and reinforce unit knowledge, ensuring a solid foundation for each student.
d. Advance (Stratified Teaching and Skill Advancement): Following personalized practice, teachers group students by proficiency levels and assign tasks suited to their readiness and interests. Students engage in independent thinking while collaborating with group members to transfer and expand their knowledge and enhance their English skills.

Analysis of GenAI's Empowering Role in Elementary English Review Lessons. GenAI responds to prompts by creating new content and ideas in the form of dialogues, stories, images, videos, and music. This intelligent and interactive capability empowers personalized teaching in elementary English review lessons and supports transitioning to a "Teacher-Student-Machine" triadic structure. Specifically, GenAI contributes in the following ways:

a. Student diagnosis and personalized review tasks: GenAI analyzes student data and uses knowledge bases, skill repositories, and learning evaluation standards to generate personalized review tasks. These themed tasks include activities such as spelling, fill-in-the-blank exercises, and sentence structure practice.

b. Precise guidance and differentiated review plans: GenAI provides precise guidance and tailored review plans to address the unique learning needs of students at various proficiency levels.
c. Context creation and diverse review resources: To enrich the learning experience, GenAI creates diverse review materials, including images, music, and videos, that align with students' learning needs and review themes.
d. Meaningful construction and thematic dialogue practice: Drawing on student data, GenAI engages students in interactive dialogues centered around unit themes. These dialogues guide students in expressing their ideas and practicing listening, speaking, reading, and writing skills, thereby fostering comprehensive and personalized development.
e. Data-driven evaluation and review effectiveness: GenAI provides immediate feedback to dynamically adjust teaching strategies by evaluating students' vocabulary, grammar, and perspectives during interactions. This supports both formative and summative assessments in elementary English review lessons.

The Preliminary Construction of the RCPA Model. This study focused on core competencies in elementary English, such as language proficiency, cultural awareness, and quality of thinking. Through an in-depth analysis of the large unit review lesson process, the study integrated GenAI tools like "ChatGLM" to enhance teaching quality. Additionally, intelligent technologies such as Knowledge Graphs, Dot-matrix Pen, and Digital Humans were utilized to represent knowledge networks, collect student data, and stimulate learning interest. This innovative approach led to the construction of a preliminary GenAI-empowered human-machine collaborative RCPA teaching model for elementary English review lessons, as illustrated in Fig. 3.

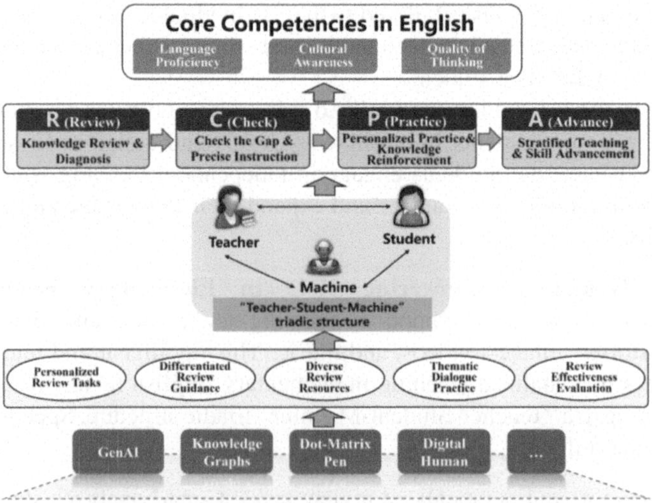

Fig. 3. The preliminary human-machine collaborative RCPA teaching model.

Model Refinement

The first round of action research: The first round of action research targeted 42 third-grade students in Class A at T. Elementary School, focusing on the theme of "Personal Information" over seven class periods. The aim was to develop core competencies in language proficiency, cultural awareness, and quality of thinking by implementing the human-machine collaborative RCPA teaching model.

a. Plan: The plan involved designing engaging English review lessons and applying the human-machine collaborative RCPA teaching model to stimulate students' interest in learning English and transform classroom activities.
b. Action. The lessons followed the established English review process, incorporating the human-machine collaborative RCPA teaching model. Data from an online interactive platform and real-time classroom feedback were used to promptly identify and address any issues, as shown in Fig. 4.
b. Observation. Through observation during this round, the following issues were identified: (1) The benefits and integration of GenAI were unclear, and its role in the teaching process was not effectively highlighted; (2) The specific roles and activities of teachers, students, and AI were not well-defined, failing to establish a cohesive "Teacher-Student-Machine" structure.
d. Reflection. The initial implementation of the human-machine collaborative RCPA teaching model was too general. In the next round, it was crucial to clearly define the role of GenAI and optimize the RCPA model based on the initial findings and student learning conditions.

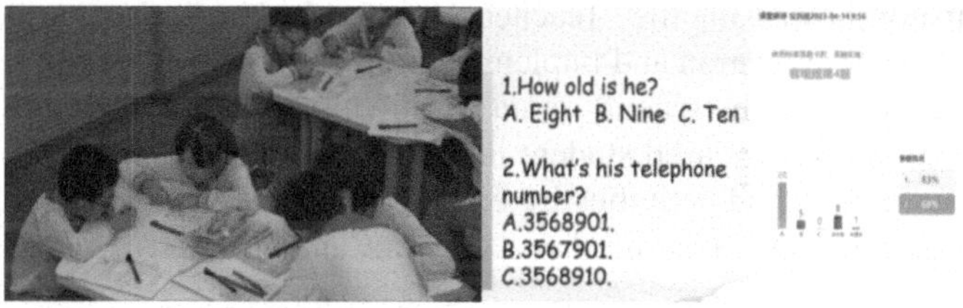

Fig. 4. Mastering the learning situation based on real-time data.

The Second Round of Action Research: Building on the first round of action research, the second round targeted 42 fourth-grade students in Class A at T. Elementary School. The unit was structured in LUT over six lessons, focusing on the theme of "Celebrations" and traditional Chinese festivals. These lessons aimed to develop core competencies such as language proficiency, cultural awareness, and quality of thinking. The primary goal of this round was to optimize and validate the human-machine collaborative RCPA teaching model for elementary English review lessons.

a. Plan. Building on the first round's experiences, the second round focused on enhancing the role of GenAI in the teaching process, embodying the "Teacher-Student-Machine" triadic structure, and optimizing the teaching model.

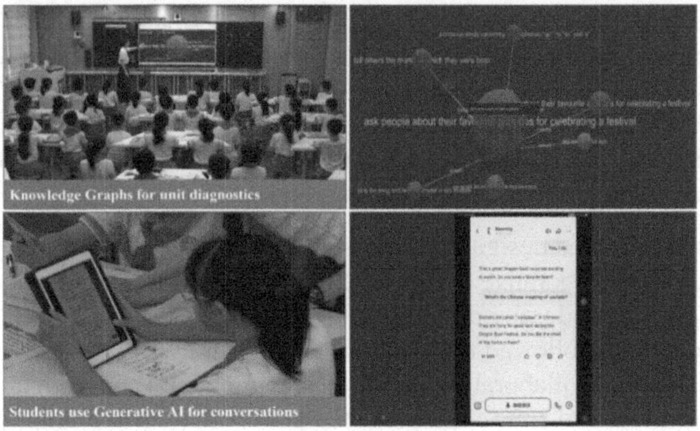

Fig. 5. Knowledge Graphs and GenAI empowering precise instruction.

b. Action. The lessons adhered to the elementary English review process, incorporating data from online interactive platforms and real-time classroom feedback. Knowledge Graphs aligned instruction with core competencies and diagnosed student learning conditions. GenAI was integrated to enhance instructional precision, thereby contributing to the development of a human-machine collaborative RCPA teaching model. The implementation process is shown in Fig. 5.
c. Observation. Observations from this round indicated that the optimized human-machine collaborative teaching model facilitated students' knowledge construction and enhanced their collaborative learning abilities. It also clarified GenAI's specific role, effectively showcasing the "Teacher-Student-Machine" triadic structure.
d. Reflection. The optimization and implementation of the human-computer collaborative teaching model in the second round clarified the direction for further improvement. This approach increased student interest and enhanced collaborative learning abilities through targeted activities by emphasizing GenAI's empowering role and fostering collaboration among teachers, students, and machines.

4.3 Modifications and Improvements

Through two rounds of teaching practice iterations, this study incorporated a detailed teaching process and implementation to further refine the GenAI-empowered elementary school English review lessons (Fig. 6).

The innovative RCPA model aimed to enhance core competencies in elementary school English, including language proficiency, cultural awareness, and quality of thinking. This model uniquely emphasized the teaching process of review lessons, integrating GenAI and other digital technologies to establish a dynamic "Teacher-Student-Machine" triadic structure in the classroom. Knowledge Graphs connected student learning data in the Review stage and illustrated the unit's knowledge network. Using this information, teachers crafted individualized learning paths, while students refined their self-review strategies and revisited unit content. This stage facilitated students' integration of concepts from both broad and specific unit perspectives, promoting knowledge accumulation and the establishment of a strong knowledge framework. In the Check stage, the

Dot-matrix Pen gathered detailed student learning data. Teachers utilized these data to guide GenAI in delivering precise instruction, fostering collaborative learning between humans and machines while encouraging autonomous self-study. Through this process, students revisited misconceptions and better understood their thought processes. In the Practice stage, GenAI created personalized review tasks and provided a variety of review materials. Teachers tailored these tasks to individual learning needs, engaging students in discussions centered around personalized topics to assess their dialogue skills. Finally, in the Advance stage, teachers guided students on values and beliefs, while GenAI produced differentiated review tasks and assessed student performance, thereby facilitating knowledge transfer and enhancement.

The GenAI-empowered RCPA teaching model for elementary school English review lessons excelled in emphasizing the core competencies of large units and the unique attributes of review lessons. It effectively used GenAI and digital technologies to streamline workload, enhance quality, and boost efficiency, showcasing a seamless integration of GenAI into the classroom through the "Teacher-Student-Machine" triadic structure.

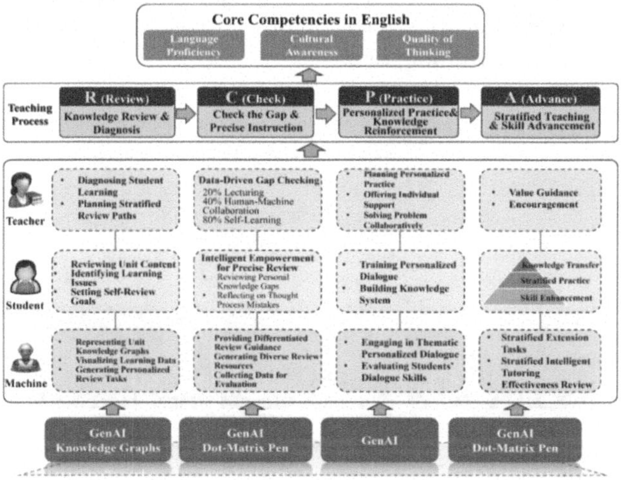

Fig. 6. The GenAI-Empowered RCPA teaching model for elementary school English review lessons.

5 Evaluation of Application Effects

To assess the effectiveness of the RCPA teaching model, this study focused on a cohort of 46 fourth-grade students from Class B at T. Elementary School. The AVA Intelligence Classroom Analysis System and 360AI Browser were employed to analyze the review lesson on "Celebrations". Participants' confidentiality was maintained by concealing their personal information. They were informed that participation was voluntary and that they could withdraw at any time. The results demonstrated that the RCPA model notably stimulated interest in English learning, enhanced core competencies in the subject, and transformed classroom teaching activities.

5.1 Stimulating Interest in English Learning

The intelligent analysis system measured students' head-up rates, providing an indication of their attention and engagement in classroom activities [11]. Figure 7 demonstrated that the overall head-up rate was relatively high, with an average of 48.4%. Notably, students were particularly focused during the 17th to 20th minutes. However, periods of head-down behavior, indicating reduced attention, were observed from the 21st to the 28th minute and again from the 32nd to the 41st minute. Upon reviewing the actual teaching segments, these observations aligned with the specific teaching strategies. From the 17th to the 20th minute, teachers randomly selected students to demonstrate and interact with GenAI, capturing the students' attention. Between the 21st and 28th minutes, students engaged with GenAI for personalized practice, reinforcing their knowledge. From the 32nd to the 41st minute, students completed differentiated tasks and advanced their learning with GenAI. During these two latter stages, students transitioned from passive listening to active participation in personalized or group human-machine collaborative dialogue training. Overall, students consistently maintained high levels of attention during both teacher-led periods and student activities. This finding demonstrates that integrating GenAI in elementary school English lessons effectively increased student engagement and interest in learning English.

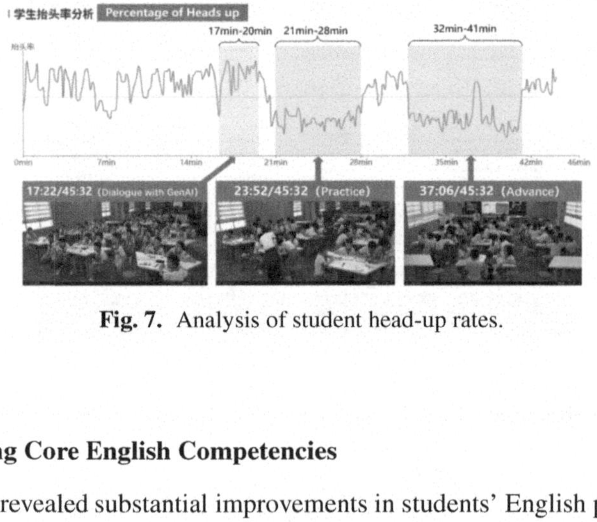

Fig. 7. Analysis of student head-up rates.

5.2 Advancing Core English Competencies

The study also revealed substantial improvements in students' English proficiency, cultural awareness, and quality of thinking through the RCPA teaching model. As shown in Fig. 8, students engaged in dialogues with GenAI about the Dragon Boat Festival, successfully completing tasks such as describing the festival and discussing its customs. Their ability to articulate these outcomes fluently in English to their peers demonstrated tangible improvements in their language proficiency. Further video analysis in the 360AI Browser showed that elementary school students engaged in human-machine collaborative learning with GenAI to explore the cultural significance of the Dragon Boat Festival. They discussed festival customs with their peers, deepening their understanding and appreciation of Chinese traditional culture, and ultimately enhancing their cultural awareness. Furthermore, students' quality of thinking was enhanced as they demonstrated the ability to understand, analyze, critique, evaluate, and create. After completing tasks, students used GenAI to assess the accuracy of AI suggestions and critically integrate its revision advice, thereby sharpening their thinking capabilities. Overall, the results demonstrated that the GenAI-empowered RCPA teaching model effectively enhanced core competencies in elementary school English, including language proficiency, cultural awareness, and quality of thinking.

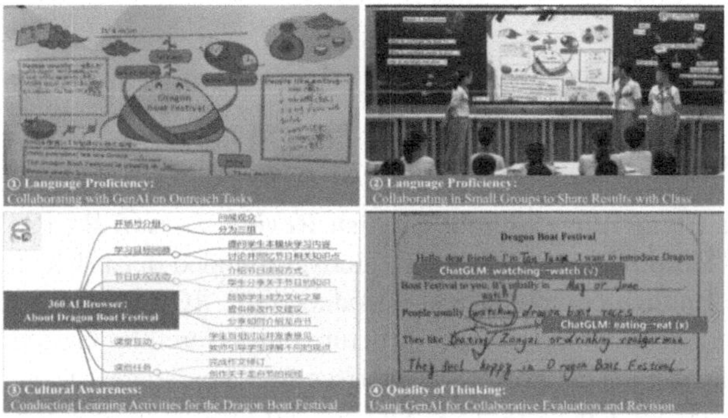

Fig. 8. The effect of advancing English core competencies.

5.3 Transforming Classroom Teaching Activities

The real-time dynamic AI system for the S-T data analysis revealed changes in teacher-student interactions during classroom activities. As illustrated in Fig. 9, this lesson was classified as a "hybrid" teaching mode, with a Teacher Behavior ratio (Rt) of 0.48 and a Teacher-Student interaction Frequency (Ch) of 0.27, indicating that 48% of behaviors were teacher-led and there were frequent transitions between teacher and student behaviors. Teacher-led activities such as guidance, explanations, and questioning accounted for 46% (21 min) of the class time. In contrast, student-led activities, including answering, independent thinking, reading aloud, and collaborative exploration with GenAI, accounted for 54% (25 min). This balanced interaction between teachers and students closely adhered to the golden classroom time distribution ratio. The results indicated that integrating GenAI into elementary school English review facilitated collaborative learning, personalized practice, and differentiated group reviews, significantly reducing workload, enhancing quality, and transforming classroom dynamics.

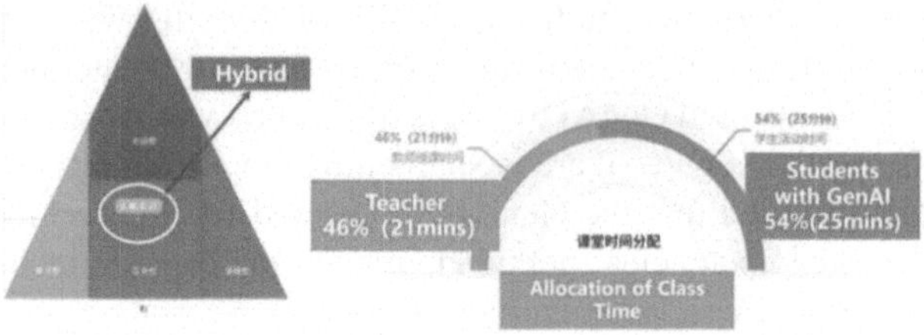

Fig. 9. Analysis of classroom teaching activities.

6 Conclusions and Discussion

6.1 Conclusion

Introducing the Elementary School English Review Lesson RCPA Teaching Process and GenAI Empowering Role. This study, grounded in LUT instructional design focused on core English competencies, introduced the RCPA teaching process within the "Teacher-Student-Machine" triadic structure empowered by GenAI for elementary English review lessons. This process comprises four key stages: Review (Knowledge Review and Diagnosis), Check (Check the Gap and Precise Instruction), Practice (Personalized Practice and Knowledge Reinforcement), and Advance (Stratified Teaching and Skill Advancement). GenAI assumes a pivotal role in this teaching model by generating personalized review tasks, providing differentiated guidance, creating diverse resources, guiding thematic dialogues, and evaluating student outcomes, thereby markedly enhancing teaching effectiveness.

Constructing the GenAI-empowered RCPA Human-machine Collaborative Teaching Model. This study aimed to enhance core competencies in elementary English (e.g., language proficiency, cultural awareness, and quality of thinking) by emphasizing the distinct features of the review lesson process. The effective integration of GenAI and other digital technologies established the "Teacher-Student-Machine" triadic structure, leading to the construction of the GenAI-empowered RCPA teaching model.

Validating the Effectiveness of the GenAI-empowered RCPA Human-machine Collaborative Teaching Model. Using the AVA Intelligence Classroom Analysis System and 360AI Browser, this study examined the application of the GenAI-empowered RCPA teaching model in elementary English review lessons. Classroom practice analysis revealed that it effectively sparked student interest, enhanced core English competencies, and transformed classroom activities.

6.2 Discussion

Integrating the GenAI-empowered RCPA human-machine collaborative teaching model represents a promising approach to elementary English review lessons. The RCPA model has demonstrated potential in effectively cultivating students' English proficiency levels, cultural awareness, and quality of thinking, which are core competencies essential for their development. However, as the sample size of this study was relatively small, the observed effects of the GenAI applications and the universality of the model require further validation through larger-scale studies in diverse educational settings. Additionally, the successful implementation of GenAI-empowered teaching requires adequate digital literacy among both teachers and students, which should be developed through comprehensive training programs to equip educators and learners with the necessary skills to use these technologies effectively. Moreover, the challenge of fully integrating the "Teacher-Student-Machine" triadic structure into classroom instructional design and teaching practice remains. Future research should focus on developing strategies to seamlessly incorporate this triadic structure, ensuring that each component effectively contributes to the learning process. Despite these challenges, we believe that with the continued advancement of educational digitalization, GenAI will have a profound and far-reaching impact on elementary English education and beyond, fostering an enriched and more engaging learning environment.

Acknowledgments. Supported by the project "Research on Digitized Construction and Scenario-based Application of Curriculum Empowered by Knowledge Graphs" sponsored by the 2023 Guangdong Undergraduate Colleges Teaching Quality and Teaching Reform Project Construction Project.

References

1. Shen, B.: Exploration of English review strategies for elementary schools based on unit main scenario diagram. Teach. Adm. **8**, 41–44 (2023)
2. Wang, L.: Types and use of thinking maps in elementary English review teaching. Teach. Adm. **3**, 110–112 (2019)
3. Qinyu, W.: Research on elementary English unit review lessons oriented by core literacy: taking Shanghai Oxford English 3A M3U1 Revision as an example. Campus English **37**, 191–192 (2021)
4. Llin, F.: On the structure of English review lessons in elementary schools based on thinking process. Res. Mod. Basic Educ. **3**, 209–213 (2021)
5. Lee, J.H., Dongkwang, S., Wonjun, N.: Artificial intelligence-based content generator technology for young English-as-a-foreign-language learners' reading enjoyment. RELC J. **54**, 508–516 (2023)
6. Dong, W., Daohua, P., Soonbae, K.: Exploring the integration of IoT and Generative AI in English language education: smart tools for personalized learning experiences. J. Comput. Sci. **82**, 102397 (2024)
7. Zhang, R., Yi, Q., Ye, L.: An empirical study on human-machine collaborative MOOC learning interaction empowered by Generative AI. In: 2023 International Symposium on Educational Technology (ISET), Ho Man Tin, Hong Kong, pp. 116–120 (2023)
8. Xi, Y., Qiuchun, L., Fangyuan, Z., Xuru, Y., Ziru, O.: The practice of 'MOOC Plus Theme Lecture–Artificial Intelligence Tutoring–Project Practice (MOOC+TAIP)' model to improve AI literacy of preschool teachers. In: 2023 Twelfth International Conference of Educational Innovation through Technology (EITT), Fuzhou, China, pp. 106–112 (2023)
9. Qiming, M.: Large unit instructional design from the perspective of learning center. Curric. Teach. Mater. Method **3**, 52–58 (2024)
10. Youru, X., Jia, L.: Classroom instructional design based on deep learning in intelligent age. E-educ. Res. **5**, 73–80 (2020)
11. Kejia, Z., Feng, T., Bo, W., Fei, S., Tao, L.: Learning status real-time collection and dynamic analysis method based on hybrid intelligence. Mod. Educ. Technol. **7**, 108–117 (2023)

Design and Development of an LLM-Empowered Arguing Pedagogical Agent with a Personalized Avatar, Movements, and Voice

Qingtang Liu, Yunfan Zuo(✉), Ruiqiu Ye, Yubei Chang, Xing Zhang, and Linjing Wu

Faculty of Artificial Intelligence in Education, Central China Normal University, Wuhan, China
liuqtang@mail.ccnu.edu.cn, zuoyunfan@mails.ccnu.edu.cn

Abstract. The rapid advancement of large language models (LLMs) has ushered in new changes in education. To address the limitations of existing arguing chatbots in adaptability and response quality, we developed an LLM-empowered innovative tool, aimed at enhancing students' argumentation skills and critical thinking. We provided the LLM with a personalized avatar, movements, and voice. Student feedback was gathered through questionnaires and interviews. The results indicate that the debate-based pedagogical agent delivers a positive interactive experience, boosting student engagement and encouraging them to consider issues from multiple perspectives. In conclusion, we examined its limitations and proposed directions for future research.

Keywords: Pedagogical Agent · Large Language Model · Debate-Based Learning · Critical Thinking

1 Introduction

With the rapid advancement of modern artificial intelligence technology, the field of education is experiencing unprecedented transformations [1]. Leveraging artificial intelligence (AI) tools to support teaching has become a popular topic in the education sector [2], particularly in argumentation instruction, where AI demonstrates significant potential and application prospects.

In educational technology research, debate is recognized as an effective teaching strategy, extensively studied and applied. Debate not only boosts student engagement but also significantly enhances critical thinking and argumentation skills [3]. The arguing chatbot has emerged as a novel technology [4]. Simulating a debate environment, the arguing chatbot leverages the computer's ability to provide instant feedback, reducing anxiety and feelings of isolation [5]. This low-pressure setting [6] helps cultivate students' argumentation skills and critical thinking [7]. Empirical studies have shown that interactive learning with AI-integrated chatbots positively impacts students' learning outcomes [8].

Q. Liu et al. (Eds.): EITT 2024, CCIS 2600, pp. 16–30, 2026.
https://doi.org/10.1007/978-981-95-1525-7_2

Previous studies on arguing chatbots indicate that machine responses must balance adaptability and consistency [9]. Retrieval-based methods [4], though providing better response consistency, cannot encompass all specific replies in their database, making it challenging to adapt to the diverse arguments posed by different students. Generative methods based on Recurrent Neural Networks (RNN) [9] face limitations in capturing long-distance dependencies, leading to less consistent responses for longer dialogue contexts compared to the current Transformer-based LLMs [10].

This study, based on social agent theory [11], aims to help students simulate debate environments, enhancing their argumentation skills and critical thinking. We have given the LLM a personalized avatar and an appropriate voice, making it more suitable for debate teaching scenarios. Its system architecture is depicted in Fig. 1. This innovative tool allows educators to avoid the issues of limited adaptability and poor response quality often found in traditional chatbots, providing students with a more personalized environment.

This paper explores the design and implementation of an arguing pedagogical agent and the feedback from its users. The remainder of the paper is structured as follows: Section 2 reviews relevant literature, assessing the strengths and weaknesses of existing research. Section 3 details the design and implementation methods of the agent. Section 4 analyzes the results from user experience surveys and interviews. Finally, the paper summarizes the research findings and suggests directions for future research.

2 Related Work

2.1 Argumentation Learning in Human-Computer Debates

The application of debate as a teaching strategy in classrooms has been extensively studied. This approach necessitates that students construct arguments, collect evidence, and counter opposing viewpoints. These activities require a comprehensive understanding of the topic, rigorous evaluation of information, and sound judgment [12]. Research indicates that debate can stimulate critical thinking, enhance engagement, and improve classroom performance [13].

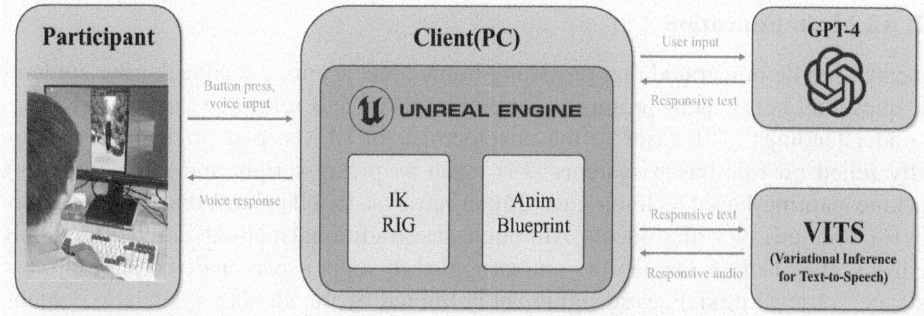

Fig. 1. Overview of our system design.

Recently, advancements in AI and natural language processing have shown that chatbots have significant potential in developing students' critical thinking and argumentation skills. Rakshit, Bowden, Reed, Misra, and Walker [4] developed Debbie, a chatbot designed specifically for debates, which helps users enhance their argumentation skills by retrieving relevant arguments from a dialogue database. Similarly, Wambsganss, Soellner, Koedinger, and Leimeister [14] created ArgueBot to help students improve the structure and logic of their arguments by identifying argumentative elements in their writing and providing feedback. Introducing chatbots into classroom debates has also proven beneficial. Guo, Zhong, Li, and Chu [7] found that students supported by the chatbot ArgueMate during classroom debates were generally positive about the tool, believing it stimulated their enthusiasm and interest in learning. These studies suggest that human-computer debates, both inside and outside the classroom, can significantly develop students' analytical and critical thinking skills, offering a stress-free [6] and enjoyable learning environment.

Nevertheless, there are still many areas where current chatbots need further improvement. Enhancing the guidance provided by chatbots to students remains a crucial challenge. The aim of debating with chatbots is to offer students different perspectives and improve their argumentation skills. However, in one study, students often preferred using their arguments over the chatbot's suggestions during debates [7]. This finding implies that chatbot responses need to be more practical and relevant. Additionally, existing chatbots primarily communicate through text, which may be less persuasive. For broader educational applications, chatbots also require greater adaptability and faster response times [4].

While some research presents a negative view of debates, noting that students may feel anxious when faced with opposing views and be reluctant to disagree [15], discomfort in learning is essential [16]. Appropriate teaching strategies can alleviate these negative emotions and boost students' motivation and engagement [6].

Face-to-face debates can foster deeper discussion and understanding [17], as students can directly ask questions, clarify viewpoints, and immediately adjust their arguments. Our pedagogical agent, designed with a human-like appearance, creates a sense of presence similar to that of in-person debates. Moreover, it incorporates the latest LLM to deliver higher-quality responses.

2.2 LLMs in Education

Research has demonstrated that providing immediate, accurate feedback after students ask questions helps them promptly correct mistakes and reinforce correct behaviors or understandings [18]. Prior to the development of LLMs, pedagogical agents typically relied on rule-based systems [19] (such as preset scripts and dialogue trees), machine learning-based technologies [20], or retrieval-based [4] text dialogue technologies to communicate with students. Although these traditional methods could offer timely feedback, they lacked adaptability and struggled to support personalized learning. For instance, retrieval models were significantly limited by the chatbot's database content, making it challenging to adapt to new user responses [9].

The emergence of LLMs has introduced an extremely effective method that provides precise and timely feedback to students while overcoming the inflexibility of traditional approaches. LLMs can create personalized, immersive interactive learning environments, aiding students in better understanding and mastering knowledge [21]. Since the introduction of GPT-3 [22], more studies have explored the application of LLMs in education, such as language learning [23] and programming [24]. Additionally, researchers have explored the use of LLMs to assist students in classroom debates [7].

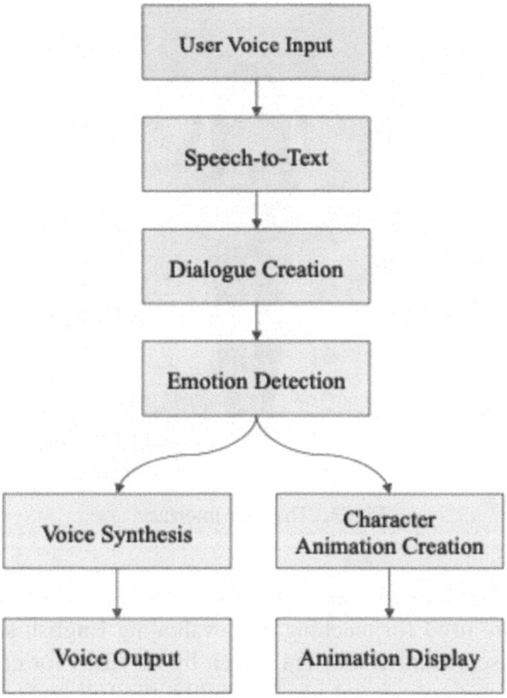

Fig. 2. The system workflow.

While LLMs have demonstrated significant potential in education, most current research focuses on text-based interactions. Therefore, our agent enhances an LLM with a friendly, approachable avatar and incorporates a voice channel, transforming student interactions from text-based to voice-based. This approach is more intuitive for students in debate scenarios and aligns more closely with the initial concept of an "intelligent learning companion."

2.3 Design of Pedagogical Agent's Avatar

From the early AVATALK system [25] to modern 3D humanoid agents capable of expressing facial and body movements [26], pedagogical agents have seen significant evolution and development.

Extensive research has been conducted to determine the most effective type of avatar for pedagogical agents. A prevalent view is that interactive systems featuring human-like avatar can enhance students' learning performance [27, 28]. Further studies indicate that young female agents are perceived as more knowledgeable, offering a friendlier learning experience and higher engagement [29].

Fig. 3. The user interface.

In the industry, mature products like AnimatoCall's Call Annie and NetEase Youdao's Hi Echo are used for teaching and evaluating English speaking skills [30]. These applications use virtual characters driven by an LLM for conversational teaching but typically feature only facial images, unlike the full-body images preferred in academic research. Most academic studies favor full-body 3D images for virtual agents [31], likely due to the influence of social agent theory [11]. In educational settings, full-body images of teachers enhance students' sense of social presence, draw more attention, and improve classroom engagement [32] as they communicate more non-verbal cues (e.g., body language, posture, and movements), thereby improving interaction and communication between the agent and students.

We utilized VRoid Studio, known for its high modeling precision, to create humanoid pedagogical agents and used Unreal Engine 5 (UE5), a tool widely employed in game development, to design their movements.

2.4 Prompt Engineering

LLMs with their remarkable abilities in content generation and real-time feedback, are rapidly becoming embedded in educational settings [33]. Effective interaction with an LLM requires users to input a guiding statement, known as a "prompt" or "prompt engineering", which helps the model understand the question and tailor its response to the specific task. As a key factor in optimizing user-LLM interaction, prompts must be carefully crafted to match various usage contexts.

A well-structured prompt generally includes five elements: an instruction, examples, an output format, a role specification, and additional information [34]. The instruction defines the task and outlines what the model is expected to do. Examples offer reference cases for the model—such as the widely used "Chain-of-Thought" prompting method [35], which illustrates how to break down problems step by step. While this few-shot approach is highly effective in fields like mathematics and symbolic reasoning, it is less suitable for the open-ended question-and-answer scenarios often encountered with pedagogical agents [36]. The output format specifies how the model should present its response—whether as plain text, markdown, or JSON—allowing the output to be adapted for different applications. Although current research suggests that assigning roles does not necessarily improve LLM performance [37], these models excel at simulating well-known personas and leveraging cultural cues. For instance, having the model take on a persona such as Mahatma Gandhi can subtly convey moral principles without explicitly stating them [38]. Other studies have explored framing the LLM as a "Socratic tutor" [39], using multi-turn dialogue to foster students' critical thinking skills.

3 Implementation of the Pedagogical Agent

In this study, we developed a pedagogical agent system designed to engage in debates with students. The system is built on a standard computer running Windows 10, equipped with an Intel Core i9 CPU, 128 GB of RAM, and an Nvidia GeForce RTX 3090 graphics card.

The overall architecture of our system is illustrated in Fig. 1. Our design philosophy centers on user-system interaction, leveraging a suite of advanced models to enhance the system's capabilities. At its core, the system is built upon a conventional Automatic Speech Recognition (ASR), LLM, and Text-to-Speech (TTS) pipeline, the workflow as illustrated Fig. 2. The system employs Paraformer [40] to transcribe the user's speech into text. This text is then sent to GPT-4 via API. Once GPT-4 provides a reply, the system uses BERT [41] to recognize the emotion in the text and then uses a pre-trained VITS model to synthesize speech. Users then receive voice responses from the agent. Users can interact with the system via a single button press.

3.1 Avatar Design

The system interface is shown in Fig. 3. We used VRoid Studio to create a young girl agent with a cartoonish look and a friendly expression. After rigging the model in UE5, we developed six types of actions across three categories: hands behind back, arms crossed, hand on chin, hands on hips, finger tapping, and arms akimbo. Specifically, during the listening mode, the agent randomly performs either hands behind back or arms crossed actions while waiting for the user to speak, as illustrated Fig. 4(a) and Fig. 4(b). While awaiting GPT-4's response, the agent performs a hand-on-chin action to signify thinking, as shown Fig. 4(c). During debates, the agent randomly performs hands-on-hips, finger tapping, or arms akimbo actions to display confidence, as illustrated Fig. 4(d), Fig. 4(e),

(a) Hands behind back (b) Arms crossed (c) Hand on chin (d) Hands on hips (e) Finger tapping (f) Arms akimbo

Fig. 4. Character action design.

and Fig. 4(f). Despite using technical measures to reduce the perceived delay due to the LLM's reasoning process, we deliberately designed the hand-on-chin thinking action to provide a brief interval while GPT-4 generates a response. This interval not only reinforces the agent's contemplative state but also enhances user experience, making interactions feel like a conversation with a real learning companion, thus improving immersion and engagement. We used a frontal view of the pedagogical agent in UE5's viewport to increase credibility [42]. Credibility is particularly important in classroom debates, as higher credibility enhances students' acceptance of the agent's arguments.

Research shows that adding a voice component to pedagogical agents, compared to text-only presentations, significantly improves learning outcomes [43]. Therefore, we trained a voice model using VITS [44] to match the appearance of a young female. Compared to using OpenAI's built-in audio synthesis, the VITS approach offers more flexibility and diverse voice options. Additionally, with BERT's support, our agent includes emotional cues [45], capable of expressing emotions such as excitement, happiness, and calmness. These emotions trigger corresponding facial expressions and tones, with some emotions accompanied by specific action sequences.

3.2 Prompt Design

The effectiveness of LLMs in performing specific tasks significantly hinges on the quality of the prompts [46]. Previous studies have shown that LLMs tend to exhibit a propensity to please humans [47], which can be problematic in debate contexts. System-level prompts are an effective solution to this problem. The GPT-4 API offers three types of role definitions, with the "system" message used to set the initial context and guide the model's behavior throughout the conversation.

In the system prompt, we define GPT-4's role as "a wise learning companion proficient in debating." We then outline the task process: "1. First, ask the student to select a debate topic. 2. Once the topic is chosen, assign the student to either the affirmative (support) or negative (oppose) side, while you take the opposite stance. 3. Conduct a structured debate by presenting arguments, rebuttals, and responses." We also specify the rules that GPT-4 must adhere to during the debate: "The following rules must be followed: 1. Your responses should be natural and conversational, avoiding robotic language, and most importantly, maintaining a respectful and educational tone. 2. Do not display a tendency to please humans during the debate. 3. Evaluate whether the student seems frustrated or confused based on their responses. If the student appears to be in a negative state, provide necessary intellectual or emotional support [16], such as offering appropriate constructive feedback and guidance. 4. Encourage students to provide evidence to support their points and guide them to critically evaluate the opponent's arguments. 5. If a student's question falls outside your knowledge base, be honest and suggest they consult relevant resources." Finally, we guide how the model should initiate: "Start by asking the student: 'What topic would you like to debate?' and then proceed based on their selection."

3.3 Reduce Feedback Delay

When GPT-4 receives a user's message, it takes some time to generate a response. After the system receives this response, it first uses the BERT model for emotion classification and then the VITS model for speech synthesis. This process introduces a significant delay, resulting in a less favorable user experience. To address this issue, we implemented a streaming output for GPT-4's responses, allowing partial responses to be returned immediately without waiting for the complete response. We then split the sentences based on punctuation marks in the received reply. Specifically, the system detects punctuation marks (such as periods, exclamation marks, or question marks) or newline characters and immediately sends the partially generated content to the BERT model for emotion classification. The responses carrying emotional information are then passed to the VITS model for speech synthesis and promptly output to the user. Essentially, the reply-generation process in the system is almost synchronized with the output to the user. Our strategy of providing partial information quickly improves the overall response speed.

4 User Evaluation

We recruited ten educational technology students from a normal university to participate in our user experience activity. Before starting the activity, researchers explained how to use the system to all participants. Each participant then engaged in a 5–10 min-long debate with our agent, with the debate topics chosen freely by the participants. After completing their debate session, each participant was asked to fill out a questionnaire immediately. We invited five participants for semi-structured interviews once the entire activity was concluded.

We conducted a user evaluation analysis using a 12-question questionnaire. Figure 5 displayed a subset of these questions. The questionnaire was designed according to the Technology Acceptance Model (TAM) [48] and the User Experience Questionnaire (UEQ) [49], covering four dimensions: usefulness, ease of use, satisfaction, and enjoyment. Responses were measured on a five-point Likert scale, with "1" representing "strongly disagree" and "5" representing "strongly agree." The questionnaire exhibited good reliability (Cronbach's = 0.856).

4.1 Questionnaire Analysis

Ten university students specializing in educational technology participated in a user experience survey. They debated topics such as "Are the benefits of AI in education greater than the drawbacks?", "Is traditional classroom teaching better than online teaching?", "Do mobile phones bring people closer or drive them apart?", and "Should elementary school students use electronic devices?" These discussions held significant value. The final analysis of the questionnaire result is shown in Table 1.

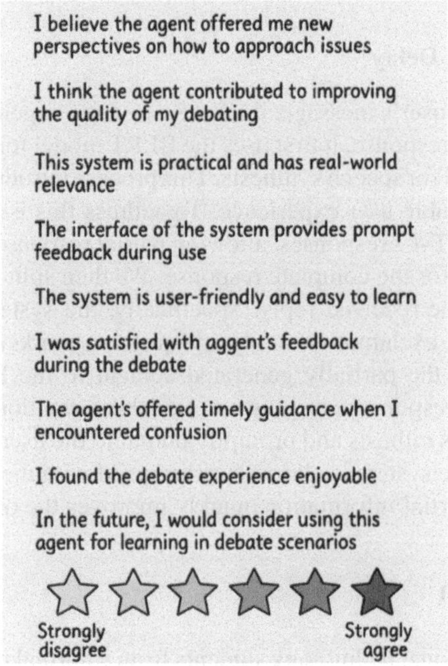

Fig. 5. Part of the questionnaire.

Although the sample size (N = 10) limits the generalizability of these quantitative findings, the consistent high ratings, coupled with qualitative data from interviews, provide valuable preliminary insights into user perceptions of the LLM-empowered pedagogical agent. The system received high ratings on multiple fronts, indicating that users were generally satisfied with the overall functionality and experience of our agent system. The highest score was for "The system is easy to learn and use," (M = 4.7, SD = 0.48), which can be attributed to the intuitive single-button interaction and the clear visual cues provided by the agent's avatar and actions. This ease of use is crucial for adoption in educational settings where complex setups can be a barrier. Similarly, high marks for "I believe this agent can help improve the quality of debates" (M = 4.5, SD = 0.53) and "This system is practical and meaningful" (M = 4.6, SD = 0.52) indicate that users perceived direct educational value, likely stemming from the LLM's ability to generate diverse arguments and the engaging multi-modal presentation which included voice and synchronized avatar movements. The strong agreement on "I believe this agent can provide me with new perspectives on issues" (M = 4.6, SD = 0.70) further suggests that the agent successfully stimulated critical thinking by exposing students to viewpoints they might not have considered, a core objective of debate-based learning and a strength of well-prompted LLMs designed to explore multiple facets of a topic. The positive attitude towards future use (M = 4.5, SD = 0.53) and recommendations (M = 4.6, SD = 0.52) underscores the potential of such systems in educational contexts.

However, a few users experienced confusion, particularly when they tried to interrupt the agent to speak but the system did not respond. This inability to handle natural turn-taking interruptions, common in human conversation, was a notable point of friction. It suggests that while the agent's speech output is fluid, the input processing and dialogue management need enhancements to support more dynamic and responsive interactions. For a debate agent, the ability to gracefully handle interruptions or allow users to interject is critical for a truly interactive experience and presents a clear direction for future system refinement, possibly through more sophisticated speech activity detection, barge-in capabilities, and dialogue state tracking. Ultimately, all users completed their debates on specific topics with the developers' assistance, indicating that the core functionality was achievable despite these interactional hiccups.

Table 1. Survey Results Analysis

Survey Dimension	Specific Content	Mean	Standard Deviation
Usefulness	Beneficial for learning	4.2	0.92
	Perspectives	4.6	0.70
	Improve response	4.5	0.53
	Practical significance	4.6	0.52
Ease of use	Easy to operate	4.5	0.71
	Easy to use	4.7	0.48
Satisfaction	Feedback	4.1	0.57
	Creativity	4.3	0.67
	Guidance	4.4	0.84
	Experience	4.3	0.82
Enjoyment	Willing to use again	4.5	0.53
	Recommend to others	4.6	0.52

4.2 Interview Analysis

We conducted semi-structured interviews with five out of ten students who participated in the user experience survey. The interviews revealed that most users had a positive attitude towards the system, appreciating its innovation and practicality. One user mentioned,

"This system is very interesting, offering a novel form of teaching." Another user added, "This is my first time experiencing a debate-style Pedagogical Agent, and I am impressed by the performance of the current LLMs in debates." Almost all users felt happy while using the system, and some were even pleasantly surprised by the agent's unconventional viewpoints.

However, some users pointed out flaws they found while using the system. One user remarked, "The emotional shifts of the pedagogical agent during debates seemed abrupt. Sometimes, it would transition from a very low tone to a very excited one in the next sentence, which felt strange." This issue may arise from the BERT model we used for emotion classification being overly sensitive to certain keywords or sentence structures, leading to rapid and sometimes contextually jarring emotional swings. Such abruptness, as users indicated, could occasionally detract from the agent's perceived naturalness and credibility, potentially disrupting the immersive learning experience. While the intention was to create an expressive agent, future iterations should focus on refining the emotion detection and expression pipeline, perhaps by incorporating smoothing algorithms for emotional state transitions, contextual awareness in emotion selection, or by training the emotion model on more nuanced conversational debate data to ensure more contextually appropriate and fluid affective displays.

Furthermore, a user with debate experience observed, "Although the agent's responses appear reasonable at first glance, they are too broad for debate scenarios and lack specificity. The agent rarely provides in-depth evaluations or explanations of viewpoints." This insightful feedback highlights a current limitation of generalized LLMs when applied to specialized tasks like academic debate, which demands not just breadth but significant depth, logical precision, and the ability to deconstruct and build upon specific argumentative points. This indicates that debates with LLMs, despite their fluency and knowledge range, still fall short of the nuanced, targeted, and deeply analytical exchanges often seen in human-to-human debates. Problems such as insufficient depth of discussion and lack of targeted arguments can diminish user engagement and limit the development of advanced argumentation skills, particularly for experienced debaters. This underscores the importance of the proposed future work on integrating knowledge bases and employing Retrieval-Augmented Generation (RAG) [50], which could potentially ground the LLM's responses in more specific evidence and allow for more focused and substantively rich counter-arguments and evaluations, thereby better aligning the agent's capabilities with the pedagogical goals of debate education and addressing user expectations for deeper engagement.

5 Conclusion

This study presented the design, implementation, and initial user evaluation of an LLM-empowered arguing pedagogical agent featuring a personalized avatar, movements, and voice. Our primary goal was to move beyond traditional text-based chatbots by creating a more engaging, embodied agent for face-to-face debate practice, aiming to enhance students' argumentation skills and critical thinking. The multi-modal design, integrating visual cues from the avatar's actions and affective expressions with synthesized voice, was intended to foster a stronger sense of presence and social interaction, which are crucial for effective pedagogical engagement. User experience evaluations, though preliminary, indicated that the agent successfully fostered a positive interactive experience, effectively engaged students' positive emotions, and was perceived as a useful tool for learning, offering a flexible and stress-free environment.

Despite these promising results, this study identified several limitations that pave the way for future research and system refinement. While the LLM provided diverse arguments, users, particularly those with debate experience, expected higher relevance, greater depth, and more specific evaluations in the agent's responses. This necessitates the integration of knowledge bases and the employment of RAG, alongside more sophisticated prompt engineering. System latency, despite efforts using streaming output, remained a concern due to the speech synthesis model, highlighting the need to explore more lightweight TTS models optimized for real-time debate scenarios. A key area for improvement is the system's current inability to gracefully handle user interruptions during its speech output, which affected the perceived naturalness of the dialogue; enhancing dialogue management to support barge-in capabilities and more fluid turn-taking is therefore essential. Furthermore, users observed that the agent's emotional transitions could sometimes appear abrupt or overly sensitive, indicating a need to refine the emotion recognition and expression pipeline for smoother, more contextually appropriate, and nuanced affective displays. While the agent featured a pre-designed "personalized" avatar, true user-driven personalization options were not implemented, and exploring such features could further enhance user engagement. Finally, the limited sample size in our user study necessitates caution in generalizing the findings and calls for larger-scale, more diverse, and potentially longitudinal evaluations.

Future research will concentrate on these areas. Specifically, we aim to integrate RAG and explore advanced prompting techniques to improve argument depth and specificity, investigate lightweight TTS models to minimize latency, develop sophisticated dialogue management for interruption handling and natural turn-taking, and refine the emotional model for context-aware expressions. We also plan to explore user-configurable personalization and conduct more extensive user studies, potentially including comparative analyses. Ultimately, this research contributes to the growing field of AI in education by demonstrating a viable approach to creating more human-like and engaging pedagogical agents. We believe that continued development in these multi-modal, LLM-driven systems holds significant promise for transforming educational interactions and supporting personalized learning experiences at scale.

Acknowledgments. This work was supported by National Natural Science Foundation of China "Research on Instructional Interaction Behavior Recognition and Pattern Mining in Informationization Classroom Integrating Multimodal Data" (grant number: 62277021), the Humanities and Social Sciences Research Project of the Ministry of Education of China (grant number: 22YJAZH067), the National Experimental Base for Collaborative Innovation in Teacher Development in Central China Normal University (grant number: CCNUTEII 2021-12).

References

1. Stokel-Walker, C.: AI bot ChatGPT writes smart essays: Should professors worry? Nature (2022). https://doi.org/10.1038/d41586-022-04397-7

2. Jeon, J., Lee, S.: Large language models in education: a focus on the complementary relationship between human teachers and ChatGPT. Educ. Inf. Technol. **28**(12), 15873–15892 (2023)
3. Oros, A.L.: Let's debate: active learning encourages student participation and critical thinking. J. Polit. Sci. Educ. **3**(3), 293–311 (2007)
4. Rakshit, G., Bowden, K.K., Reed, L., Misra, A., Walker, M.: Debbie, the debate bot of the future. In: Eskenazi, M., Devillers, L., Mariani, J. (eds.) Advanced Social Interaction with Agents. LNEE, vol. 510, pp. 45–52. Springer, Cham (2019). https://doi.org/10.1007/978-3-319-92108-2_5
5. Kohnke, L.: A qualitative exploration of student perspectives of chatbot use during emergency remote teaching. Int. J. Mob. Learn. Organ. **16**(4), 475–488 (2022)
6. Holzer, A., Tintarev, N., Bendahan, S., Kocher, B., Greenup, S., Gillet, D.: Digitally scaffolding debate in the classroom. In: Extended Abstracts of the 2018 CHI Conference on Human Factors in Computing Systems, pp. 1–6 (2018)
7. Guo, K., Zhong, Y., Li, D., Chu, S.K.W.: Effects of chatbot-assisted in-class debates on students' argumentation skills and task motivation. Comput. Educ. **203**, 104862 (2023)
8. Wu, R., Yu, Z.: Do AI chatbots improve students learning outcomes? Evidence from a meta-analysis. Br. J. Educ. Technol. **55**(1), 10–33 (2024)
9. Le, D.-T., Nguyen, C.-T., Nguyen, K.A.: Dave the debater: a retrieval-based and generative argumentative dialogue agent. In: Proceedings of the 5th Workshop on Argument Mining, pp. 121–130 (2018)
10. Vaswani, A., et al.: Attention is all you need. In: Advances in Neural Information Processing Systems, vol. 30 (2017)
11. Atkinson, R.K., Mayer, R.E., Merrill, M.M.: Fostering social agency in multimedia learning: examining the impact of an animated agent's voice. Contemp. Educ. Psychol. **30**(1), 117–139 (2005)
12. Moore, T.: The critical thinking debate: How general are general thinking skills? High. Educ. Res. Dev. **23**(1), 3–18 (2004)
13. El Majidi, A., Janssen, D., de Graaff, R.: The effects of in-class debates on argumentation skills in second language education. System **101**, 102576 (2021)
14. Wambsganss, T., Soellner, M., Koedinger, K.R., Leimeister, J.M.: Adaptive empathy learning support in peer review scenarios. In: Proceedings of the 2022 CHI Conference on Human Factors in Computing Systems, in CHI 2022. Association for Computing Machinery, New York, NY, USA (2022). https://doi.org/10.1145/3491102.3517740
15. Jonassen, D.H., Kim, B.: Arguing to learn and learning to argue: design justifications and guidelines. Educ. Technol. Res. Dev. **58**, 439–457 (2010)
16. Cheng, A.Y.: et al.: Scientific and fantastical: creating immersive, culturally relevant learning experiences with augmented reality and large language models. In: Proceedings of the CHI Conference on Human Factors in Computing Systems, pp. 1–23 (2024)
17. Altay, S., Schwartz, M., Hacquin, A.-S., Allard, A., Blancke, S., Mercier, H.: Scaling up interactive argumentation by providing counterarguments with a chatbot. Nat. Hum. Behav. **6**(4), 579–592 (2022)
18. Hattie, J., Timperley, H.: The power of feedback. Rev. Educ. Res. **77**(1), 81–112 (2007)
19. Gustafson, J., Lindberg, N., Lundeberg, M.: The August spoken dialogue system. In: Sixth European Conference on Speech Communication and Technology. Citeseer (1999)
20. Rach, N., Langhammer, S., Minker, W., Ultes, S.: Utilizing argument mining techniques for argumentative dialogue systems. In: D'Haro, L., Banchs, R., Li, H. (eds.) 9th International Workshop on Spoken Dialogue System Technology. LNEE, vol. 579, pp. 131–142. Springer, Singapore (2019). https://doi.org/10.1007/978-981-13-9443-0_12 N. Rach, S. Langhammer, W. Minker, and S. Ultes, "Utilizing argument mining techniques for argumentative dialogue

systems," in 9th International Workshop on Spoken Dialogue System Technology, Springer, pp. 131–142.(2019)
21. Nikolopoulou, K.: Generative artificial intelligence in higher education: exploring ways of harnessing pedagogical practices with the assistance of ChatGPT. Int. J. Chang. Educ. **1**(2), 103–111 (2024)
22. Brown, T., et al.: Language models are few-shot learners. Adv. Neural. Inf. Process. Syst. **33**, 1877–1901 (2020)
23. Guo, K., Li, Y., Li, Y., Chu, S.K.W.: Understanding EFL students' chatbot-assisted argumentative writing: an activity theory perspective. Educ. Inf. Technol. **29**(1), 1–20 (2024)
24. Phung, T., et al.: Automating human tutor-style programming feedback: Leveraging GPT-4 tutor model for hint generation and GPT-3.5 student model for hint validation. In: Proceedings of the 14th Learning Analytics and Knowledge Conference, pp. 12–23 (2024)
25. Hubal, R.C., Frank, G.A., Guinn, C.I.: AVATALK virtual humans for training with computer generated forces. In: Proceedings of the Ninth Conference on Computer Generated Forces. Institute for Simulation & Training, pp. 16–18, May 2000
26. Moon, J., Ryu, J.: The effects of social and cognitive cues on learning comprehension, eye-gaze pattern, and cognitive load in video instruction. J. Comput. High. Educ. **33**(1), 39–63 (2021)
27. Yılmaz, R., Kılıç-Çakmak, E.: Educational interface agents as social models to influence learner achievement, attitude and retention of learning. Comput. Educ. **59**(2), 828–838 (2012)
28. Wang, Y., Gong, S., Cao, Y., Fan, W.: The power of affective pedagogical agent and self-explanation in computer-based learning. Comput. Educ. **195**, 104723 (2023)
29. ter Stal, S., Tabak, M., op den Akker, H., Beinema, T., Hermens, H.: Who do you prefer? The effect of age, gender and role on users' first impressions of embodied conversational agents in eHealth. Int. J. Human-Computer Interact. **36**(9), 881–892 (2020)
30. Wan, Y., Moorhouse, B.L.: Using Call Annie as a generative artificial intelligence speaking partner for language learners. RELC J., 00336882231224813 (2024)
31. Dai, L., Jung, M.M., Postma, M., Louwerse, M.M.: A systematic review of pedagogical agent research: similarities, differences and unexplored aspects. Comput. Educ. **190**, 104607 (2022)
32. Guo, P.J., Kim, J., Rubin, R.: How video production affects student engagement: an empirical study of MOOC videos. In: Proceedings of the First ACM Conference on Learning@ Scale Conference, pp. 41–50 (2014)
33. Zhang, L., Xu, J.: The paradox of self-efficacy and technological dependence: unraveling generative AI's impact on university students' task completion. Internet High. Educ. **65**, 100978 (2025)
34. Schulhoff, S., et al.: The Prompt Report: A Systematic Survey of Prompt Engineering Techniques. arXiv Prepr.: arXiv240606608 (2024)
35. Jason, W., et al.: Chain-of-thought prompting elicits reasoning in large language models. Adv. Neural. Inf. Process. Syst. **35**, 24824–24837 (2022)
36. Sprague, Z., et al.: To CoT or not to CoT? Chain-of-thought helps mainly on math and symbolic reasoning. arXiv Prepr.: arXiv240912183 (2024)
37. Zheng, M., et al.: When "A Helpful Assistant" is not really helpful: personas in system prompts do not improve performances of large language models. In: Findings of the Association for Computational Linguistics: EMNLP 2024, pp. 15126–15154 (2024)
38. Reynolds, L., McDonell, K.: Prompt programming for large language models: beyond the few-shot paradigm. In: Extended Abstracts of the 2021 CHI Conference on Human Factors in Computing Systems (CHI EA 2021), pp. 1–7. ACM (2021)
39. Zhang, L., et al.: SPL: a Socratic playground for learning powered by large language model. arXiv Prepr.: arXiv240613919 (2024)
40. Gao, Z., Zhang, S., McLoughlin, I., Yan, Z.: Paraformer: fast and accurate parallel transformer for non-autoregressive end-to-end speech recognition. arXiv Prepr.: arXiv220608317 (2022)

41. Devlin, J., Chang, M.-W., Lee, K., Toutanova, K.: Bert: pre-training of deep bidirectional transformers for language understanding. arXiv Prepr.: arXiv181004805 (2018)
42. So, C., Jung, K.: Approachability and credibility of virtual character faces: the role of the horizontal viewing angle. Hum. Factors **66**(5), 1450–1474 (2024)
43. Mayer, R.E.: Multimedia learning. Psychol. Learn. Motiv. **41**, 85–139 (2002)
44. Kim, J., Kong, J., Son, J.: Conditional variational autoencoder with adversarial learning for end-to-end text-to-speech. In: International Conference on Machine Learning, pp. 5530–5540. PMLR (2021)
45. Ye, R., et al.: Design of emotional cues of pedagogical agents in a virtual eyeball learning system: a pilot study. In: 2023 Twelfth International Conference of Educational Innovation Through Technology (EITT), pp. 7–12. IEEE (2023)
46. Zhou, Y., et al.: Large language models are human-level prompt engineers. arXiv Prepr.: arXiv221101910 (2022)
47. Perez, E., et al.: Discovering language model behaviors with model-written evaluations. arXiv Prepr.: arXiv221209251 (2022)
48. Davis, F.D., Bagozzi, R.P., Warshaw, P.R.: User acceptance of computer technology: a comparison of two theoretical models. Manag. Sci. **35**(8), 982–1003 (1989)
49. Laugwitz, B., Held, T., Schrepp, M.: Construction and evaluation of a user experience questionnaire. In: Holzinger, A. (ed.) HCI and Usability for Education and Work, USAB 2008. LNCS, vol. 5298, pp. 63–76. Springer, Berlin, Heidelberg (2008). https://doi.org/10.1007/978-3-540-89350-9_6
50. Lewis, P., et al.: Retrieval-augmented generation for knowledge-intensive NLP tasks. Adv. Neural. Inf. Process. Syst. **33**, 9459–9474 (2020)

Student Emotion Recognition Research Based on Deep Learning Techniques: An Example of Primary and Secondary School Classrooms

Yimei Huang[1], Taojie Xu[1], Wei Deng[1](✉), Siyao Chen[2], and Qingtang Liu[1]

[1] Hubei Key Laboratory of Digital Education, Central China Normal University, Wuhan, China
sdengwei@mail.ccnu.edu.cn

[2] School of Economics, Wuhan University of Technology, Wuhan, China

Abstract. Under the guidance of the student-centered education philosophy, understanding students' emotions and needs is the key to improving the quality of teaching and promoting the development of personalized education. However, classroom emotion recognition faces challenges such as large numbers of people and occlusion in natural classroom environments. For this reason, this study develops student emotion recognition algorithms adapted to the classroom environment. First, the YOLOv8 model was improved by introducing the MCC module and the WIoUv3 loss function to enhance the detection accuracy and efficiency of the model. Second, a multi-channel emotion recognition network (MultiEmoNet) was developed to improve the accuracy of emotion recognition by combining facial expression, skeletal information, and environmental information. The experimental results show that the improved model performs well in various indexes and can better adapt to the classroom environment, with a view to promoting the development of classroom emotion recognition technology, providing a scientific basis for personalized teaching strategies, and enhancing the effectiveness of teaching.

Keywords: deep learning · object detection · emotion recognition · primary and secondary school classrooms

1 Introduction

Understanding and responding to the emotional needs of students in an educational setting is key to improving the quality of teaching and the student learning experience. Emotions play an important role in behavior guidance, communication, and willingness prediction [1]. With the development of artificial intelligence and machine learning technology, emotion recognition has become a hot research topic. Automatic recognition of students' emotional states helps teachers adjust their teaching strategies to students' emotional and cognitive needs, thus promoting teaching quality and students' overall development.

As part of affective computing, emotion recognition technology automatically identifies emotional states by analyzing individual physiological or non-physiological signals,

Q. Liu et al. (Eds.): EITT 2024, CCIS 2600, pp. 31–45, 2026.
https://doi.org/10.1007/978-981-95-1525-7_3

including facial expressions [2], voice [3], behavior [4], and heart rate [5]. This technology not only improves the level of personalized teaching, but also helps educators obtain more comprehensive student information and meet student needs. Studies have shown that different emotional states can support or inhibit the learning process [6]. Compared with traditional questionnaires, automatic emotion recognition technology provides a more objective means of evaluating emotional states. By monitoring facial expressions and voices in real time, it provides an empirical basis for studying the relationship between emotions and learning [7–9].

However, the application of emotion recognition technology in primary and secondary school classrooms faces challenges. The classroom environment is complex, and factors such as the position and posture of students and teachers, as well as classroom lighting and background noise affect the accuracy of emotion recognition. In addition, object occlusion will hinder the capture of key features. Although previous studies have improved the accuracy and robustness of emotion recognition by integrating multiple information sources such as facial expressions and body movements [10, 11], further development needs to consider the importance of scene background information.

This research is dedicated to developing high-precision emotion recognition methods applicable to classroom contexts. Specific objectives include: (1) Developing deep learning-based object detection algorithms to improve detection accuracy and efficiency in complex environments; and (2) Constructing a multi-channel emotion recognition network to achieve accurate emotion analysis. This initiative aims to promote the development of classroom emotion recognition technology, provide a scientific basis for personalized teaching strategies, and enhance the effectiveness of teaching.

2 Literature Review

2.1 Classification of Student Emotions in Classroom

Emotion, a multidimensional and complex theoretical concept in the field of psychology, has yet to be defined in a unified academic manner. Psychologists such as Ekman and Izard proposed six basic emotions: happiness, sadness, anger, fear, disgust and surprise. These emotions are regarded as important mechanisms for humans to adapt to the external environment and are reflected in physiological reactions and behaviors [12, 13]. These basic emotions are important references in facial expression recognition technology.

Although basic emotion classification provides a framework, it does not fully adapt to the uniqueness and dynamics of the classroom environment. Researchers such as Qinying Yu and U. Ayvaz have adjusted and expanded these emotion classifications based on the specific needs of the classroom, proposing more specific categories such as concentration, confusion, and fatigue to better adapt to teaching scenarios [14, 15]. In addition, some researchers have explored behavior-based emotion classification, indicating that emotional expressions in the classroom may differ from those in daily life situations. Kosti, Alvarez and Recasens analyzed emotions through psychological vocabulary, constructed a library of 400 emotional words, and classified these words into 26 categories, which improved the accuracy of matching teaching emotional expressions [16].

In summary, although basic emotion classification provides guidance for emotion research, classroom emotion classification needs to adopt a more specific and teaching-situation-adapted approach, which requires researchers to consider the universal characteristics of emotions and the unique environment of specific teaching activities and student interactions during the classification process.

2.2 Emotion Recognition Technology in the Classroom

Classroom emotion recognition is one of the key areas of current educational science research, and its technological innovation helps to improve the accuracy of analyzing and responding to students' emotions. Mehrabian and Ferris [17] revealed that in interpersonal communication, non-verbal information (especially facial expressions) plays a dominant role, accounting for up to 55% of communication, highlighting the importance of non-verbal cues in emotion recognition. Subsequent related research has mainly focused on facial expressions [18]. However, recent studies have also begun to explore the potential value of gestures and body postures [19]. Birdwhistell [20] pointed out that 65% of the information in a conversation is transmitted through non-verbal signals, which emphasizes the complexity of capturing true emotional states through traditional methods.

The fusion of body posture and facial expression data has been shown to enhance the accuracy of emotion recognition. Gunes and Piccardi [10] developed a method to combine these data sources to refine emotion classification through fine-grained physical expressions. In addition, Coulson [19] and Kosti, Alvarez, Recasens, and Lapedriza [21] used context-aware deep learning networks to analyze individuals and environments in static images and achieved recognition of discrete emotional states [22, 23].

In classroom settings, researchers have applied these techniques to develop models that can classify a wide range of emotions, from concentration to confusion, making the classification more adaptable to educational settings [24]. Techniques that extract facial features using Gabor wavelet transforms and classify using multilayer perceptron have been shown to effectively distinguish between positive and negative emotional states [25]. Advances in machine learning, particularly the use of convolutional neural networks, have been used to improve the robustness and accuracy of classroom emotion recognition under varying classroom conditions, such as varying lighting and occlusion [26]. These studies highlight the need for emotion recognition techniques to continue to adapt to the dynamic and often unpredictable nature of classroom environments, ensuring that emotion analysis is both comprehensive and accurate.

3 Related Work

3.1 YOLOv8 Network Model

YOLOv8 was released by Ultralytics in January 2023. It supports visual tasks such as object detection, segmentation, pose estimation, tracking and classification. It provides five versions: YOLOv8n (nano), YOLOv8s (small), YOLOv8m (medium), YOLOv8l (large) and YOLOv8x (extra-large), which are suitable for different computing needs.

The YOLOv8 model architecture consists of two parts: Backbone and Head (Fig. 1). The Backbone structure is responsible for feature extraction and consists of three modules: CBS, C2f and SPPF. The CBS module combines convolution, batch normalization and SILU activation functions. The C2f module consists of a convolution module and a residual block. The SPPF module enhances the ability to capture multi-scale information through convolution, maximum pooling and feature concatenation. The Head structure combines Feature Pyramid Networks (FPN) and Path Aggregation Network (PAN) to enhance the detection ability of objects of different scales. FPN upsamples deep feature maps and fuses them with shallow feature maps. PAN further enhances feature fusion through downsampling and upsampling. The detection head contains feature maps of three sizes, suitable for input images of different sizes, and contains independent category recognition and position positioning output branches.

YOLOv8 uses the Complete IoU (CIoU) loss function to comprehensively consider the overlapping area, center point distance, and aspect ratio of the bounding box to improve prediction accuracy. However, it still has shortcomings in dealing with dense objects, occlusion and multi-scale detection in video images. Based on these shortcomings, this paper will improve the YOLOv8 model.

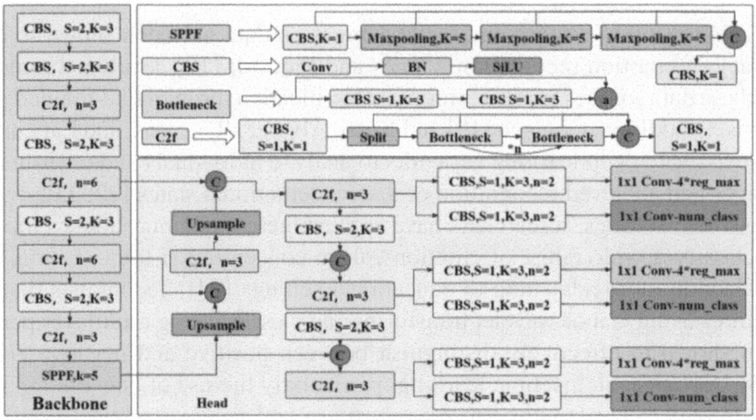

Fig. 1. YOLOv8 network framework.

3.2 EfficientNetv2 Network

EfficientNetv2 model has the characteristics of light weight, high speed and high accuracy; based on these characteristics, this study adopted this model to extract image features. EfficientNet was proposed by the Google team in 2019, through the "compound scaling factor" method to optimize the depth, width, and image resolution to realize efficient computation and accurate results. The model includes seven versions (B0~B7), which differ in width, depth, and input resolution.

EfficientNet combines the design inspiration of MobileNetV2 and ResNet. The structure includes 9 stages and is mainly composed of 16 MBConv blocks. MBConv blocks achieve a balance between performance and efficiency through deep separable convolution. EfficientNetv1-B7 achieved 88.4% top-1 accuracy in the ImageNet classification task, demonstrating its excellent classification ability. EfficientNet uses deep separable convolution to divide traditional convolution operations into channel-by-channel convolution and point-by-point convolution, reducing parameters and computational costs while effectively extracting features. The SE module processes features through global average pooling and multilayer perceptron, learns the importance of each channel, strengthens useful features, and reduces the impact of unimportant features.

Although EfficientNetv1 has excellent performance, it has a significant number of parameters and slow inference speed, making it unsuitable for fast response scenarios. To this end, the Google team proposed EfficientNetv2 in 2021. The new architecture is superior to previous models in training speed and significantly reduces the number of parameters. EfficientNetv2 uses the Fused-MBConv module to replace the MBConv module in the shallow network part to improve processing speed (Fig. 2), and it adopts a non-uniform scaling strategy to adjust according to the contribution and computational cost of each stage to make the network more efficient.

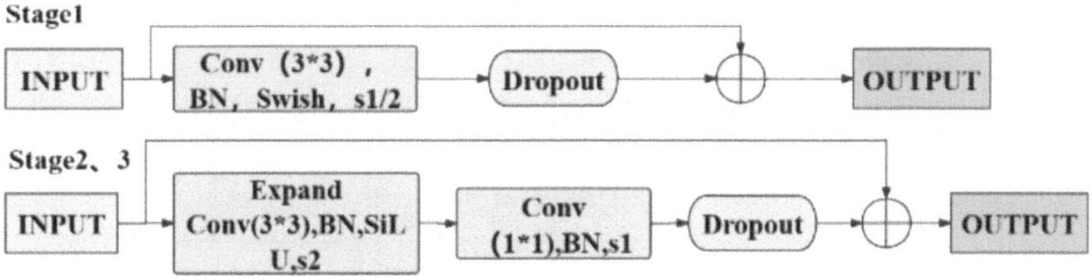

Fig. 2. Adjusted modules in EfficientNetv2.

4 Methodologies

4.1 Student Object Detection Algorithm Based on YOLOv8-Improved

To optimize the performance of YOLOv8 in object detection, this study proposes a Multi-Group Cheap Convolution (MCC) module and a WIoUv3 loss function to replace the existing module and improve the loss calculation method so as to improve detection accuracy and efficiency.

Compared with YOLOv5, YOLOv8 changes the C3 module to the C2f module on CSPLayer, which enhances the feature processing capability while maintaining low computational complexity. In order to further improve performance, this study proposes the MCC module to replace the C2f module. The specific steps of the MCC module (Fig. 3) are as follows:

- The input tensor is split into two parts, x_group and x_cheap, each occupying half of the channels.
- The x_group part is processed by group, and each group applies a convolution layer with different convolution kernel sizes (such as 3 × 3, 5 × 5) to capture features of different scales.

- The processed x_group is concatenated with the unprocessed x_cheap via the channel dimension.
- A 1 × 1 convolutional layer is applied to integrate all features and output the final feature tensor.

The calculation of the MCC module is shown in (1):

$$MCC(X) = CBS\left(Concat\left(CW_{k_1*k_1}(x_1), \cdots CW_{k_n*k_n}(x_n)\right)\right), \tag{1}$$

where $X = [x_1, x_2 \ldots ..x_n]$ represents the input features, k_i is the kernel size, C represents the convolution operation, and CBS is a combination of convolution, batch normalization, and SILU activation functions.

Compared with the traditional C2f module, the MCC module reduces the number of parameters by approximately 25%, and the overall model parameters are reduced by nearly 20%, significantly improving network efficiency and inference speed.

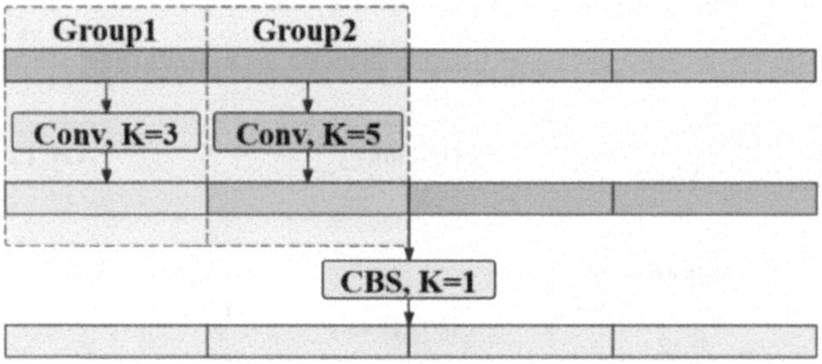

Fig. 3. MCC module diagram.

In addition, YOLOv8 uses the CIoU loss function by default, which takes into account the overlap, center point distance, and aspect ratio of the bounding box, but it is insufficient when dealing with high-quality and low-quality samples. Based on this consideration, this paper proposes the WIoUv3 loss function to replace CIoU. WIoUv1 combines distance attention to effectively distinguish high-quality and low-quality samples, reduce the penalty for low-quality samples, and improve the model's generalization ability (2), (3).

$$\gamma_{0i} = \frac{1}{n} \sum\nolimits_{k=1}^{n} \gamma_{0i}(k), \tag{2}$$

$$R_{WIoU} = exp\left(\frac{\left(x - x_{gt}\right)^2 + \left(y - y_{gt}\right)^2}{\left(W_g^2 + H_g^2\right)^*}\right), \tag{3}$$

WIoUv2 introduces a monotonic focusing coefficient to reduce the contribution of simple samples to the loss value and improve classification (4), (5), (6).

$$L_{WIoUv2} = L_{IoU}^{\gamma*} L_{WIoUv1}, \quad \gamma > 0, \tag{4}$$

$$\frac{\partial L_{WioUv2}}{\partial L_{IoU}} = L_{IoU}^{\gamma *} \frac{\partial L_{WioUv1}}{\partial L_{IoU}}, \quad \gamma > 0, \tag{5}$$

$$L_{WIoUv2} = \left(\frac{L_{IoU}^{*}}{L_{IoU}} \right)^{\gamma} L_{WIoUv1}, \tag{6}$$

WIoUv3 constructs non-monotonic focusing coefficients and dynamically adjusts gradient gains to solve the problem of slow convergence in the late stage of training (7), (8).

$$\beta = \frac{L_{IoU}^{*}}{L_{IoU}} \in [0, +\infty], \tag{7}$$

$$L_{WIoUv3} = rL_{WIoUv1}, r = \frac{\beta}{\delta \alpha^{\beta - \delta}}, \tag{8}$$

through the MCC module and WIoUv3 loss function, the object detection performance of the YOLOv8 model in complex scenes is significantly improved.

4.2 Multi-channel Emotion Recognition Algorithm in Classroom

To achieve efficient emotion recognition, this study proposes a Multi-channel Emotion Recognition Network (MultiEmoNet), which combines facial expressions, background information, and skeleton information for comprehensive analysis and which uses the lightweight and efficient EfficientNetv2-S model for feature extraction.

Facial expression recognition has stronger feature similarity than general image classification tasks. This study uses the EfficientNetv2-S model pre-trained on the VGGface2 dataset and removes the last fully connected layer for multi-channel fusion. To optimize intra-class and inter-class variations, a center loss function is used. In order to focus on important areas of the face, the Active Appearance Model (AAM) is attached after the feature extraction network; its structure is shown in Fig. 4. The AAM module includes a spatial attention module and a channel attention module, which effectively captures the interaction between features and enhances the response to key areas (Fig. 5).

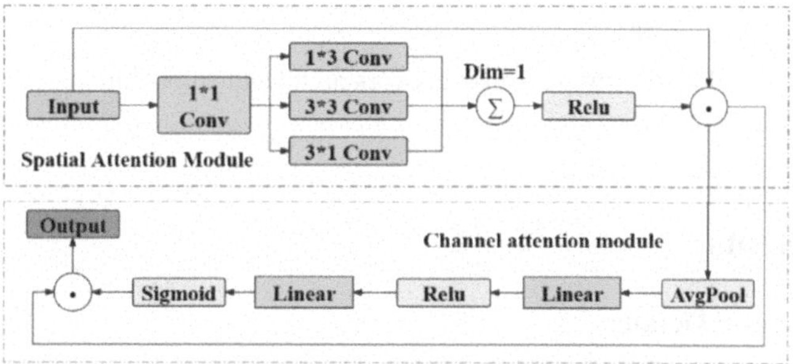

Fig. 4. AAM module structure diagram.

Multimodal fusion improves the accuracy of emotion recognition by comprehensively processing information from different channels. This paper adopts a feature layer fusion method that combines three features: facial expression, skeleton information, and background information. The MultiEmoNet network architecture is shown in Fig. 6. Each channel is extracted through independent Convolutional Neural Networks (CNN), and the AAM module is introduced into the facial image channel to enhance the feature expression of key areas of facial expression. The features output by each channel are sent to the feature fusion unit and combined into a comprehensive feature vector. The fused features are nonlinearly transformed and processed through the fully connected layer to perform the final classification task.

This architecture improves classification accuracy by integrating information from different data sources and is suitable for emotion recognition tasks in complex situations.

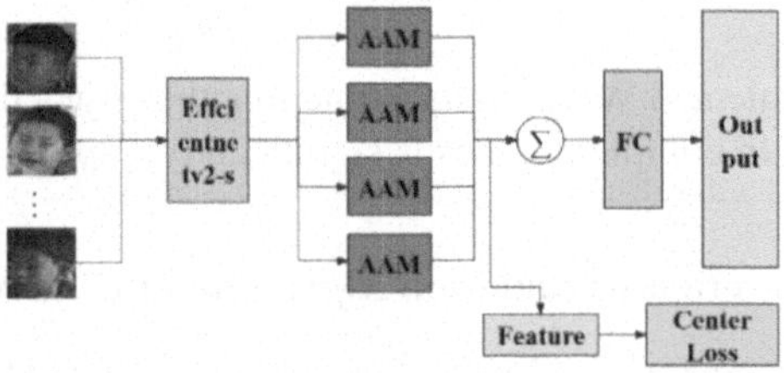

Fig. 5. FaceFeatXtractor network architecture diagram.

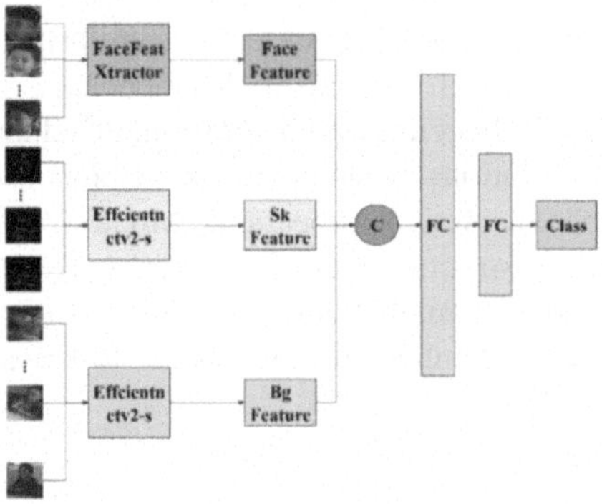

Fig. 6. MultiEmoNet network architecture diagram.

5 Experiments

5.1 Experimental Details

In order to cope with the challenges of crowded classrooms and diverse lighting conditions, this study selected the CrowdHuman dataset for model training. The CrowdHuman dataset contains approximately 25,000 images and is suitable for human detection in complex classroom environments. Each image contains an average of approximately 22.6 human instances, annotated with the head bounding box, the real human area bounding box, and the complete human bounding box (Fig. 7). The experimental environment configuration was as follows: Windows operating system, i7-12700KF CPU, 32 GB memory, RTX3080 10G GPU, CUDA11.6 computing platform, Python development language, PyCharm development IDE, and PyTorch deep learning framework.

Fig. 7. Example diagram of the CrowdHuman dataset annotation [27].

Object detection performance evaluation was primarily based on detection accuracy and speed. Accuracy metrics included Precision, Recall, Precision-Recall Curve, Average Precision (AP), and mean Average Precision (mAP). Speed metrics included Forward Propagation Elapsed Time, Frames Per Second (FPS), and Floating Point Operations (FLOPS). The binary confusion matrix (Table 1) was used to evaluate the model performance.

Table 1. Dichotomous confusion matrix.

Condition	Positive	Negative
True	TP	FN
False	FP	TN

In this study, 100 videos of primary and secondary school classrooms were collected, each lesson including both teacher perspective and student perspective, totaling 200 video files to ensure sample diversity and representativeness. The videos had a resolution of 1280 × 720 pixels and a frame rate of 25 frames per second, covering the subjects of Maths, Language, Science and English.

One frame of image was extracted every 15 s, and human body detection was performed using the YOLOv8-Improved model. A total of 33,481 images were annotated and classified into Doubt, Engagement, Fatigue, Disconnection, Pleasure, and Peace (Table 2). Fatigue and Disconnection were classified as negative emotions, Pleasure and Engagement as positive emotions, and Peace and Doubt as neutral emotions. The YOLOv8-Pose model was used to extract skeleton features. The model output the coordinates and confidence of 17 key points (Fig. 8). Key points with low confidence (such

Table 2. Number of Labeled Images in Emotion Dataset.

Emotion Classification	Number
Doubt	3857
Engagement	5987
Fatigue	4083
Disconnection	7696
Pleasure	3779
Peace	8078
Total	33,481

as knees and ankles) were removed, and the key points were drawn on a black canvas to generate a skeleton diagram (Fig. 9).

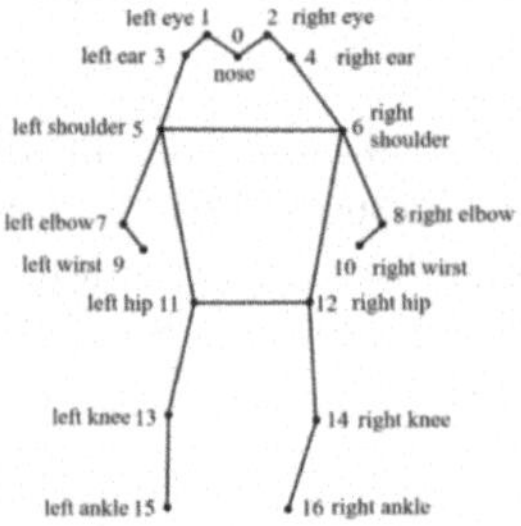

Fig. 8. Example diagram of human body key points.

To overcome difficulties in facial detection in natural classroom scenes, a "head" detector trained on the CrowdHuman dataset was used. Images were preprocessed using interpolation upsampling and mosaics data augmentation methods to improve detection accuracy, eventually generating the facial information dataset Student-Head.

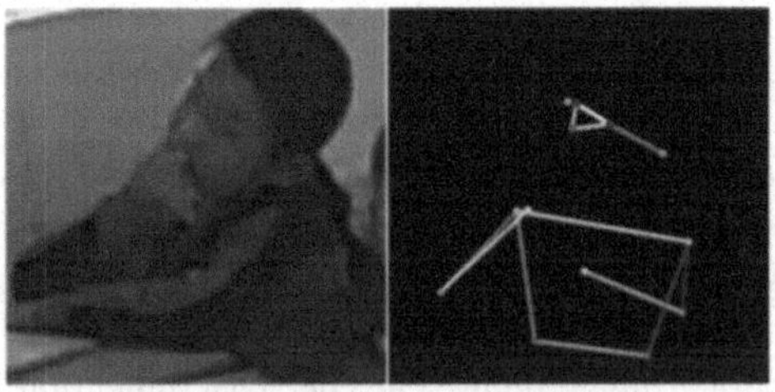

Fig. 9. Example of bone generation.

5.2 Experimental Design

The object detection experiment converted the CrowdHuman dataset into the COCO format that the YOLOv8 model could train. 15,000 training images and 4,370 validation images were prepared. All images were adjusted to 640 × 640 resolution and data augmentation was performed. The data augmentation strategies included mosaics, Blur, and Gray. Mosaic data augmentation enhanced the model's ability to identify the main objects in complex backgrounds by splicing multiple training images into a large image. Blur data augmentation used Gaussian blur to simulate motion blur in reality and improve the model's ability to handle blurred images. Grayscale data augmentation converted color images to grayscale images to help the model focus on image structure and texture. The training hyperparameters were set as follows: image size 640 × 640, load pre-trained weights, 100 training rounds, optimizer SGD, initial learning rate 0.01, final learning rate 0.001, SGD momentum 0.937, momentum decay 0.0005, and warm-up round 3.0.

The Student-Full, Student-Head, and Student-Sk datasets were used in the classroom multi-channel emotion recognition experiment as inputs for background information (Bg), face information (Face), and skeleton information (Sk), respectively. Each set of input data contained these files with the same name. If there was no corresponding information picture, a black picture was used instead. Image preprocessing only included standardization and random flipping. The training and validation data were divided into 8:2 ratios. The Adam optimizer was used during training, and the initial learning rate was set to 0.001, which was halved every three rounds, and a total of 200 Epochs were trained. The image format was PNG, 24-bit color depth, size 443 × 443 or 616 × 347, resolution 220 × 220, and progressive scanning.

5.3 Experimental Results

SSD, Faster-RCNN (Resnet+FPN), YOLOv7x, YOLOv8x, and YOLOv8x-Improved were evaluated on the CrowdHuman dataset. YOLOv8x-Improved performed best in accuracy (0.853), recall (0.772), mAP50 (0.862), and mAP50-95 (0.611), and the model size was reduced to 109M, GFLOPS was reduced to 232.2, and the processing speed reached 48 frames per second (Fig. 10).

On the student classroom dataset, YOLOv8x-Improved also performs well, with Precision of 0.802, Recall of 0.852, mAP50 of 0.863, and mAP50-95 of 0.572, which is better than YOLOv8x and YOLOv8x-COCO (Table 3).

The MultiEmoNet model performs best after combining Bg, Face and Sk information. The FaceFeatXtractor network has an accuracy of 0.721 when using face information alone, which is better than models such as Resnet50. When using skeleton information, EfficientNetv2-S has an accuracy of 0.601, which is better than traditional machine learning methods. When using background information, EfficientNetv2-S accuracy is 0.813. The MultiEmoNet model that combines multi-channel information has an accuracy of 0.841 in the Face+Sk combination and 0.914 in the Face+Sk+Bg combination, which significantly improves the accuracy of emotion recognition (Fig. 11). The classroom emotion prediction results are as shown in Fig. 12.

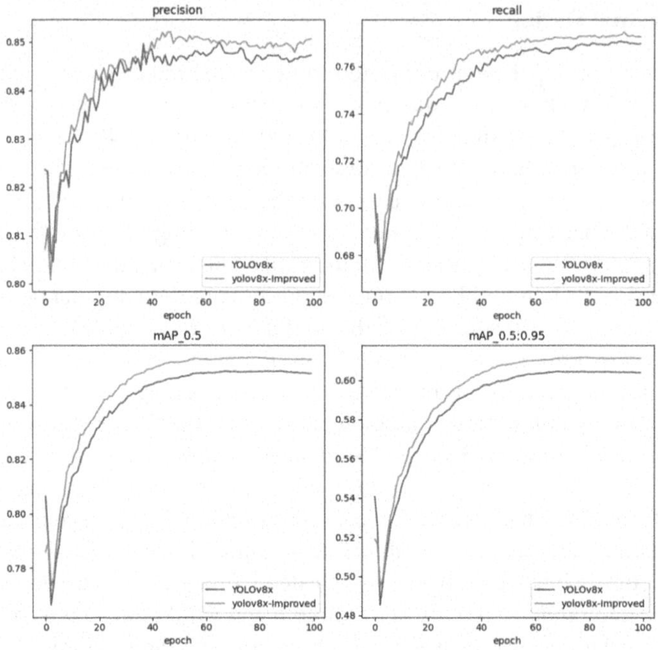

Fig. 10. Comparison of YOLOv8x-Improved and YOLOv8x training results.

Table 3. Comparison of Results on the Custom Student Classroom Object Detection Dataset.

Algorithm	Precision	Recall	mAP50	mAP50-95
YOLOv8x-COCO	0.744	0.698	0.77	0.479
YOLOv8x	0.781	0.831	0.851	0.566
YOLOv8x-Improved	**0.802**	**0.852**	**0.863**	**0.572**

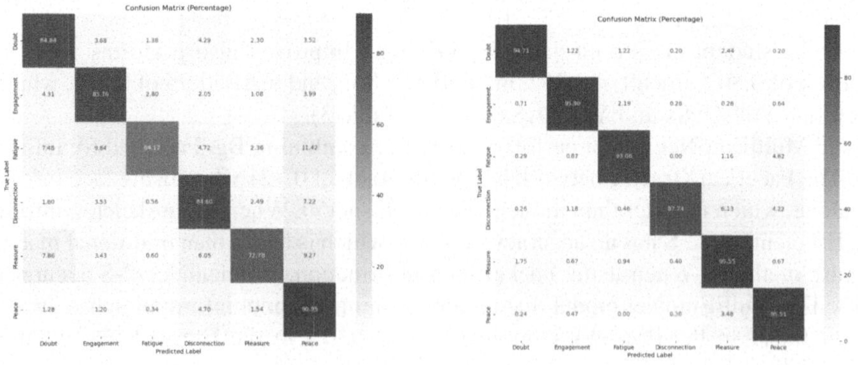

MultiEmoNet(Face+Sk) MultiEmoNet(Face+Sk+Bg)

Fig. 11. Confusion matrix.

Fig. 12. Emotion prediction.

6 Summary and Outlook

This study of student emotion recognition in primary and secondary school classrooms uses the YOLOv8-Imporved model and the MultiEmoNet model, accurate detection and emotion recognition of students in classrooms with a large number of people and severe occlusion are achieved [28]. The main contributions include: introducing the WIoUv3 loss function and the self-developed MCC module to improve the YOLOv8 model and enhance detection accuracy and speed; constructing a multi-channel classroom emotion dataset, combining facial expressions and skeleton information to provide a multi-dimensional information source; proposing the MultiEmoNet model, which improves the accuracy and generalization ability of emotion recognition by integrating facial, skeleton, and background information.

Despite its accuracy, the complexity and computational requirements of the MultiEmoNet model are high which affects the real-time processing performance, and the system has not been widely deployed. Future research will focus on model optimization and lightweight, efficient cloud deployment and cross-scenario applications to improve real-time performance and applicability, meet the needs of a wide range of users, and promote the development of emotion recognition technology in the field of education.

Acknowledgments. This work is supported by Open Fund of Hubei Key Laboratory of Digital Education (Grant No. F2024E01), the 2024 "AI + Examination and Evaluation" Special Teaching Reform Project of Central China Normal University (CCNU24GJ17), and the Fundamental Research Funds for the Central Universities under Grant (CCNU22JC011).

References

1. Andersen, P.A., Guerrero, L.K.: Handbook of Communication and Emotion. Elsevier, Amsterdam (1996)
2. Lasri, I., Solh, A.R., Mourad, E.B.: Facial emotion recognition of students using convolutional neural network. In: 2019 Third International Conference on Intelligent Computing in Data Sciences (ICDS), pp. 1–6. IEEE (2019)

3. Wani, T.M., Gunawan, T.S., Qadri, S.A.A., Kartiwi, M., Ambikairajah, E.: A comprehensive review of speech emotion recognition systems. IEEE Access **9**, 47795–47814 (2021)
4. De Gelder, B., de Borst, A.W., Watson, R.: The perception of emotion in body expressions. Wiley Interdiscip. Rev. Cogn. Sci. **6**(2), 149–158 (2015)
5. Nie, C.Y., Sun, H.X., Wang, J.: Application of chaos characteristics about physiological signals in emotion recognition based on approximate entropy. Appl. Mech. Mater. **543**, 2539–2542 (2014)
6. Hudlicka, E.: To feel or not to feel: the role of affect in human-computer interaction. Int. J. Hum. Comput. Stud. **59**(1–2), 1–32 (2003)
7. Barra, S., Hossain, S., Pero, C., Umer, S.: A facial expression recognition approach for social IoT frameworks. Big Data Res. **30**, 100353 (2022)
8. Karg, M., Samadani, A.A., Gorbet, R., Kühnlenz, K., Hoey, J., Kulić, D.: Body movements for affective expression: a survey of automatic recognition and generation. IEEE Trans. Affect. Comput. **4**(4), 341–359 (2013)
9. Liu, Z.T., Han, M.T., Wu, B.H., Rehman, A.: Speech emotion recognition based on convolutional neural network with attention-based bidirectional long short-term memory network and multi-task learning. Appl. Acoust. **202**, 109178 (2023)
10. Gunes, H., Piccardi, M.: Fusing face and body gesture for machine recognition of emotions. In: IEEE International Workshop on Robot and Human Interactive Communication 2005. IEEE, pp. 306–311. IEEE, ROMAN (2005)
11. Gunes, H., Shan, C., Chen, S., Tian, Y.: Bodily expression for automatic affect recognition. In: Emotion Recognition: A Pattern Analysis Approach, pp. 343–377 (2015)
12. Ekman, P., Friesen, W.V.: Constants across cultures in the face and emotion. J. Pers. Soc. Psychol. **17**(2), 124 (1971)
13. Izard, C.E.: Basic emotions, natural kinds, emotion schemas, and a new paradigm. Perspect. Psychol. Sci. **2**(3), 260–280 (2007)
14. Yuan, Q.: Research on classroom emotion recognition algorithm based on visual emotion classification. Comput. Intell. Neurosci. **2022**(1), 6453499 (2022)
15. Ayvaz, U., Gürüler, H., Devrim, M.O.: Use of facial emotion recognition in e-learning systems (2017)
16. Kosti, R., Alvarez, J.M., Recasens, A., Lapedriza, A.: Context based emotion recognition using emotic dataset. IEEE Trans. Pattern Anal. Mach. Intell. **42**(11), 2755–2766 (2019)
17. Mehrabian, A., Ferris, S.R.: Inference of attitudes from nonverbal communication in two channels. J. Consult. Psychol. **31**(3), 248 (1967)
18. Wallbott, H.G.: Bodily expression of emotion. Eur. J. Soc. Psychol. **28**(6), 879–896 (1998)
19. Coulson, M.: Attributing emotion to static body postures: recognition accuracy, confusions, and viewpoint dependence. J. Nonverbal Behav. **28**, 117–139 (2004)
20. Birdwhistell, R.L.: Introduction to kinesics:(An annotation system for analysis of body motion and gesture). Department of State, Foreign Service Institute (1952)
21. Kosti, R., Alvarez, J.M., Recasens, A., Lapedriza, A.: Emotion recognition in context. In: IEEE Conference on Computer Vision and Pattern Recognition 2017, pp. 1667–1675. IEEE (2017)
22. Lee, J., Kim, S., Kim, S., Park, J., Sohn, K.: Context-aware emotion recognition networks. In: Proceedings of the IEEE/CVF International Conference on Computer Vision 2019, pp. 10143–10152. IEEE (2019)
23. Mehrabian, A.: Framework for a comprehensive description and measurement of emotional states. Genet. Soc. Gen. Psychol. Monogr. **121**(3), 339–361 (1995)
24. Kołakowska, A., Landowska, A., Szwoch, M., Szwoch, W., Wróbel, M.R.: Emotion recognition and its applications. In: Hippe, Z., Kulikowski, J., Mroczek, T., Wtorek, J. (eds.) Human-Computer Systems Interaction: Backgrounds and Applications 3. AISC, vol. 300, pp. 51–62. Springer, Cham (2014). https://doi.org/10.1007/978-3-319-08491-6_5

25. Chelali, F.Z., Djeradi, A.: Face recognition using MLP and RBF neural network with Gabor and discrete wavelet transform characterization: a comparative study. Math. Probl. Eng. **2015**(1), 523603 (2015)
26. Li, L., Yao, D.: Emotion recognition in complex classroom scenes based on improved convolutional block attention module algorithm. IEEE Access **11**, 143050–143059 (2023)
27. Shao, S., et al.: Crowdhuman: A benchmark for detecting human in a crowd (2018)
28. Maji, D., Nagori, S., Mathew, M., Poddar, D.: Yolo-pose: enhancing yolo for multi person pose estimation using object keypoint similarity loss. In: Proceedings of the IEEE/CVF Conference on Computer Vision and Pattern Recognition 2022, pp. 2637–2646. IEEE (2022)

The Influence of GAI on the Development of Students' Higher-Order Thinking Ability: Meta-analysis Based on Relevant Empirical Studies at Home and Abroad Since 2022

Xiaonan Li, Zeyu Wu, Juan Zhang, Xiaoyu Bai, and Xiaojuan Li(✉)

Department of Education, Henan Normal University, Xinxiang, Henan, China
lixiaojuan2019@htu.edu.cn

Abstract. The rapid advancement of generative artificial intelligence (GAI) technology has had a significant impact on the field of education. In the 21st century, education places greater emphasis on fostering students' capacities and core literacies. This research aims to examine the impact of GAI applications in teaching on students' higher-order thinking ability. This study utilizes a meta-analysis approach to evaluate relevant domestic and international studies. Findings indicate that GAI can significantly enhance students' higher-order thinking ability, particularly within higher education settings. This improvement encompasses critical thinking, innovative thinking, and problem-solving ability. Moreover, the research reveals that various approaches to integrating GAI into education yield differing impacts on students' higher-order thinking ability. Through a thorough analysis and discussion of these findings, this paper offers insights and recommendations for fostering effective Student-GAI-Teacher collaboration in a man-machine learning environment.

Keywords: Generative artificial intelligence · All-round development · Higher-order thinking ability · Teaching · Enabling mechanism

1 Introduction

In November 2022, ChatGPT, developed by OpenAI, ushered in a new era, with the emergence of generative artificial intelligence (GAI) providing new opportunities for education reform. GAI drives the digital transformation of education by providing an innovative, flexible and personalized education environment, providing new solutions for achieving high quality education and personalized teaching.

With the advancement of science and technology, education focuses on the improvement of students' higher-order thinking ability. Higher-order thinking ability refers to the ability of an individual to internalize knowledge or skills and creatively apply them to real situations to solve complex problems. The introduction of 21st century skills has changed the ultimate goal of education from knowledge accumulation to ability cultivation and prompted extensive discussions regarding the applications of GAI in education.

Q. Liu et al. (Eds.): EITT 2024, CCIS 2600, pp. 46–58, 2026.
https://doi.org/10.1007/978-981-95-1525-7_4

Therefore, this study carried out a meta-analysis of relevant empirical studies (published since 2022) on GAI in domestic and international education, exploring how GAI is used to promote the development of students' higher-order thinking ability and what aspects of students' higher-order thinking ability GAI has positively impacted. The differences in GAI applications, GAI functions, and GAI intervention mechanisms across educational levels were analyzed. This paper explores how GAI effectively promotes the development of students' higher-order thinking ability, providing suggestions for practical educational applications and further development of GAI.

2 Concept Definition

2.1 Generative Artificial Intelligence

GAI often uses artificial intelligence generated content (AIGC) as a term in research. Its conceptual definition is based on a large language model (LLM) and massive database, through deep learning, feedback loops, and other technologies, to understand the input instructions, generate text, images, videos, audio, and other multi-modal feedback content, to achieve human-computer collaborative interaction, with strong situational understanding and content generation capabilities. In order to make an effective judgment on GAI in this research, an operational definition of GAI was made: GAI uses its four core competence advantages of inspiring content generation, dialogue situation understanding, sequence task execution, and program language analysis [1] to provide immediate and personalized solutions to the problems raised by students. For example, in 2024, the GPT-4o language model was used by Khan Academy to power its AI virtual assistant, Khanmigo, to help children with their math homework. GPT-4o can understand human tone and gradually guide children using Khan Academy to master problem-solving abilities.

2.2 Higher-Order Thinking Ability

Higher-order thinking ability refers to the ability of an individual to conduct in-depth analysis, synthesis, evaluation and creation in the cognitive process, which is an embodiment of core literacy. Bloom summarized the high level of cognition as "analysis, synthesis and evaluation" in the Classification of Educational Objectives, and Anderson revised it and improved it to "analysis, evaluation and creation," [2] in which he emphasized the importance of innovative thinking and ability, and more emphasis was placed on the internalization, application and innovation of knowledge instead of the original memory and understanding of knowledge. The 21st century skills required for students [3] proposed by the P21 Organization in the United States include critical thinking, communication, collaboration, and creativity (4C's for short), which are comprehensive abilities in the face of complex problems, and emphasize the internalization, application and innovation of knowledge. In addition, there are other factors in the process of problem solving (Fig. 1), such as individual self-regulation, cognition, meta-cognition, cooperation, and communication, that enhance the development of higher-order thinking ability. Based on the 4C model of core literacy in the 21st century [3], combined with the requirements of students' core literacy in the new era, this study divides higher-order thinking ability into critical thinking, innovative thinking, and problem-solving ability, and incorporates them as variables in the subsequent meta-analysis.

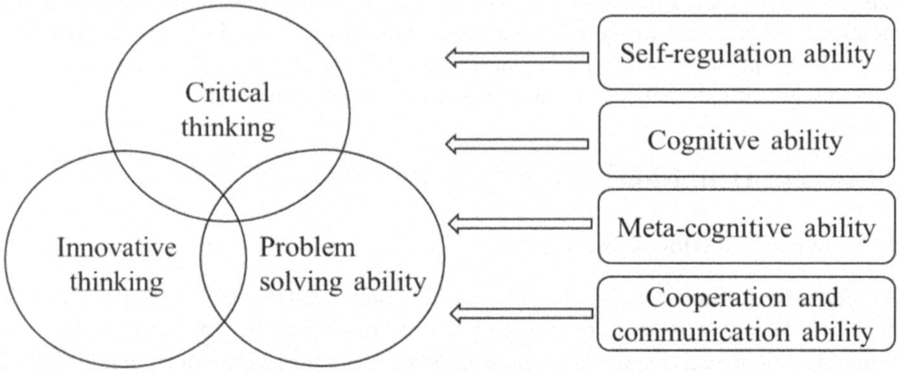

Fig. 1. The components and influencing factors of higher-order thinking ability.

2.3 The Technical Mechanism and Applied Logic of GAI Improving Students' Higher-Order Thinking Ability

Technological Mechanisms of GAI Enabling Higher-Order Thinking Ability. The ontology model of GAI is the core of its promotion of higher-order thinking ability, and its enabling of higher-order thinking ability is realized through a multi-dimensional and multi-level technical basis. First, GAI has a massive database, which provides a rich resource library for students of different learning stages and needs. Second, by monitoring and evaluating students' learning progress in real time, GAI is able to provide immediate feedback and heuristic guidance, which promotes students' critical thinking and problem-solving abilities.

Application Logic of GAI Enabling Higher-Order Thinking Ability. The technical advantages of GAI grant it unlimited potential in teaching application, and its role in the education process has attracted the attention of many scholars. Although GAI is a relatively new element in education, the unexplored relationship between teachers, students, and machines presents a new opportunity to promote the development of education. As proposed by Yang [4], the use of GAI results in three teacher-student-machine empowerment mechanisms: GAI directly empowers students, GAI-supported teachers empower students, and teachers guide students to use GAI.

Categorization of GAI Applications. 2023 is the first year of GAI. Many kinds of GAI have developed rapidly throughout the world and are being widely used in education, industry, and other fields to enhance human learning and work efficiency. The common Chinese and international forms of GAI are shown in Table 1.

Table 1. Common GAIs in China and abroad

Source	GAI Name	Large scale model	Major specialty
China	The spark language companion	Spark cognition large model	Language learning
	A word written in the mind	Wenxin large model	Chinese language processing
	Kimi AI assistant	Self-developed model	Multilingual dialogue and document processing
	Zhipu Qingyan	ChatGLM2	Multimodal processing
	Doubao	Skylark model	Text processing
	Tongyi Wanxiang	Tongyi	Image generation
	Tongyi Qianwen	Tongyi	Text processing
	IFlyZhiwen	Iflyspark cognition large model	Generate PPT outline and make PPT document
	AIPPT	Iflyspark cognition large model	Text processing and PPT generation
	IFlyspark	Iflyspark cognition large model	Text generation and language learning
	Iflytek	Iflyspark cognition large model	Speech recognition
Abroad	Alexa	Multi-parameter Alexa Teacher Model	Voice interaction and information query
	Khanmigo	ChatGPT-4o	Personalized tutoring
	Siri	LLM model and multi-channel acoustic model	Voice recognition and user command execution
	Sora	Diffusion model	Vincennes video/diagram
	ChatGPT	Unsupervised Transformer	text understanding and generation
	Midjourney	Self-developed model	Image generation

3 Research Design

3.1 Literature Retrieval

Since the launch of ChatGPT by OpenAI in 2022, the application of generative AI in education has attracted significant attention. In order to analyze this phenomenon systematically, the time frame of literature search was limited from 2022 to the present. This study used Chinese keywords (Generative Artificial Intelligence OR ChatGPT OR GenAI OR AIGC) AND (Higher-Order Thinking OR Critical Thinking OR Innovative Thinking OR Problem Solving Thinking) AND (Education OR Teaching) and English keywords (Generative Artificial Intelligence OR ChatGPT OR GenAI OR AIGC) AND (Higher-Order Thinking OR Critical Thinking OR Innovative Thinking OR Problem Solving Thinking) AND (Education OR Teaching) to search. The retrieval data came from Chinese databases Zhiwei, Vipu and Wanfang, and English databases Web of Science, EI, Springer, and Elsevier ScienceDirect.

After a preliminary search, a total of 2,555 Chinese and English literatures were obtained. The research team screened these documents to ensure that the final documents included in the analysis were of high quality and relevant. In addition, a web tool named Connected Papers was used to snowball search the literature with high topic relevance.

3.2 Literature Selection

In order to ensure the accuracy and reliability of the literature analysis, the following literature selection criteria were formulated: (a) Only the empirical research of GAI in teaching applications was included, and non-empirical research was excluded. (b) The content was highly relevant to the research topics "GAI" and "higher-order thinking ability," focusing on the study of the effect of GAI on higher-order thinking ability and excluding irrelevant topics. (c) Literature containing controlled trials was given priority, and non-controlled trials were excluded. (d) The literature focused on the analysis of the effect of GAI on the training of higher-order thinking ability, excluded the statement of no effect, and included at least one of the following: "critical thinking," "innovative thinking," and "problem solving ability." (e) The mean, standard deviation, and experimental sample size of the influence of GAI on higher-order thinking ability were given, and literature with missing data were excluded. After two rounds of selection, 15 valid samples were included, including five Chinese and 10 English papers.

3.3 Literature Coding

After the literature retrieval and screening, in order to facilitate statistical analysis, this research coded the characteristic values of the original papers involved in the literature review, including the author, date, sample capacity, subject, thinking types, period of study, function of GAI, and higher-order thinking enabling mechanism of GAI, in each piece of literature (Table 2). A total of 44 meta-analysis data were obtained in this study.

In Table 2, the sample capacity, thinking types and enabling mechanism are alphabetically coded. In the sample capacity column, C represents the sample capacity of the control group and T represents the sample capacity of the experimental group. In the column of thinking types, CT represents critical thinking, IT represents innovative thinking, and QT represents problem solving ability. In the enabling mechanism column, DS represents GAI directly empowering students, AS represents GAI assistant teachers empowering students, and IS represents teachers guiding students to use GAI.

3.4 Effect Size Calculation

The sample size for this study was small, so Hedges' g was calculated to measure the effect size in the experiment. Comprehensive Meta-Analysis3.0 software was used to calculate the effect size.

Table 2. A partial table of literature codes

Serial #	Author	Sample capacity	Subject	Thinking Types	Period of Study	Function of GAI	Enabling Mechanism
1	Li Haifeng, Wang Wei	C = 42 T = 40	Instructional design	CT IT QT	College	Support man-machine collaborative deep inquiry learning	AS
2	Zheng Lanqin et al.	C = 15 T = 15	Artificial intelligence technology	CT	College	Support conversational feedback and online learning	IS
3	Li Haifeng, Wang Wei	C = 35 T = 33	Distance pedagogy	CT IT QT	College	Provide group debate and man-machine collaborative intelligence study companion services	AS
4	Luo Heng et al.	C = 59 T = 58	Mathematics	CT	Junior high school	Teacher comments are generated	AS
5	Tzu-Yu Tai et al.	C = 29 T = 28	Oral English	QT	Primary school	Text-to-speech and automatic speech recognition (ASR) technology are available	DS
6	Kai Guo et al.	C = 60 T = 64	English writing	CT QT	College	Eva-based AI chatbots assist students in generating peer feedback	DS

4 Research Results

4.1 Publication Bias Test

The funnel plot of effect size distribution (Fig. 2) shows that the points on the funnel plot are dispersed symmetrically around the combined effect size of 0.636, which indicates that there is no publication bias. The Begg test results show $p = 0.19545$, and Egger's test results show $p = 0.33436 > 0.05$, which further indicates that there is no publication bias. Finally, a Fail-safe Number (Nfs) test was performed, and the Nfs value = 3874.0 $> 5 \times 44 + 10$ was obtained, indicating again that the effect size of unpublished studies had little impact on the total effect size of published studies. In conclusion, there is no publication bias in this study, and the samples are valid and reliable.

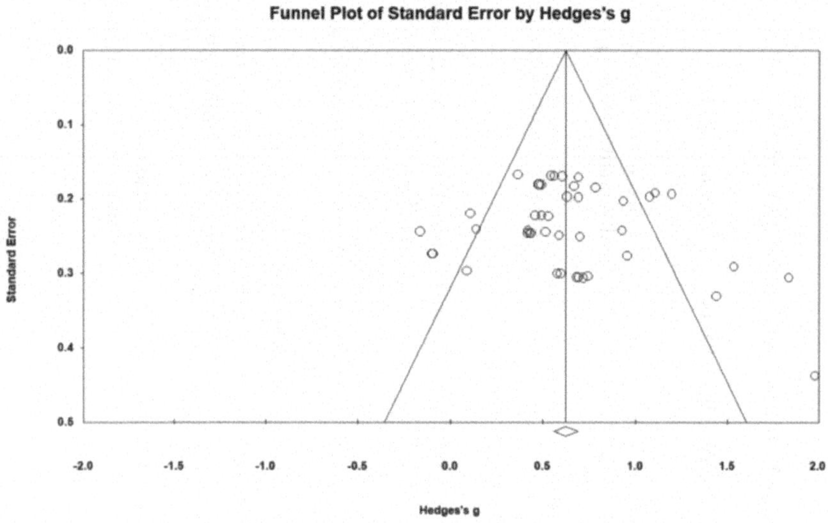

Fig. 2. Funnel plot.

4.2 Heterogeneity Test

The results of sample heterogeneity test (Table 3) show that $Q = 114.336$, $p = 0.000$, $I2 = 62.392\%$ (between 60% and 75%), indicating that there is a large heterogeneity among research results. Therefore, the random effects model was used to calculate the combined effect value.

4.3 The Overall Effect of GAI on Students' Higher-Order Thinking Ability

According to the random effects model (Table 3), the combined effect value of GAI is 0.636, attaining statistical significance ($p < 0.001$) and indicating that GAI has a positive impact on the development of students' higher-order thinking ability. According to the effect size standard proposed by Cohen [5], GAI has a moderate positive impact on the development of students' higher-order thinking ability.

Table 3. The overall effect of GAI on students' higher-order thinking ability

Model	N	Hedges' g	95%CI		2-Tail		Heterogeneity			
			Upper	Lower	Z	P	Q	df	P	I2
Fixed	44	0.625	0.691	0.559	18.581	0.000	114.336	43	0.000	62.392
Rand-om	44	0.636	0.746	0.526	11.323	0.000				

4.4 The Influence of GAI on Different Aspects of Higher-Order Thinking Ability

In order to further explore the impact of GAI on students' higher-order thinking ability, our research was conducted according to three aspects of higher-order thinking ability (Table 4). The combined effect value of critical thinking, innovative thinking, and problem-solving ability was 0.785, 0.493, and 0.503, respectively, all of which are between 0.2 and 0.8. Statistically significant p-values ($p < 0.001$) were attained, indicating that GAI had moderate positive effects on students' critical thinking, innovative thinking, and problem-solving ability.

Table 4. The influence of GAI on different aspects of higher-order thinking ability

Aspect	N	Hedges' g	95% CI		2-Tail	
			Upper	Lower	Z	P
Critical thinking	22	0.785	0.958	0.611	8.850	0.000
Innovative thinking	13	0.493	0.599	0.387	9.123	0.000
Problem solving ability	9	0.503	0.672	0.334	8.495	0.000

4.5 The Effect of GAI on Different Students

In order to further explore the difference of the influence of GAI on the higher-order thinking ability of students of different age groups, the samples were divided into three groups: college, middle school, and primary school (Table 5). Based on our literature retrieval, it is clear that most studies focus on the impact of GAI on college students' ability, and the combined effect size is 0.617 ($0.2 < ES < 0.8$), indicating that GAI has a moderately positive effect on higher-order thinking ability in college, and it is statistically significance ($p < 0.001$). At the same time, the effect sizes of middle school and primary school are 1.077 ($ES > 0.8$) and 0.956 ($ES > 0.8$), indicating that GAI has a significant degree of positive impact on students' higher-order thinking ability. However, due to the small number of application studies of GAI in primary and secondary schools and the single sample size, the actual situation may deviate from the data performance.

Table 5. The effect of GAI applications across different periods of study.

Period of study	N	Hedges' g	95%CI		2-Tail	
			Upper	Lower	z	p
College	42	0.617	0.729	0.505	10.821	0.000
Middle school	1	1.077	1.463	0.692	5.475	0.000
Primary school	1	0.956	1.497	0.415	3.461	0.001

4.6 The Effect of GAI Empowerment Mechanisms

The empowerment mechanism of GAI is divided into three dimensions: GAI directly empowers students, GAI assists the teacher in empowering students, and the teacher guides the students to use GAI (Table 6). The combined effect size of GAI directly empowering students is 0.853 (ES > 0.8), indicating that GAI directly empowering students has a large positive impact. The combined effect sizes of GAI assisting teachers in empowering students and teachers in guiding students to use GAI are 0.583 and 0.641 (both 0.2 < ES < 0.8), respectively, and attains statistical significance ($p < 0.001$), indicating that GAI-assisted teacher empowerment of students and teacher-directed student use of GAI has moderate positive impact.

Overall, it seems that GAI can promote the development of students' higher-order thinking ability under different mechanisms of action. Among them, the direct empowerment of students by GAI has the most significant effect on their higher-order thinking ability, while the indirect empowerment of students by GAI-assisted teachers has a relatively weak effect. This observation may be related to ineffective collaboration with students, the short intervention period, and students' weak subjective motivation. Considering the objective factors, this observation may be related to the small sample size and other factors.

Table 6. The effect of GAI empowerment mechanisms.

Enabling mechanism	N	Hedges' g	95% CI		2-Tail	
			Upper	Lower	z	p
GAI directly empowers students	9	0.853	1.118	0.588	6.309	0.000
GAI assisted teachers to empower students	24	0.583	0.671	0.495	13.031	0.000
Teachers instruct students in the use of GAI	11	0.641	0.916	0.366	4.566	0.000

5 Research Results and Discussion

Through a meta-analysis of 15 domestic and foreign empirical research papers on GAI and students' higher-order thinking ability published since 2022, the research conclusions are as follows:

5.1 The Effect of GAI on Different Aspects of Higher-Order Thinking Ability is Different

The analysis revealed that GAI has positive effects on critical thinking, problem solving ability, and innovative thinking. Among them, critical thinking is most influenced by GAI. The main reasons for this include students needing to engage in in-depth discussions [6, 7] on the content generated by GAI, make value-based selection judgments [8–10], and contemplate questioning strategies [11]. However, the enhancement of problem solving and innovative thinking is relatively smaller. This could be due to the insufficient connection between learning tasks and real-life situations, the lack of diversity in learning materials provided by GAI [12], and students' insufficient questioning abilities [13, 14], as well as their dependency on GAI [15], which may limit the improvement of students' problem solving and innovative thinking abilities. Concurrently, some research has explored how to avoid the "knowledge feeding" phenomenon through human-computer argumentative dialogue learning [16, 17], thereby enhancing students' higher-order thinking skills such as critical thinking and problem-solving. This approach offers research insights for GAI to facilitate the achievement of heuristic education.

5.2 Differences in the Breadth of GAI Applications at Different Stages of Education

The application of GAI in higher education has achieved remarkable results, especially in disciplines closely connected with technology, such as artificial intelligence, instructional design, and programming and educational technology. In these disciplines, teachers and students can use GAI tools to enrich the learning experience [18, 19] and improve teaching efficiency. However, the application of GAI in other disciplines of higher education still requires further research and practice to explore how to effectively integrate GAI in a wider range of subject areas. In addition, the lack of primary and secondary school resources and research ability, technical limitations, and the dependence of primary and secondary school learners on teacher guidance [20] limit the promotion of GAI in the basic education stage. However, the data in this study show that GAI has a considerable effect in primary and middle school teaching, so it is particularly important to continue studying the application and effective implementation of GAI in primary and middle school education.

5.3 There Are Differences in the Influence of GAI on Students' Higher-Order Thinking Ability Under Different Empowerment Paths

The results show that all three kinds of GAI empowerment can promote the development of students' higher-order thinking ability, but the effect of GAI directly empowering students is more significant than GAI-supported teachers empowering students and teachers guiding students to use GAI. Students can directly use GAI in their daily life, with high degrees of freedom and curiosity [21], which may make the direct empowerment mechanism more effective. However, the effect of GAI-supported teachers empowering student is relatively small, which may be because most of the selected experiments have relatively clear teaching tasks, which require both knowledge and skill learning, so the effect of improving students' advanced thinking ability is slightly weaker than that of directly empowering students. In addition, there may be reasons such as insufficient integration between teachers and GAI or insufficient initiative of students. However, the reason for the small effect of teachers' guiding students to use GAI may be due to the lack of cooperation between teachers and students, which leads to teachers' inability to correctly grasp students' situation and put forward targeted guidance, teachers' lack of proficiency in the use of GAI, and students' resistance to teachers' guidance.

6 Summary and Prospect

The findings of this study demonstrate that the application of GAI in teaching can enhance students' higher-order thinking ability. The study conducted research separately for different grade levels and empowerment modes, and based on the results suggest that the role of GAI in the "teacher-student-machine" teaching state should be comprehensively considered in the research process, so that the empowerment modes of GAI empowering teachers, GAI empowering students, and teachers using GAI to guide students can form a stable synergy. Based on this, the following suggestions are made for students, GAI, and teachers.

6.1 Improve Students' Cognitive Drive and Exert Their Agency

When students first encounter GAI, they are often attracted by its novelty and will actively try to explore the various functions of GAI. Over time, however, curiosity may fade, affecting their continued use of GAI. This phenomenon may reduce the potential of GAI as a teaching tool to develop students' higher-order thinking ability. Transforming transient curiosity into a lasting and stable cognitive drive that spontaneously generates a steady stream of power is the key to changing the teaching mode from "spoon-feeding" to "student-centered."

6.2 Deeply Explore the Advantages of GAI Heuristics Teaching to Promote the Development of Students' Higher-Order Thinking Ability

The advantage of GAI lies in its strong text understanding and natural language generation ability, which provides the technical basis for personalized guided feedback.

Its thought chain technology can realize the formation of problem chain, gradually guide students to think deeply, and effectively stimulate students' critical thinking and problem-solving ability. In addition, GAI provides personalized advice based on learning situations, stimulating students' divergent thinking and energizing innovative thinking. Moreover, in the process of case search, we find a large number of effective applications of GAI conversational robots in the field of teaching. We should think about how to develop and popularize GAI teaching aids based on heuristic teaching method to enable students to develop higher-order thinking.

6.3 Improve Teachers' Digital Literacy and Build a "Double Teacher" Teaching Model Under GAI Environment

The emergence of GAI has brought revolutionary changes to the field of education, and its innovation in teaching mode, method, and environment has brought unprecedented breakthroughs to traditional education technology. The "generative" advantage of knowledge is difficult for teachers to surpass, but GAI cannot replace the experience and emotional capabilities of teachers. Therefore, teachers should recognize their important role as facilitators, continuously improve digital literacy, and form a human-machine coordination "double teacher" teaching model with GAI, so that teachers' classroom guidance and personalized assistance provided by GAI can be combined to jointly promote the all-round development of students. In addition, GAI's multi-modal content generation technology provides teachers with rich teaching resources and innovative teaching ideas. Teachers should make full use of these technologies, adopt appropriate teaching strategies, shift their focus from "classroom" to "problem," return the classroom to students, and stimulate their subjective initiative. Through continuous learning and practical application, students will gradually improve their problem-solving ability and innovation thinking.

References

1. Lu, Y., Yu, J.L., Chen, P.H., Li, M.Y.: Educational application and prospect of generative artificial intelligence: taking ChatGPT system as an example. Chinese J. Distance Educ. **43**(4), 24–31+51 (2023)
2. Ma, S.F., Yang, X.D.: A conceptual framework of higher-order thinking. J. East China Normal Univ. Educ. Sci. **40**(11), 58–68 (2022)
3. Deng, L.: A Study of the 21st Century Skills Education Reform in the United States. Ph.D. dissertation, East China Normal University, Shanghai, China (2018)
4. Yang, Z.K., Wang, J., Wu, D., Chen, X.: Exploring the impact of ChatGPT/AIGC on education and strategies for response. J. East China Normal Univ. Educ. Sci. **41**(7), 26–35 (2023)
5. Zheng, H.M., Wen, Z.L., Wu, Y.: The appropriate effect sizes and their calculations in psychological research. Adv. Psychol. Sci. **19**(12), 1868–1878 (2011)
6. Li, H.F., Wang, W., Li, G., Wang, Y.: Intelligent midwifery teaching method: a case study of 'Intelligent Socratic Dialogue Robot' teaching practice. Open Educ. Res. **30**(2), 89–99 (2024)
7. Guo, K., Pan, M., Li, Y., Lai, C.: Effects of an AI-supported approach to peer feedback on university EFL students' feedback quality and writing ability. Internet Higher Educ. **63**, art. no. 100962 (2024)

8. Essien, A., Bukoye, O.T., O'Dea, X., Kremantzis, M.: The influence of AI text generators on critical thinking skills in UK business schools. Stud. Higher Educ. **49**(5), 865–882 (2024)
9. Essel, H.B., Vlachopoulos, D., Essuman, A.B., Amankwa, J.O.: ChatGPT effects on cognitive skills of undergraduate students: receiving instant responses from AI-based conversational large language models (LLMs). Comput. Educ. Artif. Intell. **6**, art. no. 100198 (2024)
10. Zheng, L.Q., Gao, L., Huang, Z.C.: Can chatbots based on generative artificial intelligence facilitate online collaborative learning performance? E-educ. Res. **45**(3), 70–76+84 (2024)
11. Yilmaz, R., Karaoglan Yilmaz, F.G.: The effect of generative artificial intelligence (AI)-based tool use on students' computational thinking skills, programming self-efficacy and motivation. Comput. Educ. Artif. Intell. **4**, art. no. 100147 (2023)
12. Al-Abdullatif, M.A., Al-Dokhny, A.A., Drwish, A.M.: Implementing the Bashayer chatbot in Saudi higher education: measuring the influence on students' motivation and learning strategies. Front. Psychol. **14**, art. no. 1129070 (2023)
13. Li, H.F., Wang, W.: Human-computer collaborative deep exploratory teaching mode: taking the human-computer collaborative exploratory learning system developed based on ChatGPT and QQ as an example. Open Educ. Res. **29**(6), 69–81 (2023)
14. Marek, U., et al.: ChatGPT improves creative problem solving performance in university students: an experimental study. Comput. Educ. **215**, 105031 (2024)
15. Liao, J., Zhong, L., Zhe, L., Xu, H., Liu, M., Xie, T.: Scaffolding computational thinking with ChatGPT. IEEE Trans. Learn. Technol. **17**, 1668–1682 (2024)
16. Li, H.F., Wang, W.: Human-machine arguing inquiry method: an exploration of a model for cultivating students' higher-order thinking ability supported by an argumentative intelligent chatbot. E-educ. Res. **45**(3), 106–112+128 (2024)
17. Li, H.F., Li, G.X., Tang, Y.F.: The model of human-chatbot collaborative research study supported by group Al multidimensional awareness chatbot—taking the development and teaching of 'Confucius' Big Language Model Al Chatbot as an example. Modern Distance Educ., 1–14 (2024)
18. Guo, K., Zhong, Y., Li, D., Chu, S.K.W.: Effects of chatbot-assisted in-class debates on students' argumentation skills and task motivation. Comput. Educ. **203**, art. no. 104862 (2023)
19. Tzirides, A.O.(Olnancy), et al.: Combining human and artificial intelligence for enhanced AI literacy in higher education. Comput. Educ. Open **6**, art. no. 100184 (2024)
20. Tai, T.-Y., Chen, H.H.-J.: Improving elementary EFL speaking skills with generative AI chatbots: exploring individual and paired interactions. Comput. Educ. **220**, art. no. 105112 (2024)
21. Luo, H., Liao, X.F., Ru, Q.Q., et al.: Generative AI-supported teacher comments: an empirical study based on junior high school mathematics classrooms. E-educ. Res. **45**(5), 58–66 (2024)

Design of Knowledge Graph Solutions to Enhance Resilience in STEM Education

Jian Lin(✉), Musheng He, Shanyin Mai, and Bingqian Bu

Lingnan Normal University, Zhanjiang, China
linjian@lingnan.edu.cn

Abstract. Science, technology, engineering, and mathematics (STEM) education involves a complex and diverse array of interdisciplinary knowledge structures that learners may not have in a systematic interdisciplinary framework. Therefore, when confronted with different scenarios, their manifestations of creative resilience can vary. A knowledge graph that constructs entities of interdisciplinary knowledge is advantageous for bolstering the creative resilience of learners in STEM interdisciplinary education. This manuscript presents a design scheme for a STEM knowledge graph leveraging large language model (LLM) technology, with the objective of assisting learners in improving their creative resilience and facilitating the informed transformation and enhancement of STEM education.

Keywords: STEM Knowledge Graphs · Creative Resilience · STEM Education · Large Language Models

1 Introduction

Science, technology, engineering, and mathematics (STEM) education is essential for cultivating individuals with innovative thought processes and practical competencies. As digital education progresses, the integration of interdisciplinary professional education is becoming more common [1]. Learners are faced with complex knowledge structures, and it is this complexity that elevates the cognitive barrier. Thus, engaging with challenging material necessitates enhanced resilience. This manuscript delineates resilience within STEM education as "creative resilience," which refers to a learners' capacity to manage disruptions and challenges, accomplish projects, and introduce innovations throughout the process of learning and complex problem-solving.

Knowledge graphs, by visually mapping the intricate relationships among knowledge entities, offer a comprehensive and systematic interdisciplinary knowledge network to learners, which has been extensively integrated into STEM education. Utilizing knowledge graphs enables learners to uncover both explicit and tacit linkages between procedural knowledge elements, interconnect knowledge that appears discrete, and profoundly grasp the intrinsic relationships among diverse knowledge points. This process reduces the learners' cognitive barriers and subsequently boosts their creative resilience. Concurrently, the utilization of large model technology within digital education is expanding.

Q. Liu et al. (Eds.): EITT 2024, CCIS 2600, pp. 59–72, 2026.
https://doi.org/10.1007/978-981-95-1525-7_5

To augment the creative resilience of learners in STEM education further, this paper introduces a design scheme for STEM knowledge graphs that leverages large model technology, with the goal of diminishing the learners' cognitive barriers and amplifying their creative resilience. It also aims to offer novel perspectives on the confluence of knowledge graphs with STEM education research. This study anticipates delivering robust support for the innovative practices within STEM education.

2 Review of Related Research Work

2.1 Creative Resilience Integrated with Digital Resilience

The construct of resilience is derived from the field of psychology, specifically psychological resilience, and has been progressively integrated into education to assist learners in managing academic pressures and challenges. As the digital era unfolds, challenges such as technological risks, environmental intricacies, and information surges have led to the emergence of digital resilience. This concept underscores the capacity to counter technological threats, adapt to shifting environments, and restore stability within digital contexts. It includes multiple facets such as resilience, adaptability, and recuperative power, proving to be a vital element in digital education for equipping learners with the skills to navigate digital lifestyles, bolster problem-solving, and enhance self-regulation [23]. Extant research, by integrating the educational principles of psychological resilience, has delved into fortifying learners' tenacity against academic adversities. These investigations have enriched the theoretical underpinnings of digital resilience and offered significant guidance for educational practice.

STEM education, burgeoning as a leading educational paradigm, enshrines "creativity" among its principal tenets [4]. Its objectives extend beyond nurturing future technocrats to invigorating a nation's innovative prowess and competitive edge. Cultivating resilience, by elevating learners' psychological fortitude, augments their comprehensive attributes and their adaptability to intricate global scenarios. It is evident that, within the purview of digital education, the philosophies of digital resilience and STEM education are congruent. From this congruence, "creative resilience" has emerged, delineating a novel competency framework. In essence, "creative resilience" is conceptualized as a distinct dynamic capability within STEM education, enabling learners to diminish the cognitive barriers to disciplinary knowledge and, thus, persist in innovation and the generation of novel value amidst volatile and indeterminate learning milieus.

2.2 Learner Performance in the Context of Learning Challenges in STEM Education

Within the domain of STEM education, fostering an interdisciplinary knowledge integration mindset among learners has emerged as a pivotal subject of contemporary educational inquiry. The capacity for high-caliber interdisciplinary cognitive synthesis is crucial for an individual's adaptability and innovation when confronted with intricate problems and unpredictable challenges and significantly influences the overall efficacy and quality of STEM educational initiatives. Within the context of STEM instruction, engagement in hands-on activities allows for a more profound comprehension of scientific principles and technological applications. Evidence suggests that the incorporation of design thinking into information technology curricula can markedly enhance students' innovative and design thinking proficiencies [5]. Furthermore, project-based learning (PBL) facilitates a holistic approach to project education, necessitating that learners construct maker products capable of addressing real-world issues from the ground up, thereby effecting a transition from theoretical exploration to practical implementation. This process entails the amalgamation and fine-tuning of interdisciplinary knowledge, demanding robust resilience from learners to navigate an array of challenges [6].

The profundity and scope of STEM knowledge necessitate that learners possess advanced logical and abstract reasoning capabilities, presenting a formidable challenge for some individuals. When there is a disconnect between STEM content and everyday life [7], it may lead to learner disenchantment, impacting the development of creative resilience. Moreover, the capacity for interdisciplinary integration is a significant hurdle in STEM education. Such education underscores the importance of synthesizing and applying knowledge across various domains, requiring learners to integrate diverse knowledge bases and skill sets effectively. The cultivation of this integrative skill is not instantaneous and requires extensive learning and experiential accumulation. For certain learners, the interdisciplinary integration mindset may represent an insurmountable divide, and they may frequently encounter obstacles such as tenuous knowledge retention and inadequate practical skills [6]. Particularly when tackling intricate and arduous STEM projects, learners may feel overwhelmed and even experience defeat. If unaddressed in a timely manner, such feelings of frustration can evolve into impediments to learning, adversely impacting the progression of creative resilience.

2.3 Application of STEM Education Based on Large-Scale Knowledge Graphs

In recent years, knowledge graphs have established a solid foothold in the field of education [9]. As tools for presenting structured knowledge, their role in personalized learning, curriculum design, concept mapping, and educational content recommendation systems is increasingly recognized. Thanks to advancements in information storage, computational power, and the development of intelligent algorithms, knowledge engineering has rapidly evolved towards the creation of large-scale knowledge graphs. This technology graphically depicts and connects various entities in the real world, effectively simulating societal structures. Therefore, leveraging knowledge graphs to address numerous issues in education holds special exploratory value.

The construction of knowledge graphs, as a core technology, involves several key steps: the establishment of the knowledge ontology, the identification of named entities, the mining of entity relationships, and the fusion of knowledge. Knowledge ontology is the foundational framework defining the realms and classifications of knowledge, which is crucial for building knowledge graphs. Research indicates that dynamically-generated knowledge graphs can support the design of science, technology, engineering, art, and mathematics (STEAM) learning themes [46]. Combined with concept mapping and visualization techniques to graphically represent complex concepts and their interrelationships, this strategy can produce customized knowledge graphs based on different learning objectives and contexts, thereby enhancing the effectiveness and adaptability of learning. Chen et al. [10] presented a knowledge graph system called KnowEdu, which can automatically construct knowledge graphs to support personalized teaching services and adaptive learning solutions in education, including STEM education. KnowEdu offers comprehensive knowledge management and service solutions, aiding teachers and students in organizing and understanding complex knowledge systems.

As an auxiliary tool, knowledge graphs can be applied in project-based learning in STEM education, assisting students in organizing and absorbing the knowledge and skills required for projects more effectively. With the rapid development of generative artificial intelligence, large language models (LLMs), such as GPT-3.5 and GPT-4, are causing a new surge in the field of intelligent education due to their superior performance [11]. Knowledge graphs, recognized for their strong interpretability as structured knowledge representation methods, have been widely used in various educational scenarios, such as lesson plan design, cognitive training, and intelligent question and answer (Q&A). Although LLMs possess powerful generalization capabilities, such as extraction, summarization, reasoning, and generation, they often lack factual knowledge and interpretability [9], which is undesirable in STEM education. Therefore, combining knowledge graphs with LLMs can fully leverage the advantages of both, improving the construction and reasoning processes of knowledge. This collaborative approach can effectively meet complex and variable demands while also providing more accurate, comprehensive answers and explanations, injecting new impetus into innovation and development in the field of education [12]. It is anticipated that the integration of knowledge graphs and ChatGPT will play an increasingly vital role in future educational practices.

3 Stem Knowledge Graph Overall Scheme Design

This initiative, merging LLMs with knowledge graphs, constructs a dynamic, intelligent STEM knowledge graph to bolster learners' creative resilience in STEM education. It consolidates basic interdisciplinary concepts with those related to creative resilience, employing visualization to expose their interconnectedness. An LLM is deployed to automatically curate a structured knowledge base from STEM project resources, complemented by personalized learning recommendations and real-time feedback to foster innovation and problem-solving. Ongoing enhancements maintain the graph's currency and precision, providing robust support for STEM education. The knowledge graph enables learners to identify and explore project-related concepts and their semantically-related knowledge, nurturing innovative capabilities. This paper presents the design of a STEM knowledge graph, the structure of which is depicted in Fig. 1.

3.1 The Hierarchical Nature of Creative Resilience

Bloom's Taxonomy of Educational Objectives [13] delineates a hierarchy of six learning levels: Knowledge, Comprehension, Application, Analysis, Evaluation, and Creation. These levels encapsulate the cognitive, operational, and creative facets of creative resilience, coalescing into a pyramidal framework of objectives that encompasses the

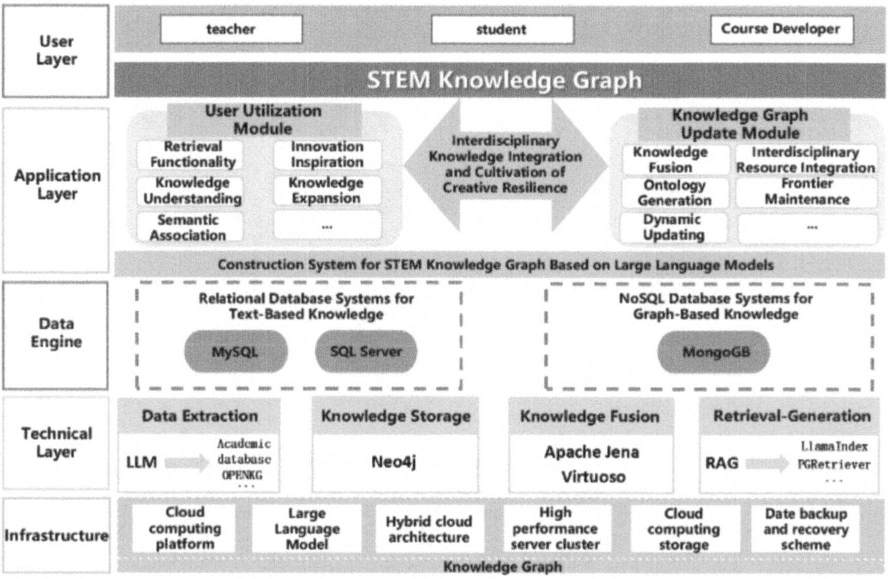

Fig. 1. Schematic diagram of the overall system design.

cognitive, operational, and creative strata. Within the cognitive stratum, STEM education deepens the apprehension of scientific inquiry and mathematical concepts, reinforcing learners' memorization and cognition of foundational theories, thus establishing a robust knowledge base for innovative thinking. In the operational stratum, STEM education integrates engineering design and technological development projects, empowering learners to apply theoretical knowledge to address intricate real-world issues, cultivating their integrated capabilities in analysis and application. At the creative stratum, STEM education spurs students to undertake innovative design and critical thinking, driving them towards novel elevations of innovative thinking through the processes of evaluation and creation. The STEM knowledge spectrum design, congruent with Bloom's Taxonomy, is intended to augment learners' creative resilience via the harmonious amalgamation of the cognitive, operational, and creative levels, instigating affirmative qualitative transformations in maker experiments (Fig. 2).

3.2 Design of the STEM Knowledge Graph Construction System Based on Large Language Models

The construction of a knowledge graph is a fundamental component of a holistic STEM educational framework, paralleling the development process of generic knowledge graphs with four key stages: knowledge extraction, knowledge storage, knowledge integration, and Retrieval-Generation. In the knowledge extraction phase, the Retrieval-Augmented Generation (RAG) technique efficiently automates the harvesting of knowledge from a wide array of academic and educational databases, managing both structured and unstructured data. Utilizing the sophisticated semantic understanding and inferential

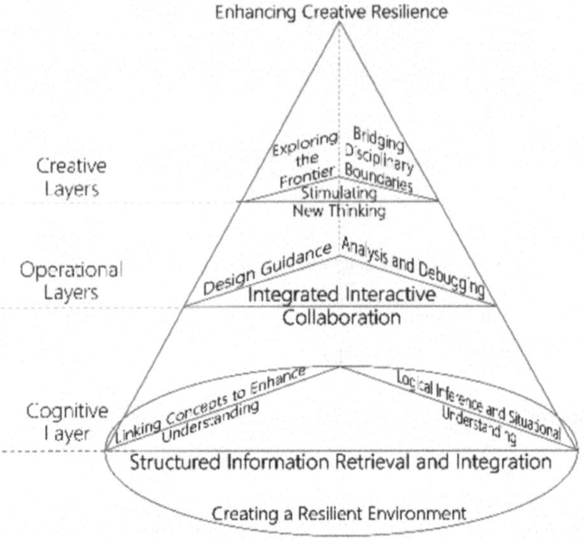

Fig. 2. Hierarchy of creative resilience.

capabilities of LLMs, the method parses data into triples [14], laying the groundwork for a preliminary STEM educational knowledge graph within the Neo4j platform.

Advancing to knowledge storage, the Neo4j framework is instrumental in storing entities, relationships, and attributes, complete with the establishment of pertinent indexes. LLMs are then engaged to add supplementary layers of relationships and attributes, thereby enriching the STEM knowledge graph's content.

Knowledge integration, underpinned by Neo4j, sees the STEM knowledge graph's logical architecture clearly divided into two core components: the schema layer and the data layer. The schema layer is responsible for defining the conceptual categories and relational framework of STEM educational knowledge, erecting a semantic scaffold through the integration of knowledge. The data layer, an embodiment of the schema layer's constructs, is carefully assembled according to the pre-established categories and relational definitions, often found within relational database management systems such as MySQL or SQL Server. The data layer's development encompasses critical processes like entity recognition and relationship formation, adhering to the conceptual and relational guidelines set by the schema layer. The synergy between the schema and data layers is mediated by the deep learning techniques of LLMs for data mapping, culminating in the STEM knowledge graph's formation within a graph database repository.

In the final stages of retrieval and generation, the RAG approach harmonizes keyword and vector searches using the PGRetriever from the LlamaIndex framework, cultivating an enriched context. The conversion of natural language queries into Cypher queries with Text-to-Cypher allows for the fusion of vector and keyword searches, effectively extracting relevant information from the STEM knowledge graph. This methodology amplifies the knowledge network accessible to learners and alleviates their cognitive load, providing a more comprehensive and efficient learning experience.

3.3 Intelligent Project Knowledge Management System

The knowledge management system is integral to STEM education, acting as a conduit between the knowledge graph and users, and serving as the central platform for innovation and creative endeavors among learners. The system is bifurcated into two core functional modules: the user engagement module and the graph maintenance module.

Within the user engagement module, learners have the capacity to interrogate the project knowledge graph for terms pertinent to their project requirements. These graphs encapsulate a multitude of knowledge points leveraged in previous projects, reflecting the technical concepts integral to the design of prior STEM initiatives. By delving into these knowledge points, learners can grasp the foundational concepts of the project and, leveraging semantic interconnections between terms, further investigate additional technical keywords associated with the STEM project. These emergent keywords broaden the learners' knowledge spectrum and indicate potential avenues for future innovation, stimulating their creative faculties.

Conversely, the graph maintenance module harnesses LLMs and machine learning algorithms to facilitate real-time updates and expansion of the project knowledge graph. As learners engage in innovative tasks, the system autonomously chronicles and amasses interdisciplinary STEM educational resources, culminating in the genesis of novel ontologies and relational constructs. These nascent knowledge points and interconnections are seamlessly integrated into the extant project knowledge graph, ensuring its perpetual enhancement and iterative progression. This paradigm guarantees that the most current knowledge and research insights in STEM projects are promptly incorporated within the knowledge graph, preserving its vanguard status and applicability.

The STEM knowledge management system, through its twin core modules of user engagement and graph maintenance, realizes the organized management and efficacious allocation of educational assets and engenders a superior and progressively evolving educational milieu for learners. It elevates the learners' capacity for self-directed learning and augments their creative resilience.

3.4 Project Learning Interaction Module

The project learning interaction module constitutes an essential element of the project-oriented knowledge graph design framework, acting as a conduit between learners and the knowledge graph and amplifying the innovative outcomes of projects through interactive and feedback mechanisms. This discussion delves into the design considerations of the project learning interaction module across several critical dimensions.

At the heart of the project learning interaction module's development is interaction design. The interface is designed to be intuitive and user-friendly, allowing for ease of use and rapid mastery by learners. To bolster the graph's accessibility and visual appeal, this design incorporates state-of-the-art graphical library technologies such as D3.js, rendering the knowledge graph into an interactive visual format. Users are provided with functionalities like searching, filtering, and dynamic path exploration, which facilitate efficient comprehension and navigation of the STEM domain's knowledge architecture. This approach empowers learners to quickly identify and access project-relevant information. The inclusion of interactive tools such as zoom and drag enhances the adaptability and ease of use of the user experience.

The project learning interaction module should integrate an intelligent recommendation system that harnesses the synergy of machine learning and knowledge graphs. By conducting in-depth analyses of learners' historical learning behaviors and ongoing project advancements, the system can proffer pertinent knowledge points, learning resources, and project examples. This form of guidance aids learners in uncovering potential domains of interest and innovative avenues, thereby invigorating their motivational and creative faculties. Additionally, the project learning interaction module is equipped to support real-time interactive and feedback mechanisms. Via an integrated Q&A platform, learners can engage in continuous communication with peers and mentors, thereby fortifying the collective creative resilience of the team. The system is endowed with an instantaneous feedback feature that supplies prompt responses and chronicles learners' behavioral data during interactions within the project-based knowledge graph or upon the completion of project milestones. This mechanism not only enriches the knowledge graph's content but also establishes a groundwork for learners' future educational endeavors.

4 Stem Learning Path Planning

Knowledge graphs, with their structured and semantically-enhanced knowledge representation, are pivotal in providing key support for personalized learning path recommendations. In STEM education, this is especially beneficial for nurturing students' innovative capabilities and problem-solving skills [15].

STEM knowledge graphs facilitate the planning of personalized learning paths by visually mapping the interconnections of knowledge points, decomposing them into a network of nodes and edges that elucidate the dependencies among concepts. This approach allows learners to discern the sequence of learning, avoiding cognitive gaps and ensuring a coherent educational progression. Moreover, these graphs are instrumental in surmounting learning impediments, providing alternative solutions through interdisciplinary knowledge networks, thus bolstering knowledge acquisition and fostering adaptive and innovative problem-solving skills.

Additionally, STEM knowledge graphs dynamically track learners' traits and advancement, tailoring learning path recommendations to align with their current knowledge and creative resilience levels. This customization enhances the efficacy of learning, bolstering learners' self-efficacy and equipping them with the confidence and tenacity to navigate challenges and engage in sustained innovation (Fig. 3).

5 Application of Stem Knowledge Graphs

5.1 Participants

This research endeavors to assess the impact of a STEM knowledge graph-assisted tool on the performance enhancement of student maker experiments through a well-crafted experimental plan. Spanning two months, the study engaged 20 learners from School L, who were randomly divided into an experimental group and a control group, each comprising 10 individuals. The experimental group underwent training on the STEM

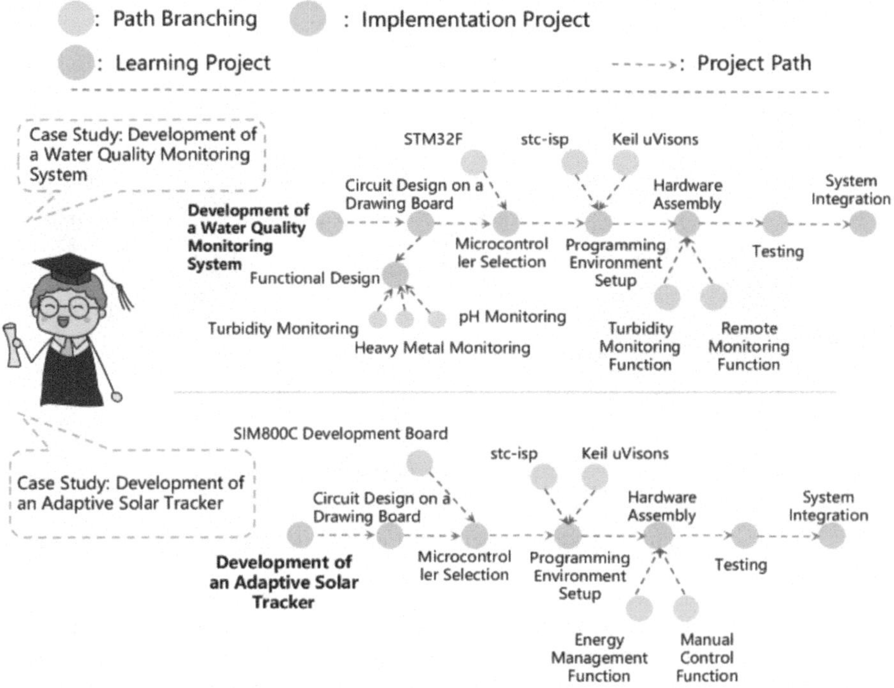

Fig. 3. Maker-oriented learning path diagram.

knowledge graph tool, which they applied in maker experiments. Performance disparities in task completion efficiency between the groups was meticulously recorded and analyzed. The outcomes will be discussed, and suggestions for enhancement will be presented. The study upholds a commitment to the protection of participant confidentiality, as well as to the integrity and security of the experimental process (Fig. 4).

5.2 Instruments

The study employed the Programming Resilience Scale for University Students (PRSUS) [16] for a quantitative assessment of learners' creative resilience, operationalized through the dimensions of "persistence (PE)," "perceived value (PV)," "difficulty cognition (DC)," and "incremental belief (IB)." Creative resilience is defined as the trait of maintaining a proactive stance in learning despite encountering setbacks in STEM education. Utilizing the integrated PRSUS scale, the study evaluates the levels of creative resilience among learners in STEM contexts.

The online survey was divided into three sections. The initial section gathered demographic information, such as gender, age, and academic standing, with a retest of scores at the conclusion to validate the survey. The second section assessed the influence of the STEM knowledge graph on maker project performance across dimensions of learning performance, innovation, and originality [17], using a 5-point Likert scale. The final section was a self-assessment of creative resilience, with participants evaluating their

Fig. 4. Maker project partial works exhibition

resilience on a 5-point scale according to the study's definition, including assessments of PE, PV, DC, and IB. The survey's structured design facilitated the collection of comprehensive data on individual creative resilience, supporting further research and practical applications.

5.3 Creating Resilience Analysis

Table 1 reveals that the mean academic performance score of the experimental group (M = 82.50, SD = 5.000) markedly exceeds that of the control group (M = 65.80, SD = 3.500), with a calculated t-value of 4.853 and a p-value of 0.001, both of which are below the threshold of 0.05. This statistical significance substantiates the impact of the STEM knowledge graph on the improvement of creative resilience.

Table 1. The impact of stem knowledge graphs on the performance of student marker projects

Variable	Group	Sample(N)	Mean	SD	t	p
Learning outcomes	Control	10	65.80	3.500	4.853	0.001
	Experimental	10	82.50	5.000		
Innovation	Control	10	19.50	1.200	5.123	0.002
	Experimental	10	26.50	2.500		
Originality	Control	10	28.70	1.700	6.200	0.001
	Experimental	10	36.20	2.100		

The findings illustrate that the experimental group achieved a notably higher mean score in interdisciplinary pedagogical knowledge compared to the control group, evidenced by a t-value of 4.809 and a p-value of 0.001. This indicates that the utilization of the STEM knowledge graph tool significantly bolsters the creative resilience of learners.

Table 2. Measurement of creative resilience

Variable	Experimental (N = 10)			CONTROL (N = 10)			Experimental vs Control
	Post	Pre	Pre-Post	Post	Pre	Pre-Post	Post
	M(SD)	M(SD)	t(p)	M(SD)	M(SD)	t(p)	t(p)
Persistence (PE)	81.29	71.57	− 4.229	81.27	71.00	− 4.875	0.008
	(1.589)	(9.081)	(0.000)	(1.554)	71.00	(0.000)	(0.993)
Perceived Value (PV)	18.10	14.71	− 7.347	16.20	14.73	− 3.2901	4.809
	(0.228)	(0.453)	(0.000)	(0.341)	(0.679)	.(005)	(0.000)
Difficulty Cognition (DC)	36.05	30.14	− 3.639	37.73	30.67	− 5.220	− 1.056
	(1.107)	(1.229)	(0.000)	37.73	(1.511)	(0.000)	(.298)
Incremental Belief (IB)	27.14	26.71	− 0.546	27.33	25.60	− 1.956	− 0.191
	(0.688)	(0.731)	(0.591)	(0.681)	(1.095)	(0.071)	(0.850)

Table 2 shows that the STEM knowledge graph positively influenced the enhancement of students' creative resilience. The analysis reveals that the post-test mean of the experimental group in PV significantly exceeded that of the control group, with a t-value of 4.809 and a p-value of 0.000. This indicates that the STEM knowledge graph effectively mitigates the challenges learners encounter with complex interdisciplinary knowledge structures, thereby diminishing cognitive barriers and enhancing creative resilience.

Both groups exhibited advancements in PE and DC, suggesting that the STEM knowledge graph aids in navigating personalized learning path issues during maker experiments. This reduction in the complexity of knowledge acquisition and deepened comprehension of STEM concepts indirectly boosts creative resilience, yet the improvements did not eclipse the impact of conventional educational approaches. In the case of IB, no significant variations were detected between the groups, implying the necessity for more extended or varied instructional strategies. Although the belief enhancement is constrained, the data hint at a tendency that the ongoing learning and practice supported by the STEM knowledge graph contribute to learners' awareness of their capacity for improvement, which is expected to positively influence creative resilience over time.

5.4 Analysis of Scale Reliability and Validity

This study employed the expert evaluation method to assess the content validity of the PRSUS scale. Based on the structure of STEM maker projects and the clear definitions of each dimension, three experts in the field of educational technology and two experts in programming education were invited to participate in the scoring work, evaluating whether all items on the scale conform to the definitions of their corresponding dimensions. Through statistical analysis of the scoring results, the consistency coefficient among the five experts is as high as 1, indicating a high level of agreement among the experts. The average score of each item ranges from 3.70 to 3.98, with a standard deviation fluctuating between 0.000 and 0.432. These data demonstrate that the PRSUS scale possesses good content validity.

Table 3. Verification of scale reliabilty

Variable	PE	PV	DC	IB
Cohen's d	0.832	0.810	0.718	0.722

To further evaluate the reliability of the scale and the actual effects of the experimental treatment, this study calculated the Cohen's d values for the control group (pre) and the experimental group (post) on four parameters: PE, PV, DC, and IB. The results presented in Tables 3 and 4 indicate that the Cohen's d values for these four parameters significantly exceed the minimum threshold of 0.2 proposed by Cohen [18] for effect size standards, thereby providing strong evidence that the parameter changes in the experimental group before and after the treatment are significant. This finding further validates the high reliability of the scale used in this study in measuring these four parameters, demonstrating good stability and consistency in the measurement results, and objectively reveals that the experimental treatment had a significant positive impact on these four parameters, successfully enhancing the subjects' levels in these areas. These indicators suggest that the scale used in this study possesses high reliability.

6 Summary

This research concentrated on the design of a STEM knowledge graph propelled by LLM technology, with the objective of strengthening learners' creative resilience. Addressing the interdisciplinary and intricate aspects of STEM education, it is essential to enhance learners' resilience amidst challenges. The study began by defining "creative resilience" and scrutinized its significance in the context of STEM education. Subsequently, it addressed prevalent challenges in STEM education, including the complex integration of interdisciplinary knowledge and the gap between theoretical understanding and practical application, which impede the growth of creative resilience. The integration of knowledge graphs with LLMs automates knowledge extraction and discerns relationships, thereby forging a dynamic and intelligent graph that lowers cognitive barriers. The experimental results indicate a marked enhancement in learners' creative resilience, with around 80.5% of participants attaining a high level. This study introduces a novel perspective for the intelligent evolution of STEM education, fostering the integration of interdisciplinary knowledge and the development of innovative competencies, thereby contributing substantial academic and practical implications.

Acknowledgments. This research was funded by the National Natural Science Foundation in China (62307023), the Ministry of Education in China Project of Humanities and Social Sciences(23YJC880063), Guangdong Province Philosophy and Social Science Planning Project (GD23XJY72), Guangdong Province Philosophy and Social Science Innovation Project Special Entrusted Project (GD22TWCXGC10).

References

1. Shen, F., Roccosalvo, J., Zhang, J., Tian, Y., Yi, Y.: Online technological STEM education project management. Educ. Inf. Technol. **28**(10), 12715–12735 (2023)
2. Tim, Y., Leidner, D.E.: Digital resilience: a conceptual framework for information systems research. J. Assoc. Inf. Syst. **24**(5), 1184–1198 (2023)
3. Qi, C., Yang, N.: Digital resilience and technological stress in adolescents: a mixed-methods study of factors and interventions. Educ. Inf. Technol. 1–47 (2024)
4. Aguilera, D., Ortiz-Revilla, J.: STEM vs. STEAM education and student creativity: a systematic literature review. Educ. Sci. **11**(7), 331 (2021)
5. Liu, S., Li, C.: Promoting design thinking and creativity by making: a quasi-experiment in the information technology course. Think. Skills Creat. **49**, 101335 (2023)
6. Lu, S.Y., Lo, C.C., Syu, J.Y.: Project-based learning oriented STEAM: the case of micro–bit paper-cutting lamp. Int. J. Technol. Des. Educ. **32**(5), 2553–2575 (2022)
7. He, Q., Wan, Z.H., Tsang, Y.F., So, W.W.M.: Learning by solving environmental problems: primary students' experiences of E-STEM activities. In: Li, Y., Zheng, Z., Song, N. (Eds.) Disciplinary and Interdisciplinary Education in STEM: Changes and Innovations, pp. 197–219. Springer Cham, Switzerland (2024)
8. Wan, Z.H., So, W.M.W., Hu, W.: Necessary or sufficient? The impacts of epistemic beliefs on STEM creativity and the mediation of intellectual risk-taking. Int. J. Sci. Educ. **43**(5), 672–692 (2021)
9. Xiao, Y., Zheng, S., Shi, J., Du, X., Hong, J.: Knowledge graph-based manufacturing process planning: a state-of-the-art review. J. Manuf. Syst. **70**, 417–435 (2023)
10. Chen, P., Lu, Y., Zheng, V.W., Chen, X., Yang, B.: Knowedu: a system to construct knowledge graph for education. IEEE Access **6**, 31553–31563 (2018)
11. Sayin, A., Gierl, M.: "Using OpenAI GPT to generate reading comprehension items. Educ. Meas. Issues Pract. **43**, 5–18 (2024)
12. Hu, X., Liu, A., Dai, Y.: Combining ChatGPT and knowledge graph for explainable machine learning-driven design: a case study. J. Eng. Des. 1–23 (2024)
13. Zhou, C., Zhang, W.: Computational thinking (CT) towards creative action: developing a project-based instructional taxonomy (PBIT) in AI education. Educ. Sci. **14**(2), 134 (2024)
14. Li, Y., Liang, Y., Yang, R., Qiu, J., Zhang, C., Zhang, X.: CourseKG: an educational knowledge graph based on course information for precision teaching. Appl. Sci. **14**(7), 2710 (2024)
15. Zhang, X., Liu, S., Wang, H.: Personalized learning path recommendation for e-learning based on knowledge graph and graph convolutional network. Int. J. Softw. Eng. Knowl. Eng. **33**(1), 109–131 (2023)

16. Fu, Q., Zhang, L.W., Hong, J.C., Dong, Y.: Development and verification of the programming resilience scale for university students. Australas. J. Educ. Technol. **37**(6), 141–155 (2021)
17. Ji, Y., Zou, X., Li, T., Zhan, Z.: The effectiveness of ChatGPT on pre-service teachers' STEM teaching literacy, learning performance, and cognitive load in a teacher training course. In: Proc. 2023 6th Int. Conf. Educ. Technol. Manage, pp. 16–22 (2023)
18. Cohen, J.: The concept of power analysis. In: Statistical Power Analysis for the Behavioral Sciences, pp. 1–17. Routledge, New York, NY (2013)
19. Ding, Z.: Empirical study on the influencing factor model of customer satisfaction of the red special collection of libraries. M.S. thesis, Nanchang Univ, Nanchang, China, (2023)
20. Loh, C.E., Gan, S., Mounsey, S.: What do children want to read? A case study of how one primary school library supported reading for pleasure. J. Libr. Adm. **62**(7), 931–945 (2022)
21. Bracken, S.S., Fischel, J.E.: Family reading behavior and early literacy skills in preschool children from low-income backgrounds. Early Educ. Dev. **19**(1), 45–67 (2008)
22. Mat-Roni, S., Merga, M.K.: The influence of extrinsic and intrinsic variables on children's reading frequency and attitudes: an exploration using an artificial neural network. Aust. J. Educ. **63**(3), 270–291 (2019)

Understanding College Students' AI Literacy in Course Essays: A Topic Modeling-Based Method

Jingxiu Huang(✉), Lixin Zhang, Yufeng Wei, and Jiajun Liao

School of Information Technology in Education, South China Normal University, Guangzhou, China
jimsow@m.scnu.edu.cn

Abstract. The development of artificial intelligence (AI) technology has made the development of AI literacy a pressing need. However, the current AI literacy assessment methods for college students are mainly traditional methods that consider reliability and hysteresis, such as questionnaires and interviews. Few studies have made use of topic modeling techniques. This study explored the AI literacy that different students demonstrated in their AI course assignments and assessed the relationships between different AI literacy topic words. The results highlight that "online," "learning," "students," and "education" are the most often used topic words in relation to AI literacy. This suggests a strong connection and possible overlap in discussions about AI literacy. In the future, these topic words could be regarded as a reference to improve AI literacy for students of different disciplines and age levels.

Keywords: AI literacy · Topic modeling · BERTopic · Topic words

1 Introduction

The introduction, advancement, and popularization of artificial intelligence (AI) technologies have greatly increased the efficiency and effectiveness of various fields and made it possible to perform duties more effectively. These technological innovations have also permeated other sectors of academia, further enhancing the effectiveness and efficiency of academic research and education [1]. In the realm of education, AI algorithms and systems are garnering considerable attention, with interest continuing to grow each year [1]. The impact of AI on college students' learning is mainly positive in terms of the potential for improved learning efficiency and performance using these tools. However, the link between the notion that AI may replace the role of the instructor in the learning process and student learning is not clear, and the correlation with other potential factors, such as time spent studying online and awareness of the advantages of the software, is weak. The impact of these factors on university students' learning is not yet fully understood, and further research is needed to determine their specific role [2].

Different scholars have used different methods of evaluating AI literacy, with more scholars favoring questionnaires and interviews [3]. However, the reliability of these

Q. Liu et al. (Eds.): EITT 2024, CCIS 2600, pp. 73–85, 2026.
https://doi.org/10.1007/978-981-95-1525-7_6

methods for evaluating AI literacy may affect the results. Moreover, there is a hysteresis in the evaluation of these methods. Therefore, emerging technologies need to be utilized to avoid these problems.

Topic modeling is a popular statistical tool for extracting latent variables from large datasets [4]. Topic modeling is a technique that utilizes groups of algorithms to reveal and annotate thematic structure in collections of documents [5]. Current topic modeling techniques mainly include Latent Dirichlet Allocation (LDA), Non-Negative Matrix Factorization (NMF), Latent Semantic Analysis (LSA), and deep learning-based methods (e.g., Generative Adversarial Networks and Variational Self-Encoders). In addition, pre-training models such as BERT have also been widely used for topic extraction. Each of these methods has its own characteristics: LDA captures topic distributions through probabilistic models, NMF extracts non-negative topic structures based on matrix decomposition, LSA reduces the text dimensionality by using singular value decomposition, and deep learning methods construct more complex topic representations with the help of neural networks. These studies provide an important framework and methodological foundation for automatic topic discovery in text data. However, despite the large amount of research that has been conducted on topic models, especially in AI literacy, there is no universally accepted topic model for the domain of education.

This study aims to extract topic words related to students' AI literacy from the textual information in assignments completed for an AI course. Kong et al. [3] described an AI literacy course for diverse university students, employing a flipped classroom approach. The study demonstrated significant improvements in students' understanding of AI concepts, suggesting the effectiveness of such interventions in fostering inclusive AI education [3].

2 Literature Review

2.1 AI Literacy

The concept of AI literacy can be traced back to 2016 when the term "AI literacy" was first defined as the ability to understand the basic technologies and concepts behind AI products [6]. With the introduction of AI literacy, scholars in China and abroad have researched its core elements and put forward their views. Many scholars have combined the components of AI literacy with Bloom's cognitive goal classification, starting from the perspective of adapting to the abilities needed for AI. Ng et al. [7] organized and summarized existing peer-reviewed literature, proposing four aspects (namely recognition and understanding, use and application, evaluation and innovation, and ethical issues) to cultivate AI literacy based on classic literature adaptation. Zhao et al. [8], based on Bloom's conclusions, divided teacher AI literacy into four dimensions and explored the correlation between each dimension. Southworth et al. [9], by integrating AI into the entire curriculum of the University of Florida, combined with Ng's model [7], constructed an AI literacy model.

Other scholars regard AI literacy as a comprehensive quality and build a framework from this perspective. Kong et al. [10] proposed a three-dimensional conceptual framework for AI literacy to promote digital citizenship. Wang et al. [11] proved through experiments that the four-structure model they proposed is sufficient to represent AI literacy. Kim et al. [12] believe that AI courses focus on achieving AI literacy based on three AI abilities: knowledge, skills, and attitudes. Some people also believe that AI literacy consists of three core parts: knowledge about AI; the ability to understand various application examples; and skills to help or teach others how to understand, imagine, develop, and apply AI [13]. Celik et al. [14] built an AI literacy model by combining concepts such as the digital divide, cognitive extraction, and computational thinking, including awareness, use, evaluation, and ethics.

In general, scholars' construction of AI literacy framework models still shows a situation of contention, and a unified index system has not yet been formed. Referring to existing research results, this paper, combined with Bloom's "Taxonomy of Educational Objectives," constructs a theoretical framework for college students' AI literacy, dividing AI literacy into six dimensions: *Knowledge*, *Thinking*, *Awareness*, *Capability*, *Application*, and *Creativity*.

2.2 Topic Modeling in Education

Topic modeling is a type of probabilistic generative model that has been widely applied in computer science in recent years, especially in text mining and information retrieval. Since the model was first proposed, it has attracted researchers' attention and widespread interest in many fields. Common topic models include LDA, Biterm Topic Model (BTM), BERTopic, etc. Topic modeling is a popular analytical tool for evaluating data [4] and been used as a statistical technique for identifying underlying themes within extensive collections of data.

With the rapid development of information technology, the field of education has experienced unprecedented changes. In this process of change, educational technologies, especially text analysis techniques, such as topic modeling, provide powerful tools for educational research and practice. In the field of education, data are often processed using topic modeling, and the sources of these data focus on the interactive content of online discussion forums and the mining of academic literature, while open-text assignments also provide a portion of valuable data resources [15]. Moreover, topic modeling has been commonly used for educational content analysis and educational assessment, as well as educational data mining [16]. Chen et al. [17] focused on *Computers & Education*, a major journal in the field of educational technology. Through a structured topic modeling analysis of 3,963 articles published in the journal between 1976 and 2018, the study revealed the topical themes of the educational technology research community and their evolution over time. Saar Kuzi et al. [18] provide insights into how traditional supervised machine learning can be extended to include richer semantic features.

Current automated scoring techniques are limited to multiple-choice, short-answer, and short-essay scoring. Research has explored ways to automate the assessment of more complex assignments, and this paper directly extends this work by exploring a finer-grained feature: topic features. Against this background, this paper attempts to answer the following two research questions:

RQ1: What topic words are represented in students' course essays across different dimensions of AI literacy?

RQ2: Are there differences in the topic words presented in students' course essays across various dimensions of AI literacy?

3 Methodology

3.1 Research Framework

The objective of our work was to retrieve textual data from the final assignments of an AI course. This textual data was then used to examine the differences in the importance of AI-related subjects across individuals with different dimensions of AI literacy. The research methodology, illustrated in Fig. 1, consists of four independent phases: data acquisition, data preprocessing, topic modeling, and topic interpretation analysis.

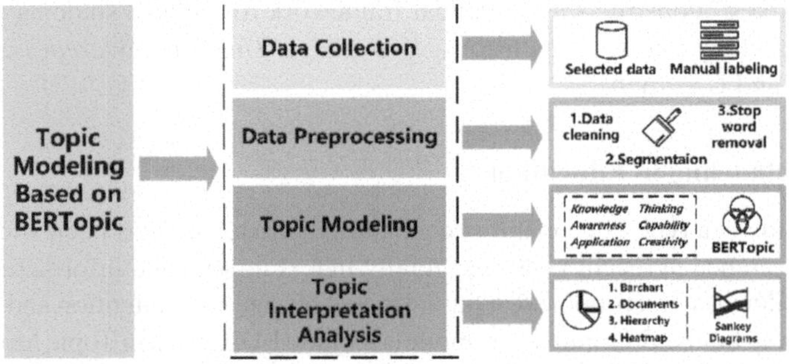

Fig. 1. Research framework.

We collected textual data from student assignments specifically focusing on the topics of *Knowledge*, *Thinking*, *Awareness*, *Capability*, *Application*, and *Creativity* related to AI. The data used in this study were derived from the textual data of assignments from students in an AI course. Subsequently, these data were fed into the BERTopic model. This technique initially used the BERT model to encode the text, converting it into vectors with a large number of dimensions. Next, the c-TF-IDF approach was employed to extract topic words from these vectors, thereby capturing the principal substance of the text. BERTopic then employed a clustering technique to group the topic words into distinct topics. These topics encompass the fundamental topic words and characteristics of AI across various dimensions.

3.2 Data Collection

From an AI course in higher education, we collected the final assignment of each student as the data source. To finish this assignment, each student wrote a 1,000-word essay about AI technology, which was submitted as a PDF or Word document. Before using the 98 essays to examine internal data, each essay was labeled with categories based on its content, including the following six dimensions of AI literacy: Knowledge (40), Thinking (28), Awareness (39), Capability (26), Application (57), and Creativity (11). The essays were divided into six groups depending on the dimensions, resulting in a total of 201 essays (with duplicates). Pandas, a Python package, was used to organize the essays into tables to facilitate later analysis.

3.3 Data Preprocessing

Before initiating further textual analysis, the data was refined to ensure the validity and reliability of the results. All the following operations were performed in Jupyter:

Data Cleaning.

Several non-relevant characters, such as numeric characters, special escape characters, and punctuation marks, were removed from the original texts.

Segmentation.

Since the data needed for topic modeling is textual, the sentences in the essays were divided into individual words. To improve the accuracy of the segmentation, we employed a Python package named jieba [19], a prevalent Chinese word segmentation tool.

Stop Word Removal.

In this step, several stop words were removed after segmentation to reduce the interference of certain characters. Stop words are words that are not directly related to the topic of the essay, such as "hello," "the," "a," "in," etc. These stop words may affect the effectiveness of the topic modeling. Therefore, we merged two authoritative lists of stop words: the Chinese stop word list and the Sichuan University Machine Intelligence Laboratory stop word list. To adopt this approach, we downloaded the two lists of stop words, imported them into Jupyter as a list, and then removed the stop words from the 201 essays.

3.4 Topic Modeling

BERTopic [20] is a topic model that utilizes clustering techniques and a class-based variant of TF-IDF to produce coherent topic representations, which can extract key topics from documents and cluster similar topics into one category. This method is particularly suitable for topic discovery and visualization of a large amount of text data. Compared with the most commonly used topic modeling approach, LDA [21], BERTopic is more adept at modeling the topic of long texts and achieves better performance. In the LDA model, each word is independent, which makes the LDA model unable to characterize context, leading to the production of missing semantics and semantic ambiguity [22]. The BERTopic model, which leverages word embeddings, compensates for the deficiencies of the LDA model based on the semantic level. BERTopic can efficiently perform word embedding tasks directly using a pre-trained model, eliminating the need to manually set the number of topics, as required by conventional topic extraction models, and circumventing the intricate process of parameter tuning.

4 Results

Table 1 reveals that the powerful themes and topic words of the six dimensions of AI literacy are *Knowledge* with three topics ("Online," "Student," "Learning," "Instruction," "Teacher," etc.), *Thinking* with two topics ("Online," "Student," "Learning," "Teacher," "Instruction," etc.), *Awareness* with two topics ("Online," "Student," "Learning," "Instruction," "Teacher," etc.), *Capability* with two topics ("Online," "Student," "Learning,"

"Instruction," "Teacher," etc.), *Application* with 5 topics ("Student," "Network," "Online," "Learner," "Interact," etc.), and *Creativity* with two topics ("Online," "Student," "Learning," "Teacher," "Instruction," etc.). Additionally, Table 1 also displays the text probabilities for each topic word. Based on the connotation of these topic words, the secondary topics were modified or paraphrased without changing the meaning.

Figure 2 consists of six sub-topic similarity matrix diagrams, which illustrate the degree of similarity between various topics. A darker hue signifies a greater similarity, indicating that the compared topic words and AI literacy dimension words share more similar content. Conversely, a lighter color suggests a lower similarity, indicating a greater difference in content between the compared topic words and AI literacy dimension words. The results show that *Awareness* is a dimension in which the two topics exhibit a strong correlation (Similarity score = 0.82). Furthermore, within the *Knowledge* dimension, Topic 0 and Topic 1 exhibited a significant correlation, with a similarity score of 0.80. Furthermore, in the dimension with the highest number of topics, *Application*, there were three pairs of closely related topics. These topics include Topic 0 and Topic 2, with a similarity score of 0.87, Topic 0 and Topic 3, with a similarity score of 0.93, and Topic 1 and Topic 3, with a similarity score of 0.82.

In order to comprehend the hierarchical structure of the topics, a hierarchical clustering technique was employed to group the relationships between topics and display the resulting clusters, as depicted in Fig. 3. The figure displays each topic on the left side, identified by a number and descriptor. The varying lengths of lines represent distinct topic categories and clustering branches, facilitating the analysis of correlations between topic words. This diagram demonstrates a clear and direct relationship between Topic 0 and Topic 1 in the *Knowledge* dimension. In the *Application* dimension, Topic 0 had a direct correlation with Topic 3.

Figure 2 and Fig. 3 were then used to consolidate analogous topics. In the *Knowledge* dimension, we integrated Topics 0 and 1 into a new topic ("Technologies Application in Online Learning"). In the *Awareness* dimension, Topics 0 and 1 were all related to how to apply AI technologies to online learning. These two topics were merged into the same topic, called "AI in Online Learning." Similarly, Topic 0 and Topic 3 of the *Application* dimension were merged into the same topic ("Learner Interaction for Online Learning").

Upon consolidating analogous topics, we extracted the substance of topics and the dispersion of topic words. A Sankey diagram illustrating the progression of AI literacy dimensions is depicted in Fig. 4. The figure illustrates that the six primary areas of AI literacy mostly focus around "teachers," "students," "teaching," "machines," and so on. In general, there is a correlation between the words related to AI literacy dimensions, indicating a significant relationship where a single topic word might represent various AI literacy dimension words. When considering research subjects, the six dimensions of AI literacy exhibit areas of overlap and distinct qualities. Within the field of AI literacy, the concepts that receive the most focus are "machine," "education," "teaching," "students," "learning," "online," "AI," and "application." Among these, the topics "online," "learning," "students," and "education" are the most commonly mentioned and are found in many different dimensions of AI literacy, indicating a strong similarity between them.

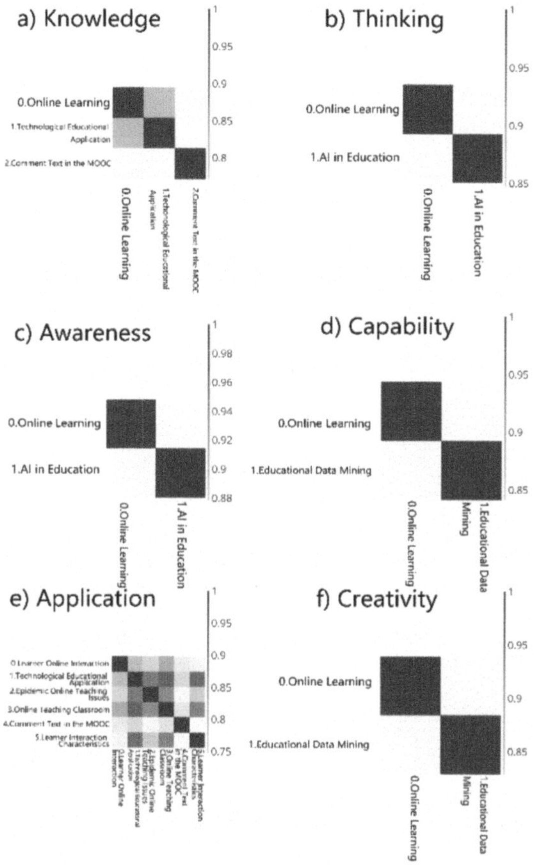

Fig. 2. Secondary topic similarity matrix diagram. In the heatmaps, the depth of the color represents the level of similarity.

5 Discussion

This study collected textual data from students' assignments in a higher education AI course and revealed the differences in the topic words of the six dimensions of AI literacy through the deployment of topic modeling. The principal findings are discussed into the following paragraphs.

In response to RQ1, the results presented in Table 1 and the figures demonstrate that all dimensions of AI literacy share a common thread of "Online," "Learning," and "Instruction." This suggests a strong interconnectivity between these topic words in the context of AI literacy. Most students often prioritized online learning in their context assignments. The assumption is supported by the fact that individuals were compelled to engage in online learning during the pandemic, which would have heightened their enthusiasm for online education [22]. Moreover, "Student" and "Teacher" demonstrate that AI literacy ultimately rests with students and teachers. There are two perspectives possible explanations for this assumption. First, the progress made in AI has highlighted

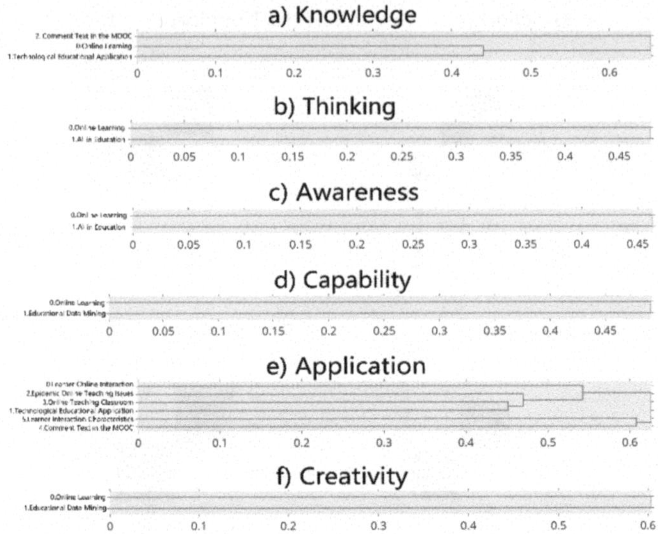

Fig. 3. Secondary topic hierarchy diagram. On the horizontal axis, smaller values indicate greater content similarity between the connected topics; larger values indicate greater divergence between the topics.

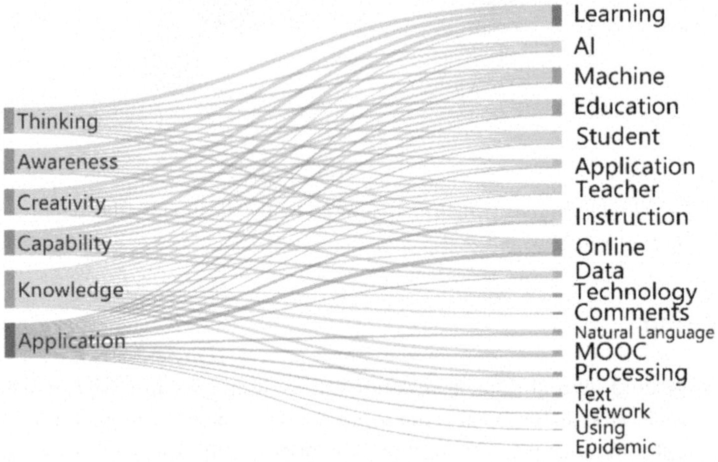

Fig. 4. Secondary topic Sankey diagram.

the need of teaching young students about fundamental AI-related knowledge, skills, values, and attitudes [23]. Secondly, teachers must have a thorough comprehension of the basic concepts and knowledge of AI through their professional development in order to be prepared for the next era of AI [24]. Furthermore, the topic words "Machine" and "AI" have a greater frequency of recurrence than other words in the list. In the realm of AI information, machine learning and the notion of AI are essential components [25].

Table 1. Text probabilities for Secondary topic word

Primary topic	Secondary topic	Top 5 topic words and corresponding probabilities				
Knowledge	0. Online Learning	Online (0.506)(在线)	Student (0.446)(学生)	Learning (0.434)(学习)	Instruction (0.416)(教学)	Teacher (0.375)(教师)
	1. Technological Educational Application	Education (0.544)(教育)	Machine (0.544)(机器)	Application (0.461)(应用)	Technology (0.459)(技术)	AI (0.429)(人工智能)
	2. Comment Text in the MOOC	MOOC (0.792)(mooc)	Natural Language (0.757)(自然语言)	Processing (0.689)(处理)	Text (0.650)(文本)	Comments (0.574)(评论)
Thinking	0. Online Learning	Online (0.576)(在线)	Student (0.542)(学生)	Learning (0.503)(学习)	Teacher (0.469)(教师)	Instruction (0.453)(教学)
	1. AI in Education	Education (0.651)(教育)	Machine (0.644)(机器)	AI (0.568)(人工智能)	Application (0.562)(应用)	Learning (0.550)(学习)
Awareness	0. Online Learning	Online (0.554)(在线)	Student (0.532)(学生)	Learning (0.502)(学习)	Instruction (0.480)(教学)	Teacher (0.463)(教师)
	1. AI in Education	Education (0.646)(教育)	Machine (0.621)(机器)	AI (0.569)(人工智能)	Learning (0.562)(学习)	Application (0.556)(应用)
Capability	0. Online Learning	Online (0.551)(在线)	Student (0.539)(学生)	Learning (0.512)(学习)	Instruction (0.441)(教学)	Teacher (0.436)(教师)
	1. Educational Data Mining	Machine (0.657)(机器)	Education (0.582)(教育)	Data (0.578)(数据)	Learning (0.539)(学习)	Technology (0.533)(技术)
Application	0. Learner Online Interaction	Student (0.342)(学生)	Network (0.333)(网络)	Online (0.327)(在线)	Learner (0.317)(学习者)	Interact(0.307)(交互)
	1. Technological Educational Application	Education (0.455)(教育)	Machine (0.437)(机器)	Application (0.402)(应用)	AI (0.396)(人工智能)	Data (0.369)(数据)
	2. Epidemic Online Teaching Issues	Instruction (0.367)(教学)	Online (0.363)(在线)	Epidemic situation (0.339)(疫情)	Problem (0.316)(问题)	Student (0.314)(学生)
	3. Online Teaching Classroom	Online (0.403)(在线)	Instruction (0.362)(教学)	Classroom (0.357)(教室)	Process (0.320)(过程)	Investment (0.313)(投入)

(continued)

Table 1. *(continued)*

	4. Comment Text in the MOOC	MOOC (0.624)(mooc)	Natural Language (0.547)(自然语言)	Processing (0.511)(处理)	Text (0.506)(文本)	Lesson (0.413)(课程)
Creativity	0. Online Learning	Online (0.568)(在线)	Student (0.552)(学生)	Learning (0.492)(学习)	Teacher (0.442)(教师)	Instruction (0.440)(教学)
	1. Educational Data Mining	Education (0.665)(教育)	Machine (0.665)(机器)	Data (0.595)(数据)	Learning (0.566)(学习)	AI (0.544)(人工智能)

Topic words like "Data," "Text," "Natural Language," and "Processing" suggest that text-based data in education might be another interest in AI. Through the process of collecting textual material, we were able to analyze it in order to obtain pertinent data, significant strategic knowledge, and insights for growth [26].

For RQ2, the results show that, in addition to "Online Learning," students with different AI literacy dimensions obtained other unique topics and topic words. According to the similarity of the topics, the dimensions could be divided into the following 3 categories: "Technological Educational *Application* and Massively Open Online Course (MOOC) Comment Interaction" (including *Knowledge* and *Application*), "AI in Education" (including *Thinking* and *Awareness*), and "Educational Data Mining" (including *Capability* and *Creativity*). Specifically, the students who showed *Knowledge* and *Application* dimensions in their assignment directed their emphasis towards a wider range of technology applications in the field of education, rather than just concentrating on the use of AI in education. This may be due to the need of integrating AI with other technologies in contemporary applications. Another potential argument is that a prerequisite for studying and using AI is an understanding of how other technologies are used in educational environments.

The topic of "MOOC Comment Interaction" was equally robust in these two dimensions, which is understandable considering that the data collected for this study was generated during the pandemic, a period when MOOCs saw a surge in popularity. Comments might be seen as a form of feedback in some situations [27]. Feedback may be provided in several different forms, such as text, voice, and video [28]. Providing feedback that contains information and explanations [29] has a beneficial effect on students' capacity to gain knowledge, since it offers ideas for improving performed tasks. Unlike *Knowledge* and *Application*, *Thinking* and *Awareness* are specifically focused on the use of AI in the educational context. Undoubtedly, it is crucial to possess an understanding of the importance of AI in education. Regarding the third category, educational data mining requires a substantial level of programming capability and critical thinking. Consequently, research has shown a strong correlation between this topic and both *Capability* and *Creativity*.

Despite the findings of this study contributing to our understanding of the relationships between different dimensions of AI literacy, there are two limitations. The first issue pertains to the research context. Some factors may have influenced the results of our study, including the limited course and students with different academic profiles. In addition, the small sample size and the analysis of coursework from only one course to determine AI literacy might have resulted in biases. Second, evaluating students' AI literacy merely based on textual data may be insufficient. The processes students use to solve real-world problems with AI in their daily lives could also serve as data sources for assessing their AI literacy. Combining textual data with behavioral data to analyze students' AI literacy would yield more scientific and explanatory conclusions.

6 Conclusion

This study utilized topic modeling techniques to provide a textual analysis of the essays, extracting topic words that appeared in the essays and researching these topic words to analyze how they reflected the AI literacy possessed by students. The study identified six dimensions of AI literacy and the representative topic words in AI literacy essays. By using BERTopic topic modeling to analyze the AI literacy topic words, the study revealed the relationships between the topic words across the six dimensions of student AI literacy and specifically examined the differences in AI literacy across various essays. The results reveal a complex yet coherent structure of interrelated topics. Notably, "online," "learning," "students," and "education" were the most frequently occurring and widely distributed topic words across AI literacy dimensions, indicating a high degree of similarity and a potential overlap in the discourse surrounding AI literacy.

Even though there were some limitations in the current study that led to experimental results that did not fully meet expectations, enhancing students' AI literacy is necessary, and using the topic words for six dimensions of AI literacy identified in this study to improve students' AI literacy is feasible. Future research should comprehensively evaluate students' AI literacy from multiple aspects, providing stronger support for future AI literacy assessment research.

Acknowledgments. This research was funded by the National Natural Science Foundation of China: 62207015, and the Humanities and Social Sciences Youth Foundation of the Chinese Ministry of Education: 22YJC880021.

References

1. Chen, L., Chen, P., Lin, Z.: Artificial intelligence in education: a review. IEEE Access **8**, 75264–75278(2020)
2. Guo, Z., Jing, T., Zhu, H., Xu, X.: Analysis of the impact of artificial intelligence on college students' learning. Front. Soc. Sci. Technol. **5**(13), 110–116 (2023)
3. Kong, S.C., Cheung, W.M.Y., Zhang, G.: Evaluation of an artificial intelligence literacy course for university students with diverse study backgrounds. Comput. Educ. Artif. Intell. **2**, 100026 (2021)
4. Vayansky, I., Kumar, S.A.: A review of topic modeling methods. Inf. Syst. **94**, 101582 (2020)
5. Kherwa, P., Bansal, P.: Topic modeling: a comprehensive review, EAI Endorsed Trans. Scalable Inf. Syst. **7**(24), e2 (2019)

6. Kandlhofer, M., et al.: Enabling the creation of intelligent things: bringing artificial intelligence and robotics to schools. In: IEEE Conference on Frontiers in Education (FIE) (2019)
7. Ng, D.T.K., Leung, J.K.L., Chu, S.K.W., Qiao, M.S.: Conceptualizing AI literacy: an exploratory review. Comput. Educ. Artif. Intell. **2**, 100041 (2021)
8. Zhao, L., Wu, X., Luo, H.: Developing AI literacy for primary and middle school teachers in China: based on a structural equation modeling analysis. Sustainability **14**(21), 14549 (2022)
9. Southworth, J., et al..: Developing a model for AI across the curriculum: Transforming the higher education landscape via innovation in AI literacy. Comput. Educ. Artif. Intell. **4**, 100127 (2023)
10. Kong, S.C., Cheung, W.M.Y., Tsang, O.: Evaluating an artificial intelligence literacy programme for empowering and developing concepts, literacy and ethical awareness in senior secondary students. Educ. Inf. Technol. **28**, 4703–4724 (2023)
11. Wang, B., Rau, P.L.P., Yuan, T.: Measuring user competence in using artificial intelligence: Validity and reliability of artificial intelligence literacy scale. Behav. Inf. Technol. **42**(9), 1324–1337 (2023)
12. Kim, S., et al.: Why and what to teach: AI curriculum for elementary school. Proc. AAAI Conf. on Artif. Intell. **35**(17), 15569–15576 (2021)
13. Deuze, M., Beckett, C.: Imagination, algorithms and news: developing AI literacy for journalism. Digit. J. **10**(10), 1913–1918 (2022)
14. Celik, I.: Exploring the determinants of artificial intelligence (AI) literacy: digital divide, computational thinking, cognitive absorption. Telematics Inform. **83**, 102026 (2023)
15. Ezen-Can, A., Boyer, K.E., Kellogg, S., Booth, S.: Unsupervised modeling for understanding MOOC discussion forums: a learning analytics approach. In: Proceedings of the Fifth International Conference on Learning Analytics and Knowledge, pp. 146–150 (2015)
16. Lee, H., Kwak, J., Song, M., Kim, C.O.: Coherence analysis of research and education using topic modeling. Scientometrics **102**, 1119–1137 (2015)
17. Chen, X., Zou, D., Cheng, G., Xie, H.: Detecting latent topics and trends in educational technologies over four decades using structural topic modeling: A retrospective of all volumes of Computers & Education. Comput. Educ. **151**, 103855 (2020)
18. Kuzi, S., Cope, W., Ferguson, D., Geigle, C., Zhai, C.: Automatic assessment of complex assignments using topic models. In: Proceedings of the Sixth (2019) ACM Conference on Learning@ Scale, pp. 1–10 (2019)
19. Zhang, X., Wu, P., Cai, J., Wang, K.: A contrastive study of Chinese text segmentation tools in marketing notification texts. J. Phys: Conf. Ser. **1302**(2), 022010 (2019)
20. Grootendorst, M.: BERTopic: Neural topic modeling with a class-based TF-IDF procedure arXiv:2203.05794 (2022)
21. Jelodar, H., et al.: Latent Dirichlet allocation (LDA) and topic modeling: Models, applications, a survey. Multimedia Tools Appl. **78**, 15169–15211 (2019)
22. Lu, C., Zhu, L., Xie, Y., Xu, W., Zhao, Y., Cao, Y.: Analysis of hot topics and evolution of research in world-class agricultural universities based on BERTopic. Appl. Math. Nonlinear Sci. **9**(1) (2024)
23. I. Chein.: The environment as a determinant of behavior. J. Social Psychol. **39**(1), 115–127 (1954)
24. Dai, Y.: Dual-contrast pedagogy for AI literacy in upper elementary schools, Learn. Instruct. **91**, 101899 (2024)
25. Du, H., Sun, Y., Jiang, H., Islam, A.Y.M., Gu, X.: Exploring the effects of AI literacy in teacher learning: an empirical study. Humanit. Soc. Sci. Commun. **11**, 559 (2024)
26. Wong, G.K., Ma, X., Dillenbourg, P., Huan, J.: Broadening artificial intelligence education in K-12: Where to start?. ACM Inroads **11**(1), 20–29 (2020)

27. Okoye, K., et al.: Impact of students evaluation of teaching: A text analysis of the teachers qualities by gender. Int. J. of Educ. Technol. Higher Educ. **17**, 49 (2020)
28. Jensen, L.X., Bearman, M., Boud, D.: Understanding feedback in online learning – A critical review and metaphor analysis, Comput. Educ. **173**, 104271 (2021)
29. Espasa, A., Mayordomo, R.M., Guasch, T., Martinez-Melo, M.: Does the type of feedback channel used in online learning environments matter? Students' perceptions and impact on learning, Active Learn. Higher Educ. **23**(1), 49–63 (2022)

Emotion Recognition for Learners Based on Facial Expressions and BVP Signals

Kai Yang[1,2], Yantao Wei[1,2](✉), Qi Xu[1,2], Tao Hu[1,2], and Peishu Chang[1,2]

[1] Central China Normal University, Wuhan 430070, China
yantaowei@mail.ccnu.edu.cn
[2] Digital Education Key Laboratory of Hubei Province, Wuhan 430070, China

Abstract. Recently, emotion recognition technology has gained widespread attention in education. This technology plays a crucial role in enhancing learning motivation, engagement, teacher-student relationships, and academic achievement. With the rapid development of deep learning, multimodal emotion recognition (MER) systems have emerged, capable of effectively integrating information from various data sources, such as audio, video, and text, thus providing more accurate emotion analysis. We propose an innovative hybrid modality emotion recognition model that combines facial expressions and blood volume pulse (BVP) physiological signals, employing a decision fusion strategy to generate the final emotion recognition results. Experimental validation on a self-constructed hybrid modality database revealed that the recognition accuracy of the model under single modalities (facial expressions or physiological signals) was 64.32% and 79.33%, respectively. However, when the two modalities were fused, the emotion recognition accuracy significantly improved to 84.55%. This result demonstrates the effectiveness of our model in multimodal emotion recognition tasks. Our research highlights the tremendous potential of MER technology and provides a more precise emotion analysis tool for the education, aiding in the creation of more intelligent and personalized learning environments.

Keywords: multimodal emotion recognition · education · facial expression · physiological signals

1 Introduction

Emotion recognition has become a critical research field, driven by its broad applications in human-computer interaction, healthcare, and education. Accurate recognition of human emotions can enhance user experiences, improve mental health assessments, and provide personalized educational feedback.

In educational practice, emotion recognition is applied on multiple levels. First, it can help teachers monitor students' emotional states in real-time, enabling them to adjust teaching strategies to better meet students' learning needs. For instance, when students exhibit confusion or frustration, teachers can provide timely assistance to prevent these negative emotions from impacting learning effectiveness. Second, emotion recognition

Q. Liu et al. (Eds.): EITT 2024, CCIS 2600, pp. 86–97, 2026.
https://doi.org/10.1007/978-981-95-1525-7_7

technology can be integrated into intelligent education systems, which can automatically adjust the teaching content and difficulty level by analyzing students' emotional responses, thus achieving personalized learning. Moreover, emotion recognition can aid in the early detection of students' psychological issues, such as anxiety or depression, providing the basis for timely psychological interventions.

Current emotion recognition methods mainly rely on facial expressions, voice, and physiological signals. Facial expression recognition has been studied extensively and shown promising results. However, it often faces challenges such as variations in lighting conditions, occlusion, and individual differences in expressions. On the other hand, physiological signals, especially blood volume pulse (BVP), offer valuable insights into emotional states by capturing subtle changes in the cardiovascular system. BVP signals are measured using photoplethysmography (PPG) and reflect blood flow over time. BVP signals are widely used in emotion recognition and health monitoring. However, relying solely on a single modality for emotion recognition is insufficient; it is necessary to integrate data from other modalities. Studies have shown that multimodal emotion recognition systems improve average accuracy by 9.83% compared to single-modality systems [1]. Therefore, multimodal emotion recognition systems are increasingly being discussed and researched by scholars both domestically and internationally.

This paper proposes a bimodal emotion recognition model for learners, consisting of a facial expression recognition branch and a BVP signal recognition branch. The facial expression recognition branch includes block attention and channel enhancement modules. The BVP signal branch connects shallow BVP features with deep features, enhancing the model's performance and robustness. The development and application of this model not only help to improve educational quality but also promote students' emotional well-being and overall development, further highlighting the importance and application value of emotion recognition in the education field. Our contributions include:

- Designing a novel multimodal framework for educational settings that effectively combines facial and physiological data, enabling the model to recognize learners' emotional states more comprehensively and accurately.
- Proposing the Block-wise Attention module and Channel Enhancement module to obtain more significant local features and a larger receptive field, enhancing the accuracy and robustness of facial expression recognition.
- Integrating shallow and deep features of BVP data, utilizing multi-scale features to capture temporal correlations in signals, and improving emotion recognition performance.

2 Related Work

Emotion classification is a key aspect of emotion recognition, involving the categorization of emotional states into distinct types. Accurate classification is crucial for enhancing human-computer interaction, improving psychological assessments, and offering personalized feedback in education. Current emotion recognition systems are generally divided into unimodal and multimodal categories.

2.1 Emotion Classification

Ekman [2] identified six basic emotions: happiness, sadness, anger, fear, disgust, and surprise, while Plutchik [3] proposed eight basic emotions, including trust and anticipation. Russell's [4] circumplex model uses valence and arousal to represent emotions, and Mehrabian and Russell's [5] PAD model includes pleasure, arousal, and dominance. Despite using these models in emotion analysis, they may not fully capture learning emotions in educational settings [6]. Pekrun et al. [7] classified emotions into achievement, cognitive, social, and topic emotions, introducing "confusion" as a positive learning emotion. D'Mello et al. [8] categorized learning emotions into anxiety, boredom, confusion, curiosity, frustration, and happiness. This study classified learning emotions into boredom, confusion, focus, happiness, neutrality, and drowsiness.

2.2 Student Emotion Recognition

Traditional classroom emotion recognition systems are typically unimodal, relying on a single data source. For example, Zhu et al. [9] and Cui et al. [10] utilized students' real-time facial expressions for emotion detection and analysis. The former leveraged feedback to understand students' classroom engagement and improve teaching methods, while the latter adjusted content based on students' interest or learning difficulty to maintain their motivation for learning English. Zhili and Xue [11] and Zhang and Li [12] also employed students' voice data for emotion recognition. Zhili and Xue [11] introduced new learning parameters and formulas to monitor students' attention states, while Zhang and Li [12] processed voice spectrogram features through multi-feature fusion and used an artificial neural network classifier for emotion recognition to analyze emotional characteristics in teaching. Shaik et al. [13] analyzed student reviews to label emotions, using unsupervised learning and deep learning methods to process the data and provide feedback on students' opinions and insights.

However, these methods face various challenges in practical application. Facial expression-based emotion recognition systems in real classrooms are susceptible to issues such as facial occlusion, lighting, and distance issues. Voice-based emotion recognition in educational settings encounters difficulties in data collection and noise interference. Emotion recognition systems that rely on student reviews or questionnaires lack real-time capability and may fail to reflect students' genuine emotions.

2.3 Multimodal Student Emotion Recognition

Khediri et al. [14] developed a multimodal intelligent tutoring emotion recognition system (MITERS), integrating facial, text, and voice data. The system uses deep learning techniques, including a Deep Convolutional Neural Network (DCNN) for facial expression analysis, a Bidirectional Long Short-Term Memory (BiLSTM) network for processing text emotions, and a Convolutional Neural Network (CNN) for speech emotion recognition. In experiments on the Multimodal Emotion Lines Dataset (MELD), the system achieved an accuracy of 97%. Cárdenas-López et al. [15] proposed three data fusion methods to improve emotion analysis accuracy through Convolutional Neural Networks (CNN) and CNN-LSTM models. After optimizing the Embedding Learning-Centered

Emotion Network (ELCE-Net) model with a genetic algorithm, the accuracy increased to 82%. Zhang et al. [16] built a bimodal emotion recognition system combining skin electrical signals and text, using Artificial Neural Networks (ANN) and Gaussian Mixture Models (GMM) for classification. Although it improved recognition accuracy, the feature extraction process was complex and susceptible to noise. Nandi et al.'s [17] research analyzed the emotional states of 10 students by integrating skin conductance and pulse wave signals. The multimodal approach improved classification accuracy and the F1 score to 0.9583, but the small sample size posed a risk of overfitting.

3 Methodology

3.1 Overview

As shown in Fig. 1, our network primarily consists of a facial expression recognition branch and a BVP signal recognition branch.

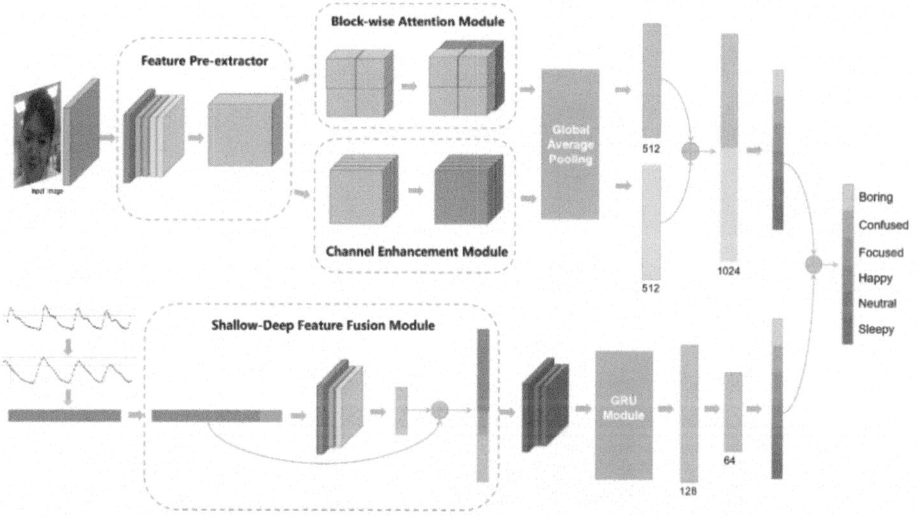

Fig. 1. Network model architecture diagram.

In the facial expression recognition branch, we first preprocess the input image by randomly cropping a 224×224 square region and standardizing it, resulting in a tensor of size (3, 224, 224). Next, a feature extractor, consisting of one 2D convolutional layer and four 5×5 convolutional blocks, extracts facial feature maps. A Block-wise Attention Module and a Channel Enhancement Module then process these feature maps. After processing, a global average pooling layer and a fully connected network are used to obtain two feature vectors, each of length 512. These vectors are concatenated and used for emotion classification.

In the BVP signal recognition branch, we start by preprocessing the raw BVP signal data, which includes outlier processing and denoising. We apply a moving average filter with a window size of 5 to smooth the BVP signal. The preprocessed data is then input into a Shallow-Deep Feature Fusion Module, which processes the data to generate a feature vector composed of original data, shallow data, and deep data. Finally, a GRU module captures temporal dependencies to produce a feature vector with stronger representational capabilities, which is then reduced in dimensionality and classified through two fully connected layers.

We jointly train the two branches, with the loss function composed of two cross-entropy losses: one from the facial expression recognition branch and the other from the BVP signal recognition branch.

$$L = \beta L_{face} + (1 - \beta)L_{bvp}, \quad (1)$$

where β is a hyperparameter used to balance the two parts. Ultimately, we fuse the emotion classification results obtained from both branches to achieve the final emotion recognition result

3.2 Block-Wise Attention Module

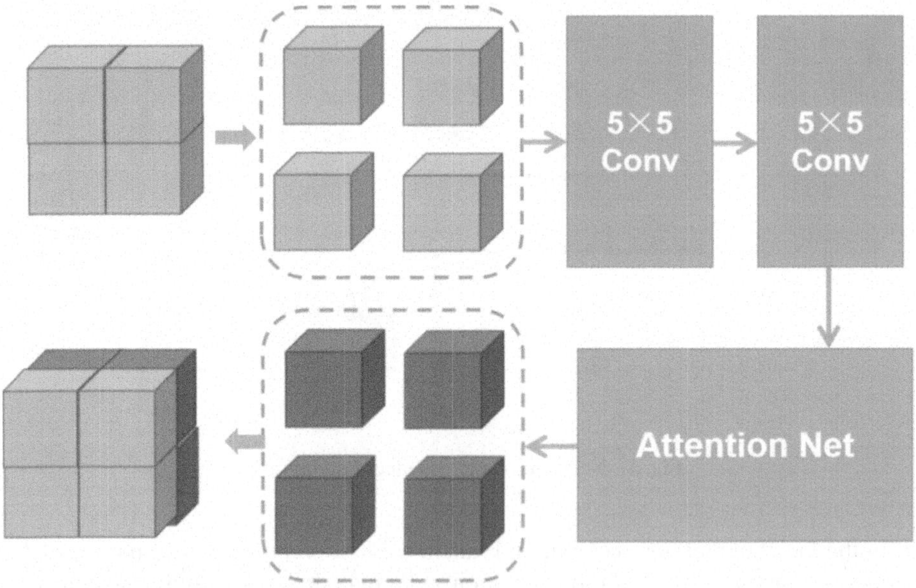

Fig. 2. Structure diagram of the block-wise attention module.

In classroom scenarios, students' faces are often occluded by glasses, hats, masks, etc. Therefore, focusing on the unobstructed local features is crucial. To capture effective local features, we uniformly divide the learned feature map F into four non-overlapping blocks. This allows each feature map to focus on its own local salient features, as shown in Fig. 2.

Specifically, the original feature map $F \in R^{\wedge(28\times28\times128)}$ (where 28 represents the width and height of the feature map, and 128 represents the number of channels) is divided into four local feature maps $f\_i \in R^{\wedge(28\times28\times128)}$, where $i \in \{1,2,3,4\}$. Each local feature map f_i is processed by two 5 × 5 convolutions followed by a Convolutional Block Attention Module (CBAM) [18]. This module consists of a channel attention module (C-channel) and a spatial attention module (S-channel). The channel attention module enhances the feature representation of different channels, while the spatial attention module extracts key information from different spatial locations. This method enhances the performance of the CNN and improves the model's perception ability without increasing its complexity.

The obtained attention weights are multiplied by each channel of the original feature map, resulting in attention-weighted channel feature maps. This emphasizes channels that are helpful for the current task and suppresses irrelevant ones. The obtained spatial attention weights are applied to the original feature map, weighting the features at each spatial location. This highlights important image regions and reduces the influence of less important ones. The local feature maps processed by the attention network are $f_i' \in R^{\wedge(14\times14\times128)}$, where $i \in \{1,2,3,4\}$. These local feature maps f_i^' are then concatenated to form an attention feature map $F' \in R^{\wedge(28\times28\times128)}$ with the same shape as the original feature map. Next, the original feature map F and the attention feature map F' are concatenated along the channel dimension to form a new feature map $F'' \in R^{\wedge(28\times28\times128)}$ for further processing.

3.3 Channel Enhancement Module

In the Channel Enhancement Module, we perform channel-wise enhancement of feature maps to improve feature representation quality. Initially, the input feature map $F \in R^{\wedge(28\times28\times128)}$is divided into four parts, $F = [f_1, f_2, f_3, f_4]$, where $f_i' \in R^{\wedge(28\times28\times32)}$, with each part containing an equal number of channels. We process these parts sequentially, applying a 5 × 5 convolution kernel to each part and enhancing the feature maps through element-wise addition, as shown in Fig. 3.

First, a 5 × 5 convolution is applied to the first feature part f_1, generating a new feature map. This convolution operation extracts new feature information from f_1. The result of this convolution is then added element-wise to the second feature part f_2, producing an updated f_2. The purpose of this step is to transfer the feature information from f_1 tof_2, thereby enhancing the feature representation in f_2. Following the pattern in (2), the same process is repeated for f_3 and f_4, where f_i' represents the feature part f_i' after the update.

$$f_i' = con_{(5\times5)}\left(f_{(i-1)}\right) + f_i \ (1 < i \leqq n) \tag{2}$$

After updating f_4, we proceed with a backward information transfer. A 5 × 5 convolution is applied to the updated f_4, resulting in a new feature map, which is then added element-wise to the updated f_3. This process aims to transfer the newly extracted features from f_4 back to f_3, further enhancing the feature representation in f_3. Repeat the process in (3) until f_1 is updated. f_i" represents the feature part f_i' after the update.

$$f_i'' = con_{(5\times5)}\left(f'_{(i+1)}\right) + f_i' \ (1 < i \leq n) \tag{3}$$

This process completes a cycle of feature map enhancement, ensuring that features across all parts are thoroughly fused and enhanced. Through these steps, we obtain a set of new feature maps that integrate the results of multiple convolution operations and element-wise additions. This ensures that both forward and backward feature subsets encompass richer scale information, allowing each part to gain a larger receptive field and more feature information, thus improving the richness and robustness of the feature representation. This method effectively enhances the feature maps' ability to represent different levels of information, thereby boosting the overall performance of the model.

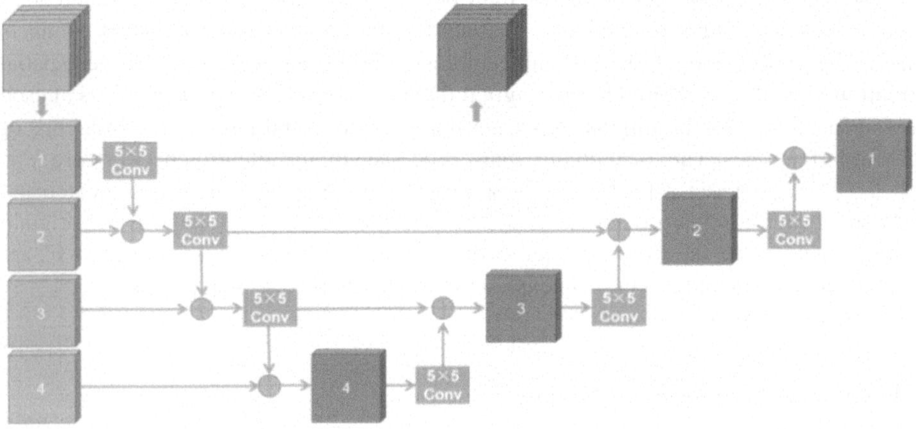

Fig. 3. Structure diagram of the channel enhancement module.

3.4 Shallow-Deep Feature Fusion Module

In emotion recognition, extracting and selecting features from physiological signal data is crucial. These features, which can be time-domain or frequency-domain, include metrics like heart rate variability (HRV), power spectral density (PSD), and wavelet transform coefficients.

Traditionally, features were manually extracted for classification. While these shallow features have performed well, manual extraction is time-consuming and error-prone, requiring extensive domain knowledge [19]. Recent studies have adopted deep learning techniques for automatic feature extraction to address these issues. For example, Poria et al. [20] used CNNs for textual features and 3D-CNNs for audio and visual features, followed by support vector machine classification. Yang et al. [21] combined CNNs and

RNNs for electroencephalogram (EEG) signal classification, achieving 91.03% accuracy. Although deep learning methods are more efficient and accurate, they may lack statistical significance and fail to capture data characteristics comprehensively.

In summary, our approach integrates both shallow and deep feature extraction methods by developing a shallow-deep feature extraction fusion module. We select features from both time and frequency domains, as detailed in Table 1.

Table 1. Selected shallow features

Domain name	Feature name	Total
Time domain	RR Interval, Root Mean Square of Successive Differences (RMSSD), Standard Deviation of Normal-to-Normal Intervals (SDNN), Max Amplitude, Min Amplitude, Mean Amplitude, Amplitude Standard Deviation, Waveform Rise Time, Waveform Fall Time, Pulse Wave Shape Index, Pulse Wave Velocity, Rise Time, Approximate Entropy, Heart Rate Acceleration	14
Frequency domain	Power Spectral Density (PSD), Low Frequency Power (LF Power), High Frequency Power (HF Power), LF/HF Ratio (Spectral Power Ratio), Peak Frequency, Spectral Bandwidth	6

First, the processed raw BVP signal is concatenated with the extracted shallow features (a total of 20), forming a vector with 100 features, which is then fed into a CNN module. After processing through three convolutional layers, a feature map with 128 channels and 156 features is generated. The average is then computed along the second dimension for each channel, resulting in 156 deep features. These deep features are concatenated with the original data and shallow features (a total of 100), forming a vector with 256 features, which is input into the gated recurrent unit (GRU) module to capture temporal dependencies, producing the final output feature vector. The entire shallow-deep feature fusion module integrates time and frequency domain feature extraction, along with CNN and GRU processing, enhancing model performance and demonstrating high adaptability and robustness in signal processing and feature extraction.

4 Experiments

4.1 Dataset

The dataset used in this study was independently constructed to simulate a learning environment, capturing learners' facial videos and BVP signal data. The participants were 58 undergraduate and graduate students from a university in central China, aged between 21 and 25 years, including 42 females and 16 males. During data collection, a frame was extracted from the video data every 4 s, and the corresponding BVP signal data was recorded for 2 s before and 2 s after the frame. All data were stored in comma-separated values (CSV) format, with each row representing a single extracted facial image and its corresponding BVP signal segment. Through frame extraction, manual screening, and annotation, a total of 8,183 facial images and their corresponding BVP signal segments were generated. The dataset quantity statistics are shown in Table 2, and the specific dataset contents are shown in Fig. 4.

Table 2. Dataset quantity statistics

	Images/BVP					
	Boring	Confused	Focused	Happy	Neutral	Sleepy
train	623	804	1592	1493	626	592
test	267	344	682	639	268	253

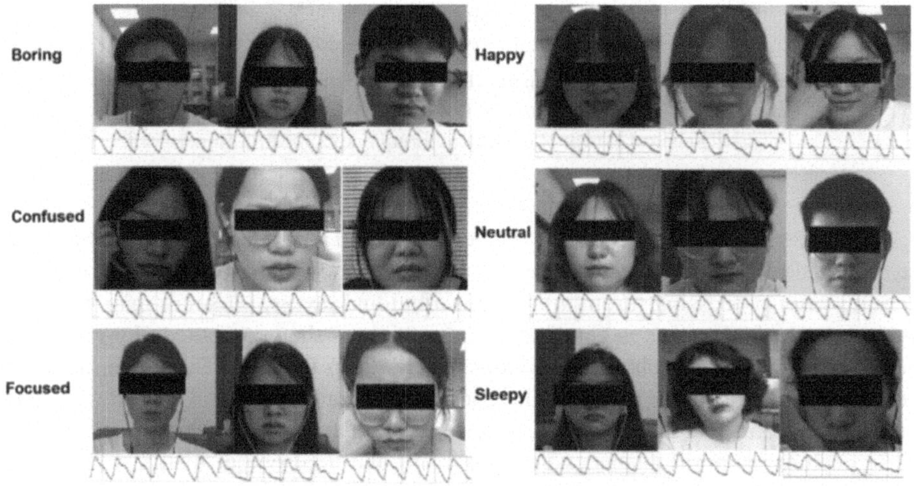

Fig. 4. Illustration of dataset contents.

5 Results

To validate the effectiveness of each component in our network, we conducted an ablation analysis on our custom dataset, focusing on the facial expression branch, the physiological signal branch, the Block-wise Attention Module, the Channel Enhancement Module, the Shallow-Deep Feature Fusion Module, and the hyperparameter β.

As shown in Table 3 and Fig. 5, the accuracy rates for using the BVP recognition module and the facial recognition module alone were 64.32% and 79.33%, respectively. Without the Block-wise Attention Module, Channel Enhancement Module, and Shallow-Deep Feature Fusion Module, the accuracy rates were 79.64%, 82.53%, and 83.42%, respectively. After fusing the two branches, the accuracy rate improved to 84.55%, with a recall rate of 84.55% and an F1 score of 84.60%. This represents increases of 20.23% and 5.22%, 20.46% and 5.22%, and 19.99% and 5.23% over the single-modal approaches, respectively. These results demonstrate the effectiveness of multimodal information fusion in emotion recognition, showing that combining information from different modalities allows the model to identify emotional states more accurately. The improvements in recall rate and F1 score further confirm the efficacy of the fusion strategy.

Table 3. Ablation experiment

Model	Accuracy	Recall	F1 Score
BVP signal recognition	64.32	64.09	64.61
facial recognition	79.33	79.33	79.37
facial recognition + BVP signal recognition (lacks a Block-wise Attention Module)	79.64	79.51	79.12
facial recognition + BVP signal recognition (lacks Channel Enhancement Module)	82.53	82.41	82.13
facial recognition + BVP signal recognition (lacks a Shallow-Deep Feature Fusion Module)	83.42	83.31	83.24
facial recognition + BVP signal recognition (ours)	84.55	84.55	84.60

To explore the impact of β on the model, we tested values ranging from 0 to 1. Figure 6 shows that the best performance is achieved when $\beta = 0.8$, indicating that the facial recognition branch is more important than the BVP physiological signal recognition branch.

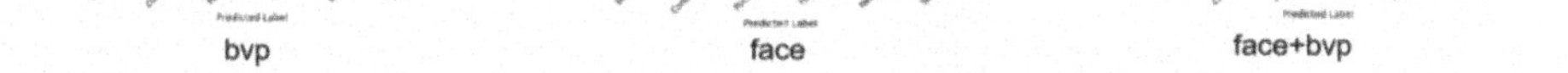

Fig. 5. Comparison of confusion matrices.

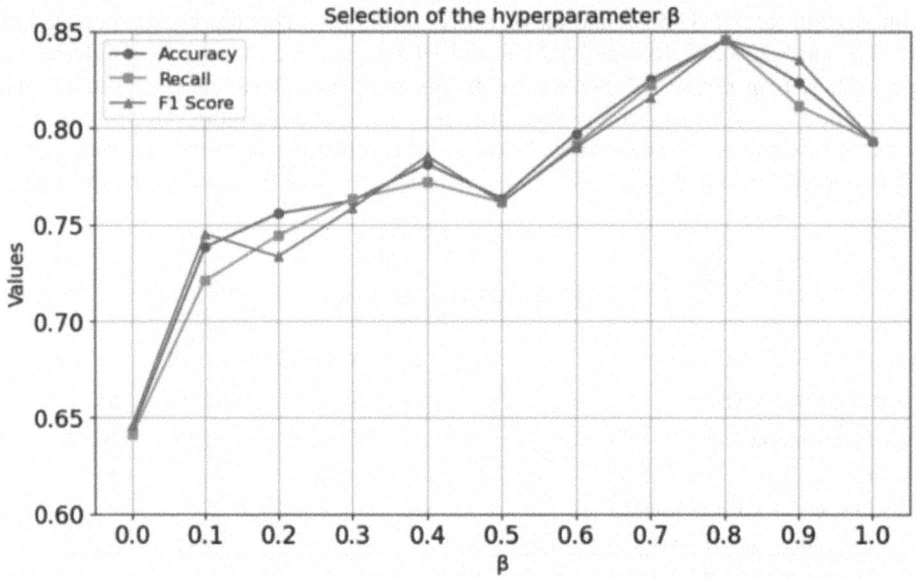

Fig. 6. Selection of the hyperparameter β.

6 Conclusion and Discussion

This study introduces a dual-modal learner emotion recognition model designed to address common issues such as facial occlusions and emotional masking in real classroom settings. The facial expression recognition branch enhances accuracy and robustness by focusing on feature-rich local regions and improving channel information interaction. The BVP signal recognition branch enriches feature information by fusing shallow and deep features, thus enhancing the reliability of physiological signals. Extensive experiments on a self-built dataset in a simulated learning environment validate the effectiveness of this network architecture. The results show that the model effectively integrates facial expressions and physiological signals, achieving superior accuracy compared to single-modal systems. However, the current model only uses BVP signals. Including other physiological signals, such as EEG, electrodermal activity (EDA), and respiratory rate (RR), which carry rich physiological and psychological information, could further enhance the model's performance and broaden its educational applications.

Acknowledgments. This work was supported in part by the National Natural Science Foundation of China under Grant 62277029, the Humanities and Social Sciences of China MOE under Grants 20YJC880100, the Knowledge Innovation Program of Wuhan – Basic Research under Grant 2022010801010274, and the Fundamental Research Funds for the Central Universities under Grants CCNU24ai018 and CCNU22JC011.

References

1. D'Mello, S.K., Westlund, J.K.: A review and meta-analysis of multimodal affect detection systems. ACM Comput. Surv. **47**(1), 1–35 (2015)
2. Ekman, P.: Are there basic emotions? Psychol. Rev. **99**(3), 550–553 (1992)
3. Plutchik, R.: Emotion: A Psychoevolutionary Synthesis. Harper & Row, New York City (1980)
4. Russell, J.A.: A circumplex model of affect. J. Pers. Soc. Psychol. **39**(6), 1161–1178 (1980)
5. Mehrabian, A., Russell, J.A.: An Approach to Environmental Psychology. MIT Press, Cambridge (1974)
6. Wei, Q., Sun, B., He, J., Yu, L.: BNU-LSVED 2.0: spontaneous multimodal student affect database with multi-dimensional labels. Signal Process. Image Commun. **59**, 168–181 (2017)
7. Pekrun, R., Goetz, T., Titz, W., Perry, R.P.: Academic emotions in students' self-regulated learning and achievement: a program of qualitative and quantitative research. Educ. Psychol. **37**(2), 91–105 (2002)
8. D'Mello, S.K., Lehman, B., Person, N.: Monitoring affect states during effortful problem-solving activities. Int. J. Artif. Intell. Educ. **20**(4), 361–389 (2010)
9. Zhu, H., Hu, P., Tang, X., Xia, D., Huang, H.: NAGNet: a novel framework for real-time students' sentiment analysis in the wisdom classroom. Concurr. Comput.: Pract. Exp. **35**(22), e7689 (2023)
10. Cui, Y., Wang, S., Zhao, R.: Machine learning-based student emotion recognition for business English class. Int. J. Emerg. Technol. Learn. **16**(5), 176–186 (2021)
11. Zhili, S., Xue, W.: Real-time assessment of student concentration based on attention and speech emotion recognition. In: 2022 7th International Conference on Intelligent Informatics and Biomedical Sciences (ICIIBMS), pp. 263–268. IEEE (2022)
12. Zhang, S., Li, C.: Research on feature fusion speech emotion recognition technology for smart teaching. Mob. Inf. Syst. 5032478 (2022)
13. Shaik, T., Tao, X., Dann, C., Xie, H., Li, Y., Galligan, L.: Sentiment analysis and opinion mining on educational data: a survey. Nat. Lang. Process. J. **2**, 100003 (2023)
14. Khediri, N., Ben Ammar, M., Kherallah, M.: A real-time multimodal intelligent tutoring emotion recognition system (MITERS). Multimed. Tools Appl. **83**, 57759–57783 (2024)
15. Cárdenas-López, H.M., Zataraín-Cabada, R., Barrón Estrada, M.L., Mitre-Hernández, H.A.: Semantic fusion of facial expressions and textual opinions from different datasets for learning-centered emotion recognition. Soft. Comput. **27**, 17357–17367 (2023)
16. Zhang, L., Ye, N., Huang, H., Wang, R.: A bimodal emotion recognition system based on skin electrical signals and text information. Comput. Syst. Appl. **11**, 103–108 (2018)
17. Nandi, A., Xhafa, F., Subirats, L., Fort, S.: MDEAW: a multimodal dataset for emotion analysis through EDA and PPG signals from wireless wearable low-cost off-the-shelf devices. arXiv (2022). https://arxiv.org/abs/2207.06410
18. Woo, S., Park, J., Lee, J.Y., Kweon, I.S.: CBAM: convolutional block attention module. In: Proceedings of the European Conference on Computer Vision (ECCV), pp. 3–19 (2018)
19. Kim, B.H., Jo, S.: Deep physiological affect network for the recognition of human emotions. IEEE Trans. Affect. Comput. **11**(2), 230–243 (2018)
20. Poria, S., Majumder, N., Hazarika, D., Cambria, E., Gelbukh, A., Hussain, A.: Multimodal sentiment analysis: addressing key issues and setting up the baselines. IEEE Intell. Syst.Intell. Syst. **33**(6), 17–25 (2018)
21. Yang, H., Han, J., Min, K.: A multi-column CNN model for emotion recognition from EEG signals. Sensors **19**(21), 4736 (2019)

Making AI Forget You: Removing Educational Data from Intelligent Education Models

Yiyu Wu[1], Yaowen Kuang[1], Xiaoyu Chen[1], and Tao Wang[2](✉)

[1] Faculty of Artificial Intelligence in Education, Central China Normal University, Wuhan, China

[2] Hubei Key Laboratory of Digital Education, Central China Normal University, Wuhan, China

tmac@ccnu.edu.cn

Abstract. The deep integration of AI and education relies on the non-intrusive collection and fusion of educational data, and while such a data-intensive research paradigm can promote the development and reform of intelligent education, the collection and analysis of educational data also brings risks of privacy infringement and unethical applications. Data destruction, as a key link in the life-cycle management of personal education data, is also an important means of protecting the ''right to be forgotten" of education participants. However, the lack of reliable mechanisms for implementing proper data destruction seriously constrains the healthy and sustainable development of intelligent education. This study introduces the concept of Machine Unlearning from computer science fields into the education field, outlines a technical model of intelligent unlearning of educational data, and explains the practical process by which educational data can be intelligently unlearned, taking facial data and graph structure data commonly as examples. The results demonstrate that the intelligent unlearning of educational data can be achieved efficiently, without retraining the original model, ensuring that the unlearned data remains unrecognizable and thus protecting education participants' "right to be forgotten." This study provides a valuable reference for optimizing the whole life cycle management of educational data and establishing more robust trust in the digital transformation of education with AI.

Keywords: intelligent education · artificial intelligence · machine unlearning · educational data · right to be forgotten

1 Introduction

Advanced artificial intelligence (AI) has become increasingly widespread in the field of education, especially models that leverage machine learning (ML) technologies to characterize educational subjects, explore educational principles, optimize educational services, and transform educational practices, thereby reshaping intelligent learning analytics.

However, as AI and ML become increasingly data-dependent, a large amount of educational data from teachers, students, and other educational participants may be at risk. Many factors, including privacy concerns, government regulations, and the need for

Q. Liu et al. (Eds.): EITT 2024, CCIS 2600, pp. 98–109, 2026.
https://doi.org/10.1007/978-981-95-1525-7_8

model generalization, are introducing a new demand: to delete data related to oneself. Specifically, laws and regulations, such as the European Union's General Data Protection Regulation (GDPR) [1] and the California Consumer Privacy Act (CCPA) [2], have been implemented to grant individuals the right to be forgotten. These laws allow users to have their private information deleted upon request. While deleting data from back-end databases satisfies regulations, it is not sufficient—ML models still retain memories of deleted data items, and membership inference attacks [3], model inversion attacks [4], and adversarial attacks [5] can reveal or reconstruct the specific contents of the deleted data. Education participants, such as graduating students, are requesting that not only data itself be deleted, but also that the impact of the deleted data item be eliminated from an already-trained learning analytic model. However, the parameters of a trained model do not always demonstrate a clear correlation with the training data. Therefore, eliminating the impact of a particular data item from an ML model is extremely challenging. It is remarkably difficult to enforce an ML model to forget a specific data item.

Currently, the education field has gradually realized the importance of educational data destruction and the need to ensure the participants' right to be forgotten. However, due to the lack of reliable implementation mechanisms and operational implementation paths, data destruction still remains at the level of framework design and management strategy. In the field of computing, "Machine Unlearning," as a new technical paradigm, can delete specific data from ML models without requiring that the models be retrained, and is expected to become an important technical path toward achieving complete data destruction. This study applies Machine Unlearning to the complex and unique field of intelligent education, introduces the concept of intelligent unlearning of educational data, and provides a path for the destruction of educational data and the protection of the right to be forgotten from a technical perspective. This approach aims to break through the privacy and ethical dilemmas faced by those employing AI technologies, with a goal of establishing sound trust mechanisms for the digital transformation of education with AI. Providing this theoretical reference and practical guidance will help educators more ethically and safely facilitate the deep integration of AI and education.

2 Educational Data and Intelligent Unlearning Models

2.1 Defining Intelligent Unlearning

Machine Unlearning refers to the process of deleting previously learned information from a ML model while ensuring that the performance of the model remains largely unchanged after the unlearning process [6], as depicted in Fig. 1. Essentially, it is the opposite of ML: while the goal of ML is to train models for identification, classification, and prediction, the purpose of Machine Unlearning is to revoke or reverse this process, ensuring that in specific scenarios, selectively removing sensitive or no longer needed information from the model is possible.

Based on the above definition, we provide an explanation of intelligent unlearning of educational data, which involves selectively and completely removing corresponding data samples, classes, and features from both the original datasets and the trained and deployed intelligent education models (such as learning analytics models, educational assessment models, learning resource recommendation models, etc.). This is done

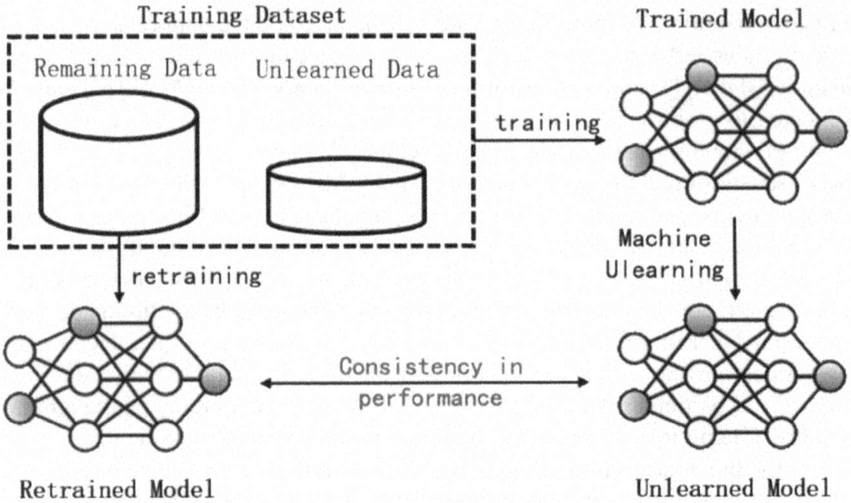

Fig. 1. A conceptual model of Machine Unlearning.

through the use of methods like model retraining, data recombination, and model modification. In response to varying levels of data unlearning requests from educational participants (students, teachers, etc.), a model may "unlearn" various types of data samples, classes, or features. The process described in this study aims to achieve accurate or approximate unlearning while ensuring that the performance of intelligent educational models remains largely unchanged. The core objectives of intelligent unlearning of educational data include:

- To safeguard educational participants' data privacy and their rights to be forgotten by effectively deleting sensitive personal data without affecting the accuracy of intelligent education model results, and by ensuring that the privacy rights of educational participants are fully respected and protected.
- To enhance the accuracy and fairness of intelligent education models and avoid instances where AI models may produce inaccurate predictions or unfair decisions due to biases or errors in historical training data. Through intelligent unlearning of educational data, we can regularly analyze the performance of intelligent education models, identify and eliminate potential biases, and thereby improve the accuracy of educational models and the fairness of educational decisions.
- To update intelligent education models in a timely manner, allowing such models to more effectively incorporate new or evolving educational environments and data, while maintaining the validity of their results. Intelligent unlearning of educational data enables the intelligent education models to adapt to new educational contexts by "unlearning" outdated and irrelevant information, promoting greater flexibility and adaptability of the models.

2.2 An Intelligent Unlearning Model for Educational Data

To clarify the technical principles of intelligent unlearning, we have constructed a technical model, as shown in Fig. 2. This model clearly outlines the standardized workflow and technical framework of intelligent education models when unlearning specific educational data. Specifically, the intelligent unlearning model primarily consists of three phases: model training, intelligent unlearning, and educational application.

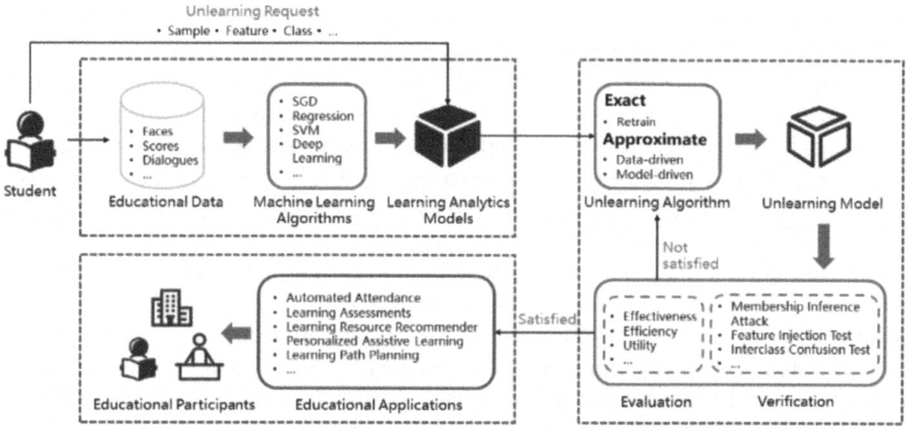

Fig. 2. Educational Data Intelligence Unlearning Technology Model.

Model Training In the model training phase, the core task is to construct intelligent education models that can meet diverse teaching needs by integrating extensive and diverse big data resources in the education field with various advanced artificial intelligence (machine learning) algorithms. This data encompasses learning behavior data (such as online study duration, click trajectories, and answering records), classroom interaction records (including teacher-student Q&A and group discussion details), and exam performance data. Through in-depth mining and analysis of this vast dataset, algorithms like deep learning and reinforcement learning are utilized to build models that can design personalized learning paths based on students' behavior patterns to boost engagement [7], predict future performance using historical data [8], and analyze emotional changes during learning via natural language processing and sentiment analysis to inform teaching strategy adjustments [9]. Crucial steps such as data preprocessing, feature engineering optimization, and algorithm fine-tuning are repeatedly verified and refined to ensure the models' robust understanding, prediction abilities, and reliability in practical applications.

Intelligent Unlearning The model utilizes accurate unlearning methods (such as retraining—though its high computational costs pose a challenge for complex models) or approximate unlearning methods (including data recombination [10] and model modification strategies [11], both of which aim for the unlearned model to be approximately indistinguishable from a retrained model in data distribution) to fulfill unlearning requests. The unlearned intelligent education model then needs to be evaluated based on different performance indicators, such as effectiveness (ensuring that the model has effectively unlearned the corresponding data according to the request), efficiency (the ability to quickly complete the unlearning process), usability (maintaining the model's accuracy for other test data), and validity (employing methods such as membership inference attacks, feature injection tests, and intra-class confusion tests to prove that the model has indeed unlearned the requested data and poses no risk of information leakage).

Educational Applications When students or other educational participants complete their course studies or are about to graduate, they may wish to delete their personal data. They can submit requests for intelligent unlearning of educational data to remove the corresponding personal data from the model. These unlearning requests involve sample unlearning (such as deleting a specific individual's data sample), class unlearning (such as deleting all facial data of a student), and feature unlearning (such as removing specific facial features from a learning sentiment analysis model). Upon receiving these requests, the owner of the intelligent education model will employ unlearning methods to delete the specified data and then evaluate and validate the unlearned model to ensure that its performance has not deteriorated and that the relevant data and its impacts have been successfully removed. If the model fails to meet these standards during evaluation, that may indicate a risk of leaking the data or information that was meant to have been unlearned, so the unlearning operation will be re-executed. Once the model passes the validation and evaluation, it will continue to provide safe and reliable intelligent education services for students, teachers, schools, and other stakeholders.

3 Two Applications of Intelligent Unlearning of Educational Data

3.1 Intelligent Unlearning for Facial Recognition Models

Intelligent education models are important for achieving intelligent and smart education. To ensure safer and more trusted usage of intelligent education models, we explore the application of intelligent unlearning to a commonly used method in intelligent education analysis—facial recognition. Facial recognition is a core module in various educational applications such as non-intrusive sign-ins, learning engagement recognition, and learning sentiment computation in current intelligent education. However, facial data is an important biometric identifier that is directly linked to personal identity and is highly sensitive. Students expect to their facial data to be deleted after they graduate or leave the classroom and no longer need to use it for any form of identity recognition.

We selected 54 individuals from the CASIA-WebFace facial dataset whose data included more than 400 face images. Each individual's face images were stored as a class based on their identity, with 60% of the images in each class were chosen for the training set; the other 40% were used as the test set. The ResNet-18 convolutional neural network, which is commonly used in image classification, was pre-trained on the training set to construct a facial recognition model. When deleting specific individuals' facial data from the training set and removing their influence in the ResNet-18 face recognition model, we applied the gradient projection unlearning method [12]. This method calculates the core gradient space of the model weights using the retained data and selectively achieves approximate unlearning of data in this space by orthogonalizing the gradient updates to the core gradient space. While updating the model weights, this process allows for the model to keep knowledge learned from the retained data. The specific process of gradient projection unlearning is as follows:

Calculation of Model Weights After the ResNet-18 facial recognition model is trained, the basis vectors and eigenvalues of the gradient space for the complete training set are calculated. By passing the training samples through each layer of the network, the output of each layer is obtained, where the input of that layer is the output of the previous layer. For convolutional layers, an input feature vector is extracted from the output of the previous layer; for linear layers, the input feature vector is equal to the output of the previous layer. Then, all input feature vectors are concatenated to form an input representation matrix. Finally, singular value decomposition is used to calculate the basis vectors and eigenvalues of the input matrix. During the calculation process, the basis vectors and eigenvalues of each layer are cached.

Construction of the Core Gradient Space Based on the unlearning request, the basis vectors of the gradient space for the remaining dataset are calculated using the cached results. The final core gradient space is constructed using the basis vectors of the input representation matrix obtained from the remaining data.

Gradient Updates and Unlearning Based on the remaining data and the model loss function, new gradients are calculated. These gradients are projected onto the core space to obtain gradient projections. By subtracting the gradient projections from the original gradients, information related to the unlearned data is removed from the remaining data.

Table 1. The experimental results of unlearning the first 5 classes using the ResNet-18 model trained on CASIA-WebFace.

Dataset	Retrained	Unlearned
Du_train	0.00%	2.75%
Du_test	43.00%	21.27%
Dr_test	91.33%	73.70%

During model training, we set the minimum batch size to 124, the initial learning rate to 0.001, the training epoch to 150, and the final learning rate to 0.0001. The facial recognition accuracy of the model obtained from the training was 90.72%. The unlearning request was to forget the first 5 classes in both the training set and test set. Table 1 presents the accuracy percentages of the ResNet-18 facial recognition model after it has been retrained using the remaining training set, and after the gradient projection unlearning process. Results are shown for the unlearned training set Du_test, the unlearned test set Du_train and the remaining test set Dr_test.

It can be clearly observed from the experimental results that retraining, as the benchmark method, involves removing the first 5 classes of facial data from the training set and then retraining the model for unlearning operations. On the Du_train dataset, the model achieves a recognition accuracy of 0%, a significant result demonstrating that the retraining-based unlearning has achieved precise effectiveness—i.e., the model has completely "unlearned" the removed specific facial data categories and cannot recognize them anymore. On the Du_test dataset, the recognition accuracy remains at 43%, which may be attributed to multiple factors: the richness and complexity of facial features themselves, with subtle differences existing among different individuals; and the high similarity of facial features within the dataset, enabling the model to recognize un-unlearned data to a certain extent based on shared feature information. On the Dr_test dataset, the recognition accuracy reaches 91.33%, indicating that after unlearning and retraining, the model maintains excellent recognition performance on this dataset—fully reflecting the effectiveness of retraining in ensuring precise unlearning of target data while preserving recognition capabilities for other data.

Comparatively, the experimental results of approximate unlearning methods (which do not retrain the model but modify parameters, etc.) exhibit different characteristics. On Du_train, the model accuracy is 2.75%—though not reduced to 0%, this value is significantly lower than the original model, demonstrating that approximate unlearning substantially reduces the model's recognition ability for removed data and achieves considerable unlearning effects. On Du_test, the accuracy is 21.27%, lower than that of the retrained model; on Dr_test, the accuracy is 73.70%, showing a noticeable performance decline compared to the retrained model's 91.33%, which highlights the insufficiency of approximate unlearning in maintaining model generalization. However, despite performance loss, approximate unlearning still enables the model to retain relatively stable recognition accuracy on remaining datasets after specific facial data unlearning. This indicates that through approximate strategies like parameter optimization and data recombination—without consuming substantial computational resources for retraining—it is possible to meet the application goals of rapid data deletion and privacy protection, providing a more flexible technical choice for practical unlearning scenarios.

In addition, by analyzing the confidence level of the unlearned model as shown in Fig. 3, entropy—an important indicator for measuring uncertainty—has a distribution that intuitively reflects the model's "confidence" in data prediction. Specifically, when the unlearned model predicts unlearned data, the generated entropy distribution curve shows a high degree of similarity to that of the retrained model. Both the central tendency of entropy values and the degree of distribution dispersion exhibit relatively consistent morphological characteristics. This means that even without using time-consuming and labor-intensive retraining, the model adjusted by approximate unlearning methods can achieve similar effects to the retrained model in terms of prediction uncertainty when facing data marked for unlearning. This consistency is not only reflected in numerical values but also in the overall trend of distribution morphology, indicating that approximate unlearning methods can achieve effects comparable to retraining in changing the model's "cognition" of specific data, providing strong data support for the efficiency and feasibility of unlearning technology in practical applications.

Therefore, by modifying specific parameters of the model, data unlearning can indeed be efficiently achieved. In various applications that rely on intelligent education models, the technique of intelligent unlearning of educational data can be adopted to remove specific data and any "data shadows" from the model, thus effectively safeguarding the participants' right to be forgotten, protecting their data privacy, and truly achieving data destruction in the full lifecycle management of educational data.

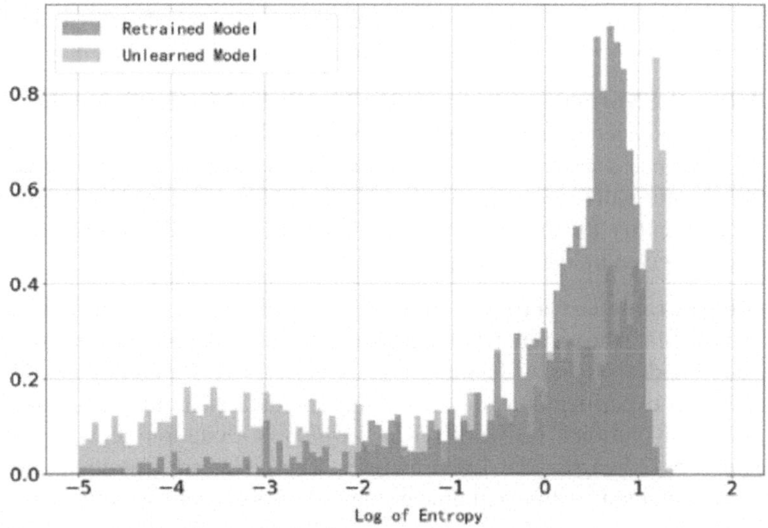

Fig. 3. Distribution of Entropy of Unlearned Data on Unlearned Model and Retrained Model.

3.2 Intelligent Unlearning for Graph Data

Learning analytics technology serves as a crucial tool for achieving personalized education, optimizing learning experiences, and driving innovations in educational practices. Among the various data types in learning analytics, graph-structured data plays a significant role in visually representing relationships among students, learning resources, learning tasks, and other elements, enabling educators to gain a deeper understanding of students' learning paths, interaction patterns, and learning achievements [13]. However, graph-structured data also contains a large amount of personal and sensitive information, such as records of students' access to and interactions with learning resources. Once students graduate or leave the learning environment, they may desire that such graph-structured data be unlearned to protect their privacy and personal information security.

We will base our approach on an innovative Graph Influence Function (GIF) method [14] that introduces an additional loss term into the traditional influence function to account for the loss of affected data neighborhoods caused by structural dependencies. In graph data, nodes are connected through edges, creating mutual dependencies. The neighborhood aggregation mechanism of Graph Neural Networks (GNNs) further enhances the similarities between linked nodes. Therefore, unlearning a certain node, edge, or feature will inevitably affect the states of its multi-hop neighbors. Taking inspiration from the idea of data perturbation in traditional influence functions, this method considers graph dependencies and derives a graph-oriented influence function. The specific process is as follows:

(1) Defining k-hop Neighbors Based on the shortest path distance (SPD) in a graph, we can define the k-hop neighbors of a node or an edge. The k-hop neighbors of a node refer to all nodes that can be reached from the node through no more than k edges. The k-hop neighbors of edge AA may refer to the set of nodes or edges that are directly connected to this edge or indirectly connected through no more than k edges.

(2) Calculating the Normalized Propagation Matrix and Determining the Affected Area Given the adjacency matrix and the corresponding degree matrix (a diagonal matrix with diagonal elements representing the degree of each node) of a graph, we can compute the normalized propagation matrix. Utilizing this normalized propagation matrix, we can identify the affected areas in different types of graph unlearning tasks associated with a given unlearning request.

(3) Estimating Parameter Changes and Achieving Unlearning For different unlearning tasks, we calculate the parameter changes, update the model parameters, and remove the influence of nodes, edges, or features.

We conducted experiments using the Cora graph dataset on the generalized graph model Graph Convolutional Network (GCN), which is a citation network consisting of 2,708 nodes, 5,429 edges, and 1,433 features. The nodes are classified into 8 categories. We implemented customized unlearning tasks appropriate for different unlearning requests and demonstrated the advantages of each in terms of model utility, unlearning efficacy, and unlearning efficiency.

In the model training configuration, we set the minimum batch size to 520, an initial learning rate of 0.001, and the training epoch to 100. To evaluate the efficacy of the unlearning method and the usability of the unlearned models, we randomly unlearned nodes, edges, and features in the models, with a unlearning ratio of 0.05. In Table 2, we provide detailed comparisons of the F1 scores (a reconciled average of precision and recall that can be used to assess the usability of a model, with values closer to 1 indicating better model performance) and runtimes (RT) for the original model, the retrained model, and the unlearned models. These measurements enable us to assess the efficiency of the unlearning method and the performance of the unlearned models.

Table 2. The experimental results of unlearning using a GCN model trained on Cora.

Unlearning request		Original	Retrained	GIF
nodes	F1	0.8472	0.7457	0.7690
	RT(seconds)	/	14.34	0.26
edges	F1	0.8443	0.8166	0.8195
	RT(seconds)	/	14.46	0.30
features	F1	0.8428	0.8361	0.8380
	RT(seconds)	/	14.60	0.25

From the experimental results, in the evaluation of three key tasks: node unlearning, edge unlearning, and feature unlearning, the F1 scores of the retrained model are 0.7457, 0.8166, and 0.8361 respectively; while the model using approximate unlearning achieves F1 scores of 0.7690, 0.8195, and 0.8380 respectively. This set of data indicates that compared with the retraining method, Graph Influence Function Unlearning (GIF) demonstrates stronger adaptability and higher precision in handling different types of data deletion requests. Overall, regardless of the type of unlearning request, the GIF method performs better, with F1 scores comparable to or higher than those of the retrained model. This implies that the GIF method can not only effectively achieve data unlearning operations but also maintain and improve model performance more effectively.

In terms of time efficiency, the GIF method shows a significant advantage. Experimental data clearly shows that the runtime of retraining is approximately 14 s, while GIF only requires 0.26 s, with the former taking about 50 times longer than the latter. In practical application scenarios, especially when facing large-scale datasets and frequent data update requirements, this huge difference in time efficiency makes the GIF method more practical and scalable for handling data unlearning tasks. Further analysis of different types of unlearning requests reveals that among node, edge, and feature unlearning requests, feature unlearning not only has higher F1 scores but also the best time efficiency. This result suggests that in the data structure, feature unlearning operations may have a more concise computational logic, or their impact on the overall model performance is more concentrated, enabling better results to be achieved in a shorter time.

To verify the actual effect of data deletion more comprehensively and rigorously, the study introduces an innovative strategy inspired by the concept of adversarial attacks. The specific operation process is as follows: First, a certain number of edges are carefully introduced into the training graph to ensure that these newly added edges connect nodes of different classes. This design aims to artificially disrupt the model's original learning pattern by injecting adversarial noise, causing performance impairment and degradation. Subsequently, these adversarial edges are set as unlearning requests, and two methods,

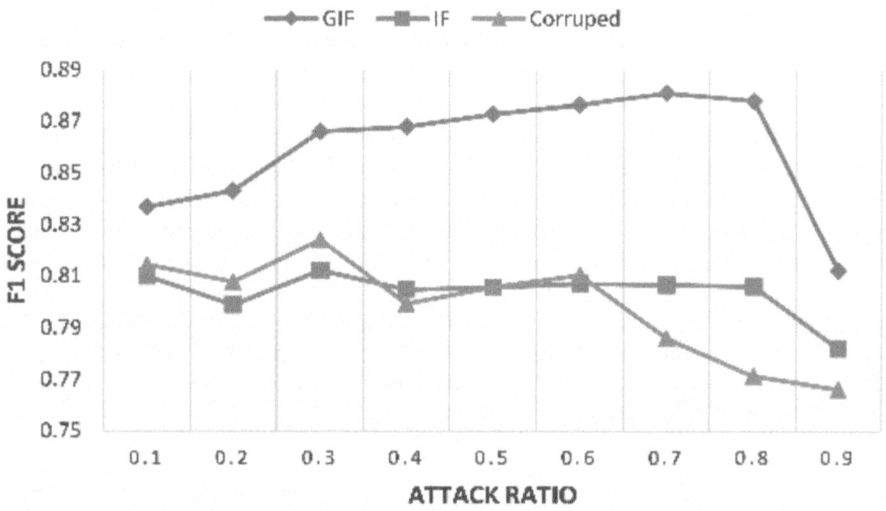

Fig. 4. Unlearned Models with Different Attack Ratios.

Graph Influence Function Unlearning (GIF) and traditional Influence Function Unlearning (IF), are used to systematically evaluate the actual performance of the model after "unlearning" these edges. Figure 4 details the F1 scores of models generated by GIF and IF under different attack ratios on the Cora dataset, while incorporating models directly trained on noisy graphs as controls.

Based on the performance of models trained on corrupted graphs, as the attack ratio increases, the model's performance declines accordingly. However, the GIF and IF methods, by unlearning adversarial edges, are able to significantly improve the model's performance, and GIF outperforms IF. Notably, GIF's performance even improves as the attack ratio increases.

Therefore, the adoption of innovative graph influence function methods indeed enables efficient unlearning of graph data. In various application scenarios that rely on graph data for learning analysis, we can leverage the intelligent unlearning method for educational data to effectively safeguard the right to be forgotten.

4 Conclusion

This study addresses the key issues of data destruction and protection with regard to the integration of artificial intelligence with educational technology and the right to be forgotten of educational participants. In the context of intelligent education applications, we have presented the concept and technical model of intelligent unlearning of education data. We have illustrated the actual process of education data destruction using two technologies commonly used in intelligent education: facial recognition technology and graph data. This study also outlines a practical mechanism and path to protect the right to be forgotten and provide secure technical support for the strategic action of education digital transformation and the development of data-driven intelligent education.

We primarily define, model, and apply a theoretical practice by which educational data can be "unlearned" or forgotten by AI models. In the future, this foundation will contribute to the development of service-oriented applications for different intelligent education applications, including solutions that integrate top-level interoperability, bottom-level interconnectivity, and algorithmic, system-level, and service-level integrations of hardware and software. We hope to complement the weakest link in the full lifecycle management of educational data, break through privacy concerns and ethical limitations that hinder the deep reform of intelligent education, promote greater trust in the digital transformation of education, and build a secure intelligent education ecosystem.

Acknowledgments. The research work reported in this paper is supported by the National Natural Science Foundation of China under Grants 62477017, 62277021 and 62277029, and the Fundamental Research Funds for the Central Universities under Grants CCNU24ai017, CCNU24ai018, CCNU22JC011 and CCNU22QN013.

References

1. Hoofnagle, C.J., Van Der Sloot, B., Borgesius, F.Z.: The European Union general data protection regulation: what it is and what it means. Inf. Commun. Technol. Law **28**(1), 65–98 (2019)
2. Pardau, S.L.: The california consumer privacy act: towards a european-style privacy regime in the united states. J. Tech. L. Pol'y **23**, 68 (2018)
3. Shokri, R., Stronati, M., Song, C. Shmatikov, V.: Membership inference attacks against machine learning models. In: IEEE Symposium on Security and Privacy (SP) 2017. pp. 3–18 IEEE (2017)
4. Fredrikson, M., Jha, S., Ristenpart, T.: Model inversion attacks that exploit confidence information and basic countermeasures. In: Proceedings of the 22nd ACM SIGSAC conference on computer and communications security, pp. 1322–1333. (2015)
5. Ren, Z., Baird, A., Han, J., Zhang, Z., Schuller, B.: Generating and protecting against adversarial attacks for deep speech-based emotion recognition models. In: ICASSP 2020–2020 IEEE International Conference on Acoustics, Speech and Signal Processing (ICASSP), pp. 7184–7188. IEEE (2020)
6. Xu, H., Zhu, T., Zhang, L., Zhou, W., Yu, P.S.: Machine unlearning: a survey. ACM Comput. Surv. **56**(1), 1–36 (2023)
7. Sümer, Ö., Goldberg, P., D'Mello, S., Gerjets, P., Trautwein, U., Kasneci, E.: Multimodal engagement analysis from facial videos in the classroom. IEEE Trans. Affect. Comput. **14**(2), 1012–1027 (2021)
8. Ludwig, S., Rausch, A., Deutscher, V., Seifried, J.: Predicting problem-solving success in an office simulation applying N-grams and a random forest to behavioral process data. Comput. Educ. **218**, 105093 (2024)
9. Tonguç, G., Ozkara, B.O.: Automatic recognition of student emotions from facial expressions during a lecture. Comput. Educ. **148**, 103797 (2020)
10. Bourtoule, L., Chandrasekaran, V., Choquette-Choo, C.A., Jia, H., Travers, A., Zhang, B., Papernot, N.: Machine unlearning. In: 2021 IEEE Symposium on Security and Privacy (SP) pp 141–159. IEEE (2021)
11. Golatkar, A., Achille, A., Ravichandran, A., Polito, M., Soatto, S.: Mixed-privacy forgetting in deep networks. In: Proceedings of the IEEE/CVF Conference on Computer Vision and Pattern Recognition pp. 792–801. IEEE (2021)
12. Hoang, T., Rana, S., Gupta, S., Venkatesh, S.: Learn to unlearn for deep neural networks: minimizing unlearning interference with gradient projection. In: Proceedings of the IEEE/CVF Winter Conference on Applications of Computer Vision, pp. 4819–4828. (2024)
13. Zheng, X.L., Huang, J., Xia, X.H., Hwang, G.J., Tu, Y.F., Huang, Y.P., Wang, F.: Effects of online whiteboard-based collaborative argumentation scaffolds on group-level cognitive regulations, written argument skills and regulation patterns. Comput. Educ.. Educ. **207**, 104920 (2023)
14. Wu, J., Yang, Y., Qian, Y., Sui, Y., Wang, X., He, X.: Gif: a general graph unlearning strategy via influence function. Proc. ACM Web Conf. **2023**, 651–661 (2023)

A Low-Consumption Teaching Role Recognition Method for Teaching Analysis

Gang Zhao[1,2], Yinan Zhang[1](✉), and Jing Wang[1]

[1] Hubei Key Laboratory of Digital Education, Central China Normal University, Wuhan, China
996zhangyn@mails.ccnu.edu.cn

[2] Faculty of Artificial Intelligence Education, Central China Normal University, Wuhan, China

Abstract. Recognizing teaching roles in classrooms is a primary work in teaching analysis. Previous work utilized speaker diarization systems to distinguish "who is speaking when" in a scene. However, in complex classroom scenes, the effect of the proposed method may be limited by an unknown number of speakers, microphone interference, etc. To construct a teaching role recognition method suitable for classroom scenes, this paper found a commonality of speakers with the same teaching identity at the acoustic level. Based on this commonality, this paper adjusted the speaker recognition task and applied it to recognize the teaching identities of different speakers in classroom scenes. A neural network named SA-TDNN was constructed to embed speakers. SA-TDNN achieved 10.09% on CN-Celeb, which is more accurate than other methods. SA-TDNN achieved 13.25% on the self-built dataset, which is more accurate than other methods. In addition, the proposed method outperforms the existing general SD system by 3.83% on the validation set of self-built dataset, which supports the feasibility of this method in the Chinese classroom.

Keywords: deep learning · acoustic feature · teaching role recognition · teaching analysis

1 Introduction

The rapid expansion of artificial intelligence and big data analysis provides more powerful technical support for teaching utterance analysis [1, 2]. In the field of education, analyzing the implicit behaviors and emotions in the utterances of teaching participants can provide data support for uncovering the rules of teaching dialogue and teachers' reflection on the teaching process. In the workflow of teaching utterance analysis, the primary step is to recognize from which teaching role the current utterance comes [3]. At present, in the actual environments such as meeting discussions, customer service dialogues, and interviews, researchers generally use speaker diarization (SD) systems (as shown in Fig. 1) to distinguish speaking roles [4]. However, due to the complex classroom environments and numerous speaking roles, the voices that may exist in the classroom include not only teachers and students but also multimedia and environmental noise. This situation leads to uncertainty as to the number of speakers in different

Q. Liu et al. (Eds.): EITT 2024, CCIS 2600, pp. 110–118, 2026.
https://doi.org/10.1007/978-981-95-1525-7_9

classrooms, so the SD system still has the following problems when applied in the actual classroom environment: (1) The SD system is based on a clustering algorithm to distinguish speaker embedding codes, so it relies on the prior knowledge of speakers. As a result, the existing SD system is difficult to adapt to the teaching analysis of the actual classroom environment. (2) This method does not work well when the input data and trained model do not match. In addition, too many speakers and long audio clip scenes will also affect the SD [5]. (3) It takes much time for the SD system to process when the total length of the input audio is too long. After observing many classroom videos, we found that teaching participants with different teaching identities had specific acoustic features when speaking in Chinese classroom environments. For example, as an individual, each teacher has unique acoustic characteristics, but different teachers have certain commonalities in characteristics such as volume and pitch. And because teachers are leaders in the teaching scene, their voices are significantly different from other teaching participants [6]. This finding inspired us to explore the feasibility of distinguishing teaching roles by a method that recognizes speaker identity through voiceprint recognition.

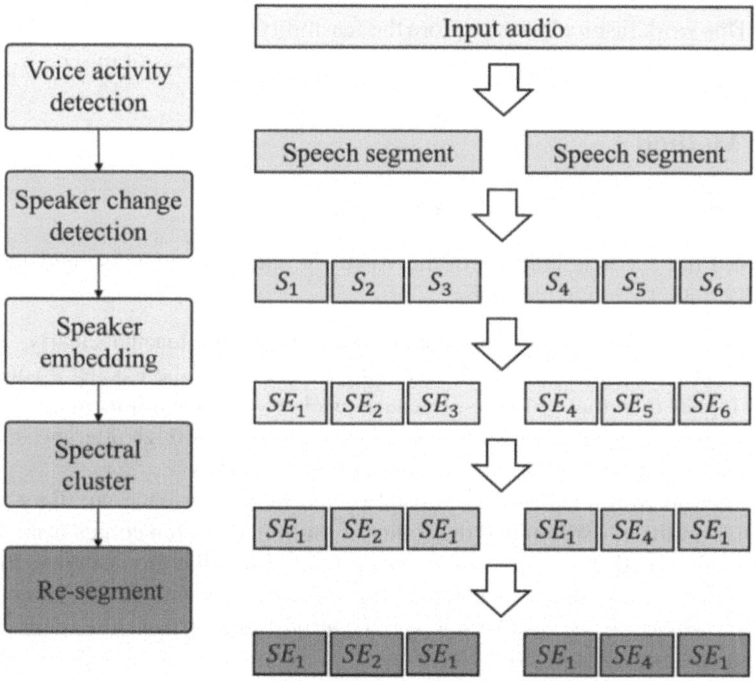

Fig. 1. The common workflow of speaker diarization system.

2 Related Work

In previous research on teaching utterance analysis, manual annotation is generally used to mark the source of a certain utterance. The manual method is efficient and accurate, but it is time-consuming and laborious because the data needs to be manually annotated and proofread, which leads to a lag in the analysis results.

Nowadays, speaker recognition tasks have been widely deployed in various fields. The speaker recognition task aims to infer who is speaking among many known speakers. In speaker recognition related tasks, researchers generally regard speakers as independent of each other, that is, each speaker is assigned a fixed label [7]. For example, Shah, Moinuddin and Khan [8] regarded dialect information as one of the clues for speaker recognition and constructed a speaker recognition system based on multi-layer perceptron. Tang, et al. [9] explored combining different levels of features extracted from gender and ethnic information for speaker recognition. Guntur, Ramakrishnan and Vinay-Kumar [10] proposed a method based on regional English accents according to the unique pronunciation or primitive behavior patterns of different regions.

In summary, these speaker recognition methods all try to find commonalities among the speaker group and embed the speaker voiceprint representation by means of deep learning. This work inspired us to explore the feasibility of finding commonalities in the utterances of speakers with different classroom identities in classroom scenes.

3 Our Method

3.1 Acoustic Features of Teaching Participants

We analyzed the acoustic features in the voices of teachers, students, student groups, multimedia, and environmental noise.

(1) **Teachers:** For the students in the classroom to hear the teacher clearly. Teachers often wear a microphone, which results in the energy of the teacher's voice being much higher than that of others. Because teachers will teach information, lead the classroom rhythm, and standardize the classroom order through utterances, the tone and pitch of teachers' utterances changes more frequently.
(2) **Multimedia:** Multimedia is often used by teachers to provide auxiliary teaching materials in the classroom and the sound of multimedia often comes from the electronic whiteboard. Therefore, the sounds from multimedia often have the following characteristics: (1) There is accompanying music in the sound. (2) The human voice from the smart device speakers is significantly different from the natural human voice generated by vocal fold vibration.
(3) **Student (single):** Students' utterances are mainly to answer teachers' questions and give feedback to teachers. At the acoustic feature level, the intonation of students is gentler than that of teachers, and the energy of the voice is generally lower.
(4) **Students (group):** The utterances of student groups usually come from collective answers to teachers' questions and classroom discussions. Due to the simultaneous inclusion of numerous speakers in this type of audio, it generally has significant characteristics such as high energy and noise.

(5) **Noise:** Noises in the classroom environment are usually manifested as the friction of desks, chairs, and chalk in the classroom and a cough in the classroom.

In conclusion, in the Chinese classroom environment, the acoustic characteristics of teaching roles all have certain commonalities. Therefore, to directly obtain the teaching roles from the acoustic features in the prediction stage, we constructed a classroom sound dataset containing teacher voices, student voices, student group voices, and multimedia and environmental noises according to the analysis of teaching roles.

3.2 Feature Preprocessing

For the audio which input to the model, Firstly, we used the speech activity detection module in the webrtcvad library in Python to split the audio into short audio clips. Subsequently, we uniformly sampled it at 16kHz and performed audio preprocessing operations such as windowing, framing, and pre-emphasis on each audio segment. Finally, Mel Frequency Cepstral Coefficient (MFCC) features were engaged through librosa audio library in Python to convert the audio from time-domain to frequency-domain. To enhance the generalization ability of the model, we masked the features using random SpecAugment [11] and TimeStretch, respectively. At the same time, we randomly added Room Impulse Responses (RIRs) [12] noise to the dataset. The processed feature can be expressed as $A_{\mathrm{MFCC}} = \{\mathrm{a}_1, \mathrm{a}_2, \mathrm{a}_3, \ldots, \mathrm{a}_\mathrm{n}\}$.

3.3 Sa-Tdnn

We designed a Self-attention Time Delay neural network (SA-TDNN) to effectively focus on both frame-level features and sequence features in teaching participants' utterances. We deployed a multi-layer one-dimensional convolutional neural network (1D-CNN) to capture all the frame-level features. A self-attention pooling (SAP) module was used to make the model focus on the valuable frame-level features, and six layers of transformers [13] were deployed to capture the sequence features in the audio. The structure of SA-TDNN is shown as Fig. 2.

The 1D-CNN receives the input MFCC features. The input channel is 256 and the output channel is 512. The multi-layer CNN extracts the frame-level features X_{frame} in the utterance from A_{MFCC}. Each CNN layer is followed by a BatchNorm-1D for feature normalization, and an activation function ReLU is deployed to map vector to output.

The SAP receives X_{frame} captured by the 1D-CNN, and the frame-level feature X_{frame} which is more closely related to the target is selected. The attention mechanism selects more valuable audio frames X_{frame} in the channel and calculates the score of each X_{frame}. The original information is then compared with the score of attention mechanism.

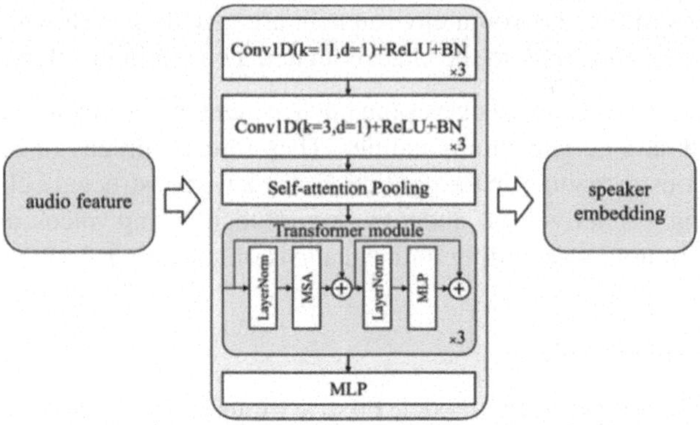

Fig. 2. The structure of SA-TDNN.

Subsequently, the fully connected layer shrinks the feature a_t from 512 dimensions to 256. Finally, the global-level sequence features of audio are extracted by transformers.

4 Experimental Results and Discussion

4.1 Experimental Datasets

The CN-Celeb [14] was used as the experimental dataset for SA-TDNN. Our proposed CVD was used to simulate the sounds that appear in the actual classroom. Two trained students were asked to label the data and Cohen Kappa coefficient were used to check for consistency in this process. The Kappa value of 0.93 indicates that the labeled data are highly consistent. The data distribution of the CVD dataset is shown in Table 1.

Table 1. The data distribution of CVD dataset.

Type of Voiceprint	Total
Teacher	1698
Student (Single)	1540
Student (Group)	770
Multimedia	522
Noise	**373**

4.2 Evaluation Metrics

In the experiments on CN-Celeb, equal error rate (EER) was adopted as the evaluation index. We used diarization error rate (DER) as the evaluation index to contrast our proposed method with the existing SD system. The SD system being used for comparison claims to achieve 8.0% DER on real conversations with 2–10 speakers.

4.3 Implementation Details

The experimental environment was an Intel i7-11700k, an NVIDIA 3080Ti, and a Pytorch 1.10.1 deep learning framework. The experimental parameters are noted in Table 2.

Table 2. Experimental parameters.

Parameter Name	Setting
Batch Size	64
Learning Rate	0.00005
Training Epoch	60

Ten-fold cross validation was adopted, and training test sets were split 8:2.

4.4 Experimental Results on CN-Celeb and CVD

We specified methods on the CN-Celeb as a comparison to SA-TDNN. These models are described as follows:

Table 3. The experimental result on CN-Celeb and CVD.

Method	**Dataset**	**EER. (%)**
E-TDNN	CN-Celeb	11.15
ResNet (Attention)	CN-Celeb	10.77
x-vector (Attention)	CN-Celeb	10.12
Ours (SA-TDNN)	**CN-Celeb**	**10.09**
Res2Net	CVD	15.33
CAM + +	CVD	14.01
Ecapa-Tdnn	CVD	13.43
Ours (SA-TDNN)	**CVD**	**13.25**

(1) **E-TDNN** [16]: Serialized attention was added to TDNN.

(2) **ResNet** with Attention Backend [17]: An attention mechanism was added to ResNet to extract audio features.
(3) **X-Vectors** with Attention Backend [18]: The scaling point self-attention mechanism was used to construct a self-attention network to extract the x-vector features of audio.
(4) **Res2Net** [19]: The multi-scale features were represented at the granularity level and the receptive field of each layer was increased.
(5) **CAM + +** [20]: A delay network based on dense connections (D-TDNN) was used.
(6) **Ecapa-Tdnn** [21]: Res2Net and squeeze-excitation blocks were used. Ecapa-Tdnn can explicitly model dependencies between channels.

In Table 3, SA-TDNN is shown to have achieved leading results for the public dataset CN-Celeb with an EER of 10.09%, which demonstrates the advance of the SA-TDNN method in Chinese voiceprint recognition tasks. The EER reached 13.25% on the self-built dataset CVD, which demonstrates that our proposed method can be applied to voiceprint recognition in the Chinese classroom environment.

Table 4. The comparison results with the SD system.

Method	EER/DER. (%)
SD	17.50
Ours	**13.67**

In Table 4, our method leads by 3.83% compared to the general SD system in actual classroom scenes. In addition, the training resources (amount of training data and average running time) required by our method are much smaller than SD system. The results indicate that our method has certain advantages in the task of teaching role recognition in Chinese classrooms, and it is easier to construct accurate models for different classes.

5 Conclusion

Recognizing teaching roles in teaching scenes is the primary step of teaching utterance analysis, and realizing automatic teaching role recognition can greatly improve the effectiveness of teaching utterance analysis. At present, the general SD system is difficult to apply to the complex classroom environment, while training SD systems dedicated to the classroom requires significant resources. Therefore, we proposed a low resource consumption teaching role recognition method and constructed a neural network named SA-TDNN. Experimental results on the public dataset CN-Celeb demonstrate the superiority of our method in the representation of voiceprint embedding. The experimental results for the self-built data set CVD show that the method has more advantages in the Chinese classroom environment. However, there are still some drawbacks in this work. If a speaker change detection module could be added, the accuracy and robustness would be further improved. In conclusion, we will consider speech content as one of the cues for teaching role recognition in the future.

Acknowledgments. This study was supported by the National Natural Science Foundation of China (Grant No. 62377020) named "Research on the Intelligent Evaluation Method of Teacher Information Technology Application Ability based on Classroom Teaching Behavior Characteristics", Fundamental Research Funds for the Central Universities named "Research on Intelligent Evaluation Technology and Application of Teachers' Teaching Ability based on Multimodal Data" (CCNU22JC011), and Open Fund of Hubei Key Laboratory of Digital Education (Grant No. F2024K02) named "Research on the Key Technology of Teachers' Digital Teaching Competence Assessment Based on the Characteristics of Teaching Behavior in the Classroom".

References

1. Järvinen, K., Kähkönen, A.L., Nieminen, P., Mäntylä, T.: Talking like a teacher: a study of pre-service teachers' voice and speech characteristics in learning and teaching situations. Educ. Sci. **14**(2), 210 (2024)
2. Lu, Y., Chen, Z., Chen, R., Shi, Y., Zheng, Q.: Research on the application framework of intelligent technologies to promote teachers' classroom teaching behavior evaluation. Front. Educ. China **18**(2) (2023)
3. Zhao, G., Zhang, Y., Chu, J.: A multimodal teacher speech emotion recognition method in the smart classroom. Internet of Things **25**, 101069 (2024)
4. Mershad, K., Said, B.: DIAMOND: a tool for monitoring the participation of students in online lectures. Educ. Inf. Technol. **27**(4), 4955–4985 (2022)
5. Park, T.J., et al.: A review of speaker diarization: recent advances with deep learning. Comput. Speech Lang. **72**, 101317 (2022)
6. Zhao, W., Ma, J., Cao, Y.: What is effective classroom dialog? A comparative study of classroom dialog in Chinese expert and novice mathematics teachers' classrooms. Front. Psychol. **13**, 964967 (2022)
7. Bai, Z., Zhang, X.L.: Speaker recognition based on deep learning: an overview. Neural Netw. **140**, 65–99 (2021)
8. Shah, S.M., Moinuddin, M., Khan, R.A.: A robust approach for speaker identification using dialect information. Appl. Comput. Intell. Soft Comput. (2022)
9. Tang, Y., et al.: Attention based gender and nationality information exploration for speaker identification. Digital Signal Process. **123**, 103449 (2022)
10. Guntur, R.K., Ramakrishnan, K., Vinay-Kumar, M.: An automated classification system based on regional accent. Circ. Syst. Signal Process. **41**(6), 3487–3507 (2022)
11. Park, D.S., et al.: Specaugment: a simple data augmentation method for automatic speech recognition. arXiv:1904.08779 (2019)
12. Ko, T., Peddinti, V., Povey, D., Seltzer, M.L., Khudanpur, S.: A study on data augmentation of reverberant speech for robust speech recognition. In: IEEE International Conference on Acoustics, Speech and Signal Processing (ICASSP), pp. 5220–5224 (2017)
13. Vaswani, A. et al.: Attention is all you need. Adv. Neural Inf. Process. Syst. 30 (2017)
14. Fan, Y., et al.: Cn-Celeb: a challenging Chinese speaker recognition dataset. In: IEEE International Conference on Acoustics, Speech and Signal Processing (ICASSP), pp. 7604–7608 (2020)
15. Kvålseth, T.O.: Note on Cohen's kappa. Psychol. Rep. **65**(1), 223–226 (1989)
16. Zhu, H., Lee, K.A., Li, H.: Discriminative speaker embedding with serialized multi-layer multi-head attention. Speech Commun. **144**, 89–100 (2022)
17. Zeng, C., Wang, X., Cooper, E., Miao, X., Yamagishi, J.: Attention back-end for automatic speaker verification with multiple enrollment utterances. In: IEEE International Conference on Acoustics, Speech and Signal Processing (ICASSP), pp. 6717–6721 (2022)

18. Zhu, Y., Ko, T., Snyder, D., Mak, B., Povey, D.: Self-attentive speaker embeddings for text-independent speaker verification. In: Interspeech, pp. 3573–3577 (2018)
19. Gao, S.H., Cheng, M.M., Zhao, K., Zhang, X.Y., Yang, M.H., Torr, P.: Res2net: a new multi-scale backbone architecture. IEEE Trans. Pattern Anal. Mach. Intell. **43**(2), 652–662 (2019)
20. Wang, H., Zheng, S., Chen, Y., Cheng, L., Chen, Q.: Cam++: a fast and efficient network for speaker verification using context-aware masking. arXiv:2303.00332 (2023)
21. Desplanques, B., Thienpondt, J., Demuynck, K.: Ecapa-tdnn: emphasized channel attention, propagation and aggregation in tdnn based speaker verification. arXiv:2005.07143 (2020)

Open Management of Research-Oriented Laboratories in Universities Based on an "AI for Science" Paradigm

Yi Liu(✉)

Hubei Key Laboratory of Digital Education, Central China Normal University, Wuhan, China
liuyi@ccnu.edu.cn

Abstract. As scientific research is transitioning from traditional paradigms to an "AI for Science Paradigm", university research-oriented laboratories are facing a series of challenges in resource allocation, instruments and equipment management, scheduling management, reservation management, and operational management. This paper proposes measures to optimize the open management of research-oriented laboratories to meet research needs based on the AI for Science Paradigm, including performance-oriented laboratory resource allocation management based on DEA, optimization of the configuration management based on AHP and interdisciplinary application management of instruments and equipment, the application of the Internet of Things (IoT) in the management of instruments and equipment, network security management based on machine learning, point-based management for usage rights, and mobile management for usage requirements.

Keywords: Open Management · Laboratory · Artificial Intelligence · Scientific Research Paradigm

1 Introduction

Research-oriented laboratories in universities are laboratories that are based on the construction of disciplines and are primarily aimed at scientific research activities. They not only undertake the teaching and research work for master's and doctoral students of their affiliated universities, but also are responsible for the implementation of scientific research projects at various levels. They serve as fundamental platforms for talent cultivation, scientific research, and technological development [1]. Open management is the core of research-oriented laboratories. Implementing open management scientifically, efficiently, and safely has important and practical significance for fully exerting the functions of research-oriented laboratories.

In recent years, there has been a significant change in the paradigm of scientific research. Turing Award winner Jim Gray divided scientific research into four paradigms: the first paradigm is the Experimental Paradigm, which is the simple observation and summarization of natural phenomena. The second paradigm is the Theoretical Paradigm, which is based on observation to establish simple mathematical models described in

Q. Liu et al. (Eds.): EITT 2024, CCIS 2600, pp. 119–130, 2026.
https://doi.org/10.1007/978-981-95-1525-7_10

mathematical language, possessing a certain degree of extensibility and universality. The third paradigm is the Computational Science Paradigm, which is based on numerical computation methods to analyze problems. The fourth paradigm is the Data-intensive Paradigm, which is based on statistical analysis of big data methods, shifting from traditional hypothesis-driven methods to the exploration of scientific data. Currently, the paradigm of scientific research is undergoing revolutionary changes and has further evolved into a fifth paradigm, the AI for Science Paradigm, which uses artificial intelligence technology and methods for research, simulation, prediction, and optimization [2, 3]. Intelligent technologies represented by artificial intelligence algorithms, big data analysis, and perception of the Internet of Things (IoT) are increasingly widely applied in scientific research. Although research-oriented laboratories in different disciplines have differences in functional positioning, they have a high degree of similarity and synchronicity in the research paradigm. The AI for Science Paradigm has posed new issues and higher requirements for the open management of research-oriented laboratories, and it is necessary to implement more vigorous reforms and innovations in the open management of research-oriented laboratories (Fig. 1).

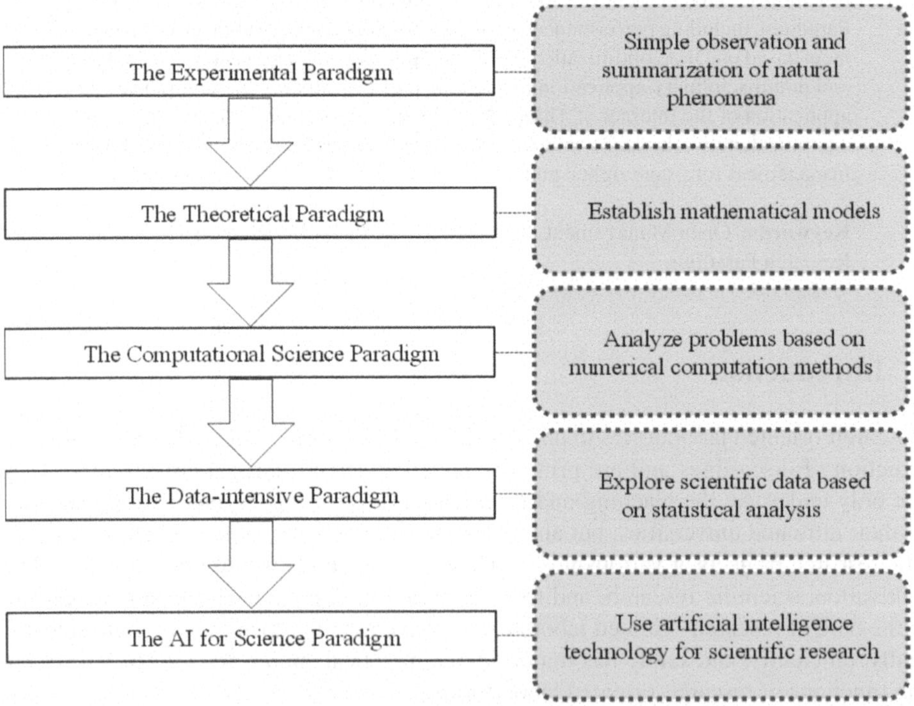

Fig. 1. Evolution of scientific research paradigms

2 Literature Review

2.1 The Open Management Method of Research-Oriented Laboratories

Open management of research-oriented laboratories includes management strategies to increase experimental time, experimental content, and experimental equipment [4]. To improve the level of research-oriented laboratory open management, it is necessary to optimize the spatial planning of research-oriented laboratories, establish an efficient scientific research team, integrate the use of laboratory resources, and improve the reservation system and sharing system for instruments and equipment of laboratories [5].

2.2 The Main Problems and Solutions in the Open Management of Laboratories

The main problems in the open management of university research-oriented laboratories are the insufficient implementation of laboratory systems, low utilization rate of laboratory resources, lack of high-quality laboratory management personnel, and shortage of laboratory operation funds. The measures to solve the above problems are to improve the laboratory management system, increase the sharing level of laboratories, provide support to laboratory management personnel in terms of salary, treatment and promotion, and establish a special fund to strengthen laboratory construction [6].

2.3 The Design of Open Management System Architecture of Laboratories

The open management system of laboratories primarily includes four modules: system login, laboratory reservation, laboratory management, and equipment inquiry [7], achieving functions such as basic data information entry and maintenance, business information management, information inquiry and statistics, laboratory reservation management, laboratory arrangement management, laboratory equipment and resource management, and laboratory safety management [8, 9].

2.4 Safety Issues in the Open Management of Laboratories

In order to elevate the level of safety management of laboratories, it is necessary to further improve the safety management system of laboratory, effectively strengthen laboratory safety inspections and supervision, enhance the emergency response capacity for laboratory accidents, establish a sound laboratory emergency management system, reasonably allocate and properly manage necessary emergency rescue facilities, establish laboratory safety accident early warning response mechanisms, formulate a comprehensive emergency work plan for laboratory safety accidents, and standardize the emergency response procedures for laboratory safety accidents [10].

Overall, based on the research findings from literature, there has been no study on the open management of research-oriented laboratories from the perspective of the AI for Science Paradigm.

3 Challenges Faced by the Open Management of Research-Oriented Laboratories Based on the AI for Science Paradigm

3.1 The Research Team's Demand for Laboratories Has Increased Significantly

AI has enhanced human-computer collaboration, leading to an increase in on-site researchers in research-oriented laboratories, extended usage time, and a significant increase in the demand for laboratory resources.

Humans participate in the training and testing of algorithm construction, continuously training and validating models in a sequential manner. This is a process of combining machine and human intelligence to achieve the best long-term results. In conventional scientific research, researchers often intervene in the research as observers. Whether it is the Empirical Paradigm, Theoretical Paradigm, Computational Science Paradigm, or Data-intensive Paradigm, humans are always observing and summarizing the objective phenomena of the physical world, using mathematical logic, theoretical concepts, formulae, and models to deal with problems. However, when this model fails in the face of complex problems and researchers turn to use AI, especially after machines have learned knowledge created by humans, researchers should integrate human intuitive experience or expert experience into the data model or computational model. In doing so, they should use human expert experience to guide the improvement of machine inefficient exploration, leverage the advantages of machine computing power and human intuition, and conduct scientific research in the form of human-machine integration to counter the limitations of machines. Thus, when scientific research teams widely use AI for scientific research, more researchers will need to carry out work in laboratories and engage in face-to-face communication and operations, leading to a sharp increase in the demand for open laboratories and creating an imbalance between laboratory usage needs and laboratory resources. In this case, it is urgently necessary for research-oriented laboratories to create better conditions for researchers, that is, to provide greater support for open management of the software and hardware environment of research-oriented laboratories.

3.2 Increasing Complexity in the Open Management of Research-Oriented Laboratories Due to AI Driving Interdisciplinary Research

The AI for Science Paradigm focuses on the interaction between humans, machines, and data, emphasizing the integration of human decision-making mechanisms and data analysis, reflecting the organic combination of data and intelligence. The multiplying effect produced by this combination enables scientific research to achieve higher efficiency, more in-depth discussions, and more accurate grasp of concepts, promoting the integration of interdisciplinary knowledge, expanding the scope of scientific research, and finding new paths to solve complex problems. Under this background, on the one hand, the number of interdisciplinary research teams in research-oriented laboratories has increased rapidly, leading to a surge in applications for laboratory space use, and the demand for laboratory usage time is no longer concentrated in the daily work period but is scattered and long-term; on the other hand, the demand for interdisciplinary application of instruments and equipment by research teams is also growing, It is necessary for the research-oriented laboratories to adopt more innovative measures in the management of large-scale instruments and equipment.

3.3 Further Enhancement of the Network Security Management Level of the Research-Oriented Laboratories

Based on the AI for Science Paradigm, the demand for cloud computing by research teams has greatly increased, putting greater pressure on the network security management of research-oriented laboratories.

Cloud computing technology utilizes cloud systems to complete massive computational tasks, improving data analysis efficiency, reducing hardware investment, and storing scientific research data safely and conveniently, reducing the probability of data damage events that may occur in traditional data storage methods, thereby effectively improving the efficiency of scientific research. At the same time, cloud computing leads to the expansion of laboratory networks at scale, the complication of network applications, and the increasing of network terminal devices. Cloud computing also makes previously closed networks and systems more open, resulting in laboratory network security risks.

3.4 The Risks Posed by Long-Term Heavy Use of Equipment to the Management of Research-Oriented Laboratories

Based on the AI for Science Paradigm, the massive data environment coupled with the widespread use of machine learning leads to instruments and equipment operating continuously for long periods at high loads, which conflicts with the normal working hours of the laboratory, thereby causing a series of operational safety management issues such as laboratory electrical safety and equipment safety.

3.5 Difficulties in Laboratory Reservation Management

Through the laboratory reservation management system, online reservations for laboratories have generally been implemented. However, there are the following problems with online reservations: First, there are many conflicts in reservation times. Due to various reservation requests having the same priority level, the system cannot process conflicting requests automatically; communication between the person making the reservation and the laboratory administrators is required to resolve the issue. For the lack of clear rules and mechanisms, in most cases, the reservation system can only accommodate requests in the order they were received; otherwise, requests can only be manually processed on-site by the laboratory administrators. This situation not only results in low efficiency and fails to allocate laboratory resources scientifically, but this system also does not favor teams with strong research capabilities and outstanding performance.

The emergence of these issues not only means that scientific research is closely following the forefront and continuously deepening, it also means that the function of the research-oriented laboratory is being recognized and valued by researchers, and the role of the research-oriented laboratory is being more and more fully utilized. It also indicates that the current open management of research-oriented laboratories urgently needs timely upgrades and optimizations from the technical and institutional aspects.

Therefore, research-oriented laboratory administrators should promptly grasp the new trends in laboratory demands, swiftly and flexibly adopt new measures to respond, and also leave room for future improvements in management methods with a forward-looking perspective.

4 Innovations in the Open Management of Research-Oriented Laboratories to Adapt to the AI for Science Paradigm

4.1 Implement Performance-Oriented Laboratory Resource Allocation Management

To resolve the contradiction between the limited laboratory resources and the ever-growing demand for their use, and to effectively allocate laboratory resources to high-level research teams, it is necessary to adopt a performance-oriented laboratory resource allocation management model: First, construct a performance evaluation index system for research teams that is centered on the quality of innovation and scientific research achievements. In the performance evaluation index system, input indicators include three types: human resources, financial resources and material resources. Output indicators include three types: scientific research benefits, social benefits, and economic benefits. On this basis, the researcher conducted an "input-output" efficiency analysis to evaluate the performance of scientific research teams, and used the evaluation results as a scientific basis for the allocation of laboratory resources.

This study took 10 research teams of the D laboratory at H University as an example and conducted a performance analysis based on the BCC model of DEA (Data Envelopment Analysis). DEA uses linear programming to evaluate efficiency, analyzing whether the inputs and outputs of decision-making units have reached an optimal state. When using DEA, the first step is to establish an input-output indicator evaluation system. Then, all the data from the decision-making units are input into the model to evaluate efficiency.

The models commonly used in DEA are the CCR model and the BCC model. The main difference between these two models is that the CCR model assumes constant returns to scale, while the BCC model assumes variable returns to scale. However, in practice, not all production units can be in the optimal production state, so the assumption of the CCR model does not match reality in most cases. Therefore, the BCC model was adopted in this study. The general formula for BCC model is as follows:

$$\begin{cases} min\theta \\ \sum_{j=1}^{n} X_j\lambda_j + S^- = \theta X_0 \\ \sum_{j=1}^{n} Y_j\lambda_j - S^+ = \theta Y_0 \\ \sum_{j=1}^{n} \lambda_j = 1 \\ \lambda_j \geq 0, S^- \geq 0, S^+ \geq 0 \end{cases}, \quad ((1))$$

where θ is the composite efficiency of the decision-making unit, Xj and Yj are input and output variables, λj represents the weights of the input and output variables, and S + and S– are slack variables.

Based on the BCC model and input-output indicators, the DEAP 2.1 software was used to conduct DEA analysis in this study. The results of the DEA analysis are shown in Table 1, and the distribution of composite efficiency is reported in Table 2.

Table 1. The composite efficiency values of the ten research teams in the D laboratory of H University

NO	Decision- Making Units	Composite Efficiency	NO	Decision- Making Units	Composite Efficiency
1	T1	1.000	6	T6	0.835
2	T2	0.818	7	T7	1.000
3	T3	1.000	8	T8	1.000
4	T4	1.000	9	T9	0.562
5	T5	0.762	10	T10	1.000

It was found that 60% of the teams in the D laboratory of H University have reached an ideal state of input and output according to Table 2. However, 10% of the teams have a lower composite efficiency score, with less research output that does not match the input in human, material, and financial resources.

Table 2. Distribution of composite efficiency

Distribution of Composite Efficiency	Number of Research Teams	Proportion
1.0	6	60%
0.8–1.0	2	20%
0.6–0.8	1	10%
0–0.6	1	10%

4.2 Optimize the Configuration Management of Instruments and Equipment in Research-Oriented Laboratories

To improve the performance of instruments and equipment, it is necessary to construct a performance evaluation index system for the use of instruments and equipment, use scientific methods for performance evaluation, and then improve the management system of instruments and equipment based on the evaluation results. The evaluation results also serve as the basis for the purchasing and operating management of instruments and equipment, aiming to optimize the allocation of instruments and equipment among different scientific research teams.

Analytic Hierarchy Process (AHP) is a decision-making method that breaks down the elements related to the decision into hierarchies such as goals, criteria, and alternatives, and then performs qualitative and quantitative analysis on this basis. This study took the D laboratory of H University as an example and applied AHP to evaluate the performance of the laboratory's instruments and equipment.

(1) **The Selection of Performance Evaluation Indicators for Instruments and Equipment.** The degree of instrument and equipment activity, research achievements, service revenue, utilization of function, maintenance and repair were selected as the first-level qualitative analysis indicators for the performance evaluation of instruments and equipment. Then, each first-level indicator was further specified to form second-level indicators that could be quantitatively analyzed. All of the indicators are shown in Table 3.

Table 3. Performance Evaluation Indicators for Instruments and Equipment

Main Goal (A)	First-Level Indicators (Bi)	Second-Level Indicators (Cj)
Performance Evaluation for Instruments and Equipment (A)	Degree of instruments and equipment activity (B1)	Instruments and equipment operating time (C1)
		Ratio of effective operating time (C2)
	Research achievements (B2)	Number of published papers (C3)
		Number of patents (C4)
		Number of awards (C5)
	Service revenue (B3)	Internal income (C6)
		External income (C7)
	Utilization of function (B4)	Usage ratio of existing functions (C8)
		Number of new functions added (C9)
	Maintenance and repair (B5)	Cost of maintenance and repair (C10)
		Duration of maintenance and repair (C11)

(2) **Determining the Weight of Performance Evaluation Indicators for Instruments and Equipment.** Twenty industry experts in the management of instruments and equipment from universities, research institutes, and government departments were selected, and a questionnaire was designed based on Table 3 for experts to fill out to

determine the weight of important indicators for laboratory instruments and equipment. Each first-level indicator was compared pairwise, and values were assigned according to their importance. The numbers 1, 3, 5, 7, and 9 were used to represent equally important, slightly important, moderately important, very important, and absolutely important indicators, respectively. The numbers 2, 4, 6, and 8 were used as intermediate values between the aforementioned judgments. By pairwise comparison analysis, scores were assigned to the relative importance of each evaluation indicator. A judgment matrix was formed by integrating expert opinions.

The eigenvector of the matrix was calculated using the arithmetic mean method to obtain the weight of the first-level indicators Wi, followed by a consistency check. Further pairwise comparisons of importance were conducted among the second-level indicators. Using the same method, the eigenvector of the matrix was calculated and a consistency check was performed to obtain the weights of the second-level indicators Wj. Finally, based on the importance weights of the first-level indicators Wi and the second-level indicators Wj, a composite weight Wk (Wk = Wj × Wj) was formed.

Table 4 shows the first-level indicators, the second-level indicators, and the weights of indicators for the performance evaluation of the instruments and equipment, as well as the composite weights calculated based on these indicators. In Table 4, it was found that A and B are the top two key indicators in terms of weight, which have significant impact on the performance evaluation of instruments and equipment. This result is consistent with the general practical experience of laboratory management.

(3) **Calculation of the Performance of Instruments and Equipment.** Based on the weights Wk of the various indicators in Table 4, combined with the scores Sk assigned by experts to each indicator, the performance of the instruments and equipment was calculated as follows:

$$P = \sum_{k=1}^{11} W_k \times S_k, \tag{2}$$

where P is the performance of the instruments and equipment, Wk is the weight of the indicators, and Sk is the score given by the experts for the indicators.

(4) **Application of the Results from the Performance Evaluation of Instruments and Equipment.** Laboratory administrators can use the results of the performance evaluation of instruments and equipment to analyze the specific reasons affecting performance. They can strengthen the procurement and operational management of instruments and equipment, thereby achieving an optimized allocation of the instruments and equipment.

4.3 Promote Interdisciplinary Applications of Instruments and Equipment

Laboratory administrators can refine the categorization of instruments and equipment, construct high-dimensional vectors for keywords and performance indicators of instruments and equipment related to interdisciplinary fields. Then by using text analysis methods and vector similarity methods in natural language processing to measure the application similarity between different instruments and equipment, laboratory administrators can assist research teams in the interdisciplinary application of instruments and equipment, seek new research themes, and achieve complementary advantages between different disciplines.

Table 4. Weights of various indicators in the performance evaluation of instruments and equipment

Main Goal (A)	First-Level Indicators (Bi)	Weight (Wi)	Second-Level Indicators (Cj)	Weight (Wj)	Composite weight (Wk)
Performance Evaluation for Instrument and Equipment (A)	Degree of instruments and equipment activity (B1)	0.398	Instruments and equipment operating time (C1)	0.875	0.349
			Ratio of effective operating time (C2)	0.125	0.050
	Research achievements (B2)	0.244	Number of published papers (C3)	0.625	0.152
			Number of patents (C4)	0.136	0.033
			Number of awards (C5)	0.239	0.058
	Service revenue (B3)	0.177	Internal income (C6)	0.407	0.072
			External income (C7)	0.593	0.105
	Utilization of function (B4)	0.126	Usage ratio of existing functions (C8)	0.750	0.094
			Number of new functions added (C9)	0.250	0.031
	Maintenance and repair (B5)	0.056	Cost of maintenance and repair (C10)	0.514	0.029
			Duration of maintenance and repair (C11)	0.486	0.027

4.4 Utilize IoT for Efficient Management of Instruments and Equipment

To fully meet the rapidly growing demands of scientific research teams based on the AI for Science Paradigm and to improve the utilization rate of instruments and equipment, IoT technology can be fully utilized in the instruments and equipment. IoT refers to the interconnection of objects to the internet through sensors, enabling interconnectivity and communication among all things, with functions such as identification, positioning, tracking, and monitoring. Instruments and equipment management systems for research-oriented laboratories based on IoT mainly consist of four parts: the user layer, the perception layer, the network layer, and the application layer.

The user layer: System administrators manage user identities and set permissions for instruments and equipment users at the user layer, including access control authorization, reservation rights, usage rights, and viewing of the status of equipment usage. Users can operate through computers or mobile phones from network port or mobile apps.
The perception layer: The perception layer collects and uploads information about instruments and equipment, the environment in which the instruments and equipment are located, and personnel operating through cameras, access control systems, electronic tags, measuring devices, and environmental sensors.
The network layer: The network layer is used to process and transmit various information obtained from the user layer and the perception layer.
The application layer: At the application layer, system administrators use the information obtained from the user layer and the perception layer to perform equipment management, operator management, and safety management.

4.5 Construct a Laboratory Network Security System Based on Machine Learning

To ensure the normal operation of the laboratory network, A laboratory network security system based on big data and machine learning should be established. The overall architecture of the system consists of three layers: the foundation layer, the network layer, and the analysis decision layer. The foundation layer stores raw network flow data and security event data for analyzing the network's security status. The network layer collects network flow data and log data from network devices and application program to provide evidence of network security events. The decision layer first performs data cleaning and normalization, then uses machine learning algorithms for feature extraction and model training that reflect the attributes and patterns of network flow. Finally, based on the well-trained machine learning model, by continuously learning and updating the model, the system can gradually improve the accuracy and efficiency of security event detection.

4.6 Explore a Points-Based Management System for Research-Oriented Laboratories

To enhance the safety management of research-oriented laboratories, a points-based assessment system for student management should be established in research-oriented laboratories. Students who obtain laboratory access qualifications can earn points by completing reservation check-ins, fulfilling duty responsibilities, and properly organizing laboratory equipment after experiments. Deductions are made for behaviors such as violating electrical safety regulations and occupying fire lanes. The points are linked to the selection of student scholarships and honorary titles, thereby encouraging students to standardize their behavior in laboratories, enhance their awareness of abiding by laboratory management regulations, and ensure the smooth operation of laboratory.

4.7 Improve the Mobile Management Functions of the Research-Oriented Laboratories

Based on summarizing the operational characteristics, demand features, and management experiences of research-oriented laboratories, a mobile management function can be added to help laboratory administrators keep track of laboratory operation dynamics at any time and handle various usage applications and special situations that occur during open laboratory hours more promptly.

5 Conclusion

The essential goal of the AI for Science Paradigm is to enhance research efficiency using AI. The AI for Science Paradigm deeply integrates AI technology with scientific research, offering significant assistance in data analysis, model building, and experimental operations. It facilitates the discovery of new knowledge and the resolution of scientific challenges and is expected to bring about more scientific innovation and breakthroughs.

By optimizing the management of laboratory resource allocation, improving the configuration and interdisciplinary application management of instruments and equipment, accelerating the application of the IoT, implementing point-based management for usage rights and mobile management for usage requirements, and strengthening network security management, research-oriented laboratories in universities can better adapt to the changes in scientific research caused by the AI for Science Paradigm. These measures can also be further promoted in other types of laboratories in universities, thereby comprehensively advancing the achievement of the goal of building first-rate universities and disciplines.

References

1. Su, X.J., et al.: Several issues in the management of open research laboratories in colleges and universities. J. High. Educ. **17**, 144–145 (2015)
2. Zhang, Y.M.: Transformation of humanities and social sciences research caused by Artificial Intelligence. J. Univ. Jinan (Soc. Sci. Edn.) **29**, 20–28 (2019)
3. Yang, J.L., Jiang, J.: Embracing a new scientific research paradigm: the scientific revolution brought by Artificial Intelligence. Sci. Soc. **13**, 11–21 (2023)
4. Yang, X.P., et al.: Exploration of the university laboratory opening management. Lab. Sci. **23**, 186–189 (2020)
5. Zhang, L., et al.: Comparison and consideration on management model of research laboratory in Chinese and Canadian universities. Res. Explor. Lab. **38**, 232–236 (2019)
6. Li, Y.D., et al.: Problems and solutions to the present situation of laboratory opening. Lab. Sci. **22**, 150–153 (2019)
7. Hong, W.X., et al.: Research on the management model in open laboratory for 'Emerging Engineering Discipline'. Exp. Sci. Technol. **18**, 153–156 (2020)
8. Lu, X.L.: Research and design of university laboratory open management system based on B/S architecture. Software **44**, 168–171 (2023)
9. Shi, Z.J., et al.: Laboratory open management system in construction of university based on process and enterprise maturity model. Res. Explor. Lab. **42**, 279–284 (2023)
10. Luo, K., et al.: Research on safety management strategies of university laboratories under the background of open sharing. J. High. Educ. **7**, 136–139 (2021)

Tech-Enhanced Curriculum and Teaching Innovations

STEM, Cultural Heritage, and Junior High School Labor Curriculum: An Example of the 3-way Luban Lock C-STEAM Project

Shiping Qiu[1](✉) and Zhizhen Zhang[2]

[1] School of Educational Technology, Faculty of Education, Beijing Normal University, Beijing, China
qiu.shiping@qq.com

[2] School of Educational, Faculty of Education, Beijing Normal University, Beijing, China

Abstract. STEM education can not only cultivate students' STEM literacy, but also promote students' inquiry and critical thinking skills. Currently, STEM education in China is mostly carried out in the form of school-based curriculum or joint-projects with external agencies, such as science centers or museums. It needs to be integrated into the national curriculum. The National Compulsory Education Curriculum Standards (2022 Edition) focus on labor curriculum as a way of furthering the cultivation of students' labor literacies. The labor curriculum is a natural field for STEM education because of its comprehensive practical nature. However, there is an urgent need to explore exactly how to implement STEM projects in regular school teaching. C-STEAM education is a STEM education model that aims at the transmission of traditional Chinese culture. In order to explore STEM education in the labor curriculum, we designed and implemented a labor curriculum project based on the 6C model of C-STEAM education. Based on a systematic analysis of the traditional cultural characteristics of the location (Beijing), the Luban Lock Project takes mortise and tenon construction (榫卯结构) as its theme, asking students to consider the cultural import of this architectural technique and to craft their own 3-way Luban Locks (三通鲁班锁) as ways of cultivating their core labor literacy. After the four-week project, students' performance, reflections, and evaluations showed that the project did promote labor literacy and that students were satisfied with this learning experience.

Keywords: C-STEAM education · 3-way Luban Lock making · junior high school · labor curriculum · cultural heritage

1 Introduction

STEM is an acronym for the four disciplines of Science, Technology, Engineering and Mathematics. STEM education in the narrow sense requires the combination of the content of the four disciplines to form an organic whole [1]. Falloon, Hatzigianni, Bower, Forbes, and Stevenson argue that STEM education emphasizes practice and interdisciplinary learning [2], and that STEM students complete practical activities through design

Q. Liu et al. (Eds.): EITT 2024, CCIS 2600, pp. 133–144, 2026.
https://doi.org/10.1007/978-981-95-1525-7_11

and problem solving. STEM education not only enhances students' STEM literacy, it can cultivate students' inquiry abilities, critical thinking abilities, and communication abilities [3]; it has thus received extensive attention from researchers. The National Compulsory Education Curriculum Standard (2022 Edition) proposes to strengthen integrated curriculum design and interdisciplinary learning [4], which is exactly the concept advocated by STEM education. At present, STEM education in China is mostly combined with other related educational concepts, such as integrated school-based curriculum or joint teaching opportunities with museums [5]. Although STEM courses are included in the curriculum of some schools, they have not been closely integrated with compulsory courses and are relatively marginalized [6]. Expanding STEM education into the regular teaching of the national curriculum is an important aspect of the in-depth development of STEM education.

The National Compulsory Education Labor Curriculum Standard (2022 Edition) separates labor from the comprehensive practice activity curriculum and makes it an independent piece of the national curriculum. Labor education aims to cultivate students' sense of labor as a key competency, urging them to perceive, admire, and appreciate the spirit of model workers the spirit of craftsmanship [4]. However, due to the relatively recent introduction of labor curriculum, practical limitations, and the influences of ideology and institutional cultures, labor education curriculum is still marginalized and lacking resources [7]. In actual teaching, the quality of the curriculum is inconsistent, which affects the implementation of labor curriculum standards [8]. Junior high school students need to properly experience metalworking, carpentry, and other projects after being exposed to basic housework. Such traditional crafts in the labor curriculum are meant to focus on real situations and help students comprehensively apply multidisciplinary knowledge to solve problems [9]. At the same time, the school curriculum of traditional crafts needs a more complex transformation in order to balance the value of craft education and other classroom teachings.

There are many proven cases and educational models for STEM education [10, 11], but fewer successful teaching cases in the independent national curriculum. Labor curriculum teachers are few and can struggle with curriculum design, which suggests that the labor curriculum needs to provide a proven model as a guide. Therefore, drawing on STEM education models in the labor curriculum is not only an extension and supplement to STEM education in schools, but also an enrichment and expansion of labor curriculum and teaching resources. In this study, we designed and developed a C-STEAM project for 7th-grade students: "Woodworking Wizard—The Technical Challenge of the 3-way Luban Lock," a puzzle based on traditional Chinese cultural inheritance and intended to cultivate the core literacy of labor within the junior high school curriculum.

2 The Philosophy of C-STEAM Education

STEAM education is an extension of STEM education, but in China there has been a lack of construction of cultural meaning for local contexts. Zhan Zehui proposes a localized STEAM education: C-STEAM for short, where C represents "Culture" [12]. This model promotes both traditional Chinese culture and national sentiments. C-STEAM education takes cultural inheritance as a primary value, and STEAM education is the means to realize cultural inheritance.

In order to implement C-STEAM education in teaching, Zhan Zehui constructed the 6C model of C-STEAM education based on contextual cognition theory and the design-based double-loop inquiry model (Fig. 1). Based on the 6C model, researchers conducted a curriculum practice of "Foshan Cantonese Opera Culture: Small Artisans of Tiangongkaiwu" [13]. The C-STEAM model and practice cases provide a reference for the program design and implementation of our study.

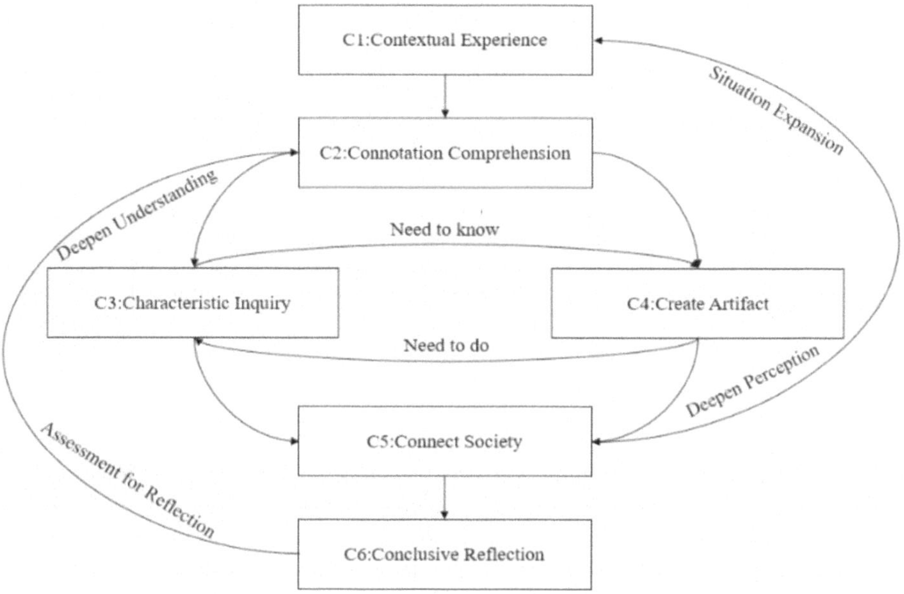

Fig. 1. The 6C Model of C-STEAM Curriculum Design for Cultural Heritage

3 Project Design of "Woodworking Wizard—The Technical Challenge of the 3-way Luban Lock"

The theme of this project is "Woodworking Wizard—he Technical Challenge of the 3-way Luban Lock," the target students are 7th grade students in Beijing, and the length of the project is four weeks. The project is described here across six aspects: cultural background, objective analysis, assessment design, project breakthroughs, activity design, and project implementation.

3.1 Cultural Background

Traditional culture is usually regional. In the field of carpentry, Beijing's buildings are deeply distinctive. Among them, mortise and tenon construction is a distinctive craft, and Beijing has a museum dedicated to mortise and tenon construction. Beijing has numerous Ming-style buildings with horizontal beams and vertical frames. Every gap between parts is connected by mortise and tenon [14]. Among the representative buildings that exemplify the mortise and tenon structure are the Imperial Palace and the Temple of Heaven.

Table 1. Curriculum Objectives for the 3-way Luban Lock Making C-STEAM Project.

	Culture	Science	Technology	Engineer	Art	Mathematics
Knowledge areas	Connotation, characteristics, and advantages of mortise and tenon construction	Manufacture of wood products	Woodworking scribing and cutting techniques	Iterative design, optimization, and improvement	Aesthetics of mortise and tenon construction	Three-view drawings of Luban Lock pieces
Abilities and literacy	Understand, identify, and inherit the cultural value of mortise and tenon construction	Practice spatial thinking skills	Execute effective wood-cutting capabilities	Demonstrate hands-on production capabilities	Improve design sensibilities, capability, and communication skills	Practice spatial thinking skills and mental rotation ability

It would be difficult to implement a direct teaching of Ming-style architecture as a woodworking product. Therefore, in order to incorporate mortise and tenon construction into the woodworking projects of the labor curriculum, the Luban Lock was chosen as the theme. The Luban Lock is an example of mortise and tenon construction that presents a manageable workload. It is also a Chinese Puzzle, which is of interest to students and ensures a high level of student participation.

3.2 Objective Analysis

The C-STEAM program focuses on the cultural aspects of mortise and tenon construction and basic craftsmanship. Students identify with the culture of mortise and tenon construction by making a 3-way Luban Lock using woodworking techniques. The teaching objectives of this project can be organized into a 2-by-6 table (See Table 1), which outlines students' knowledge, abilities, and literacy in six areas: Culture, Science, Technology, Engineering, Art and Mathematics. The specific performance outcomes are as follows:

(1) Understand the cultural connotations, characteristics, and advantages of mortise and tenon structures, (2) Rationalize the process of making a 3-way Luban Lock and the steps of its creation, (3) Read the technical drawings of the Luban Lock design, (4) Master the skill of scribing woodworking components and be able to accurately scribe according to the three-views drawing, (5) Master the essentials of an electric jigsaw and be able to saw according to the appropriate markings, and (6) Optimize the Luban Lock, establish awareness of quality production, and cultivate a spirit of striving for perfection.

3.3 Assessment Design

The labor-focused curriculum evaluates students' performance during the labor process as well as their sense of labor as a core literacy. The Luban Lock Project allows us to evaluate four aspects of students' performance: labor ability, labor habits and qualities, labor views, and labor spirit.

Labor ability requires students to select appropriate tools to produce simple woodworking craft pieces. Evaluation criteria are set for scribing and sawing the three wooden components (the basic workpiece, general workpiece and specific workpieces 1 and 2). Labor habits and qualities is a standard that requires students to work safely and consistently and to develop qualities such as conscientiousness and hard work. Teachers and students use the same evaluation indicators to assess student ability, habits, and qualities in terms of excellent (A +), good (A), fair (B), and needs improvement (C), as shown in Table 2.

The labor views criteria require that students develop a sense of respect for labor and workers, and labor spirit ideally manifests in the students' perception of and appreciation for the spirit of willingly striving for perfection. Students present their work projects in class, expressing their thoughts on the difficulties and successes they faced in making the Luban Lock. Students will appreciate their hard work and craftsmanship in the process of articulating their labor process.

Table 2. Evaluation Form for Labor Ability and Labor Habits.

Evaluation Indicators	Excellent	Good	Fair	Needs Improvement
Straight lines and precise measurement				
Straight and precise sawing of basic workpieces				
Straight and precise sawing of general workpieces				
Straight and precise sawing of specific workpiece1				
Straight and precise sawing of specific workpiece2				
Moderate elasticity with Luban Lock				
Proper and safe use of woodworking tools				
Goggles worn at all times				

3.4 Project Breakthroughs

After defining the teaching objectives for this project, we designed the following learning resources to address the teaching and learning challenges of making a Luban Lock.

Task Selection for the Luban Lock Project. As mentioned earlier, Beijing's architecture reflects expert carpentry skills that include the use of mortise and tenon joints. However, the craftsmanship involved in entire buildings is complicated and students are usually not very interested. Therefore, the Luban Lock was chosen as the project task. As a Chinese puzzle for students, the simple structure of the Luban Lock can show the complete mortise and tenon construction in a more manageable way. Teachers do not have to spend a lot of time on one-on-one instruction to ensure the implement ability of the project.

Scribing Aid List. Prior to the implementation of this project, students worked independently on the production of cable-stayed bridges. However, the raw material of the Luban Lock is a three-dimensional structure, which creates difficulties for student production. Therefore, we focused on clearly illustrating the design of the wooden components, provided the three-view technical drawings of the Luban Lock, and presented video resources showing the 3D rotational view. Teachers constructed finished versions of the Luban Lock product in advance for students to observe. The three-view drawing (Fig. 2) presents the complete production process and details of each step to help students practice interpreting three-view drawings. During formal instruction, the teachers also provide 3D animations to help students who may have difficulty visualizing sawing angles and positions.

Student Independent Learning Assessment Checklist. To help students keep the instructional objectives in mind during the project, we also provided an independent learning assessment checklist for students.

3.5 Activity Design

This C-STEAM project was designed based on the 6C model, which includes contextual experience, connotation comprehension, characteristic inquiry, creation of an artifact, connection to society, and conclusive reflection.

(1) **Contextual Experience.** This session provokes students' thinking by creating a situation and asking questions. Students are shown pictures of the Imperial Palace, the Temple of Heaven, and the Pagoda of Fogong Temple. Students are then invited and inspired to think about the raw materials of these buildings and the way their materials are connected, thus stimulating students' embodied knowledge of and experience with mortise and tenon construction.

(2) **Connotation Comprehension.** This session allows students to understand the cultural connotations of mortise and tenon construction in a step-by-step manner through video viewing. Through the appreciation of detailed pictures of existing buildings, students observe the characteristics of mortise-and-tenon joints. Teachers show a video of an earthquake demonstration of the Imperial Palace model and guide students to observe the effects in comparison to effects on other wooden structures and walls. Students think through the comparison and summarize the role and advantages of mortise and tenon construction in architecture.

(3) **Characteristic Inquiry.** This session mobilizes students to investigate the characteristics of mortise and tenon culture on their own. The project builds on students' prior knowledge and invites them to explore various applications of mortise and tenon construction. Students collect information independently after class, organize free exchanges among classmates during class, and send representatives to make public presentations.
(4) **Creation of the Artifact.** This session allows students to complete the creation and optimization of the 3-way Luban Lock. First, instructors help students to conceptualize the composition of the 3-way Luban Lock and find the similarities and differences among the three wooden components. With the introduction of an intermediate "general workpiece," the whole production process is divided into three steps: basic workpiece, general workpiece, and specific workpieces 1 and 2 (Fig. 2).

Secondly, to assist in the process of drawing lines where the wood should be cut, one teacher made up a mnemonic: "where to cut, cut to where, marking the cut." With the support of 3D video resources and drawings of the Luban Lock, the marking process is more organized. Students have already learned to use an electric jigsaw in their prior woodworking project, "Cable-stayed Bridge Making." Building on this foundation, they will learn "diagonal sawing" to accurately cut the wooden parts.

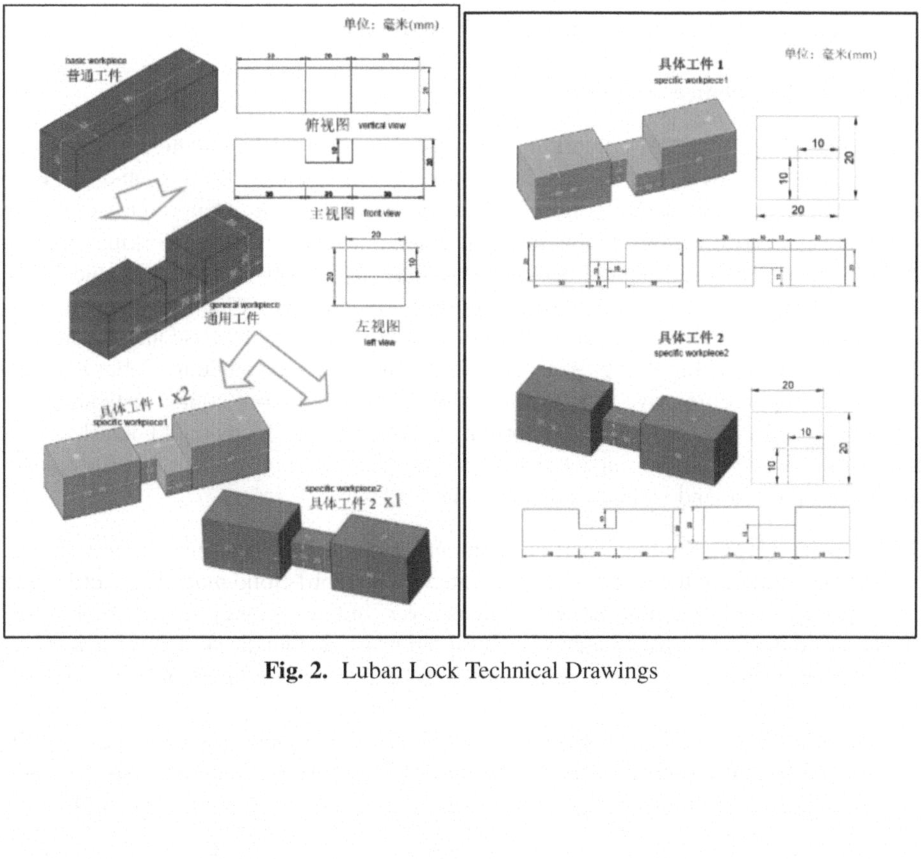

Fig. 2. Luban Lock Technical Drawings

Finally, the assembly process requires students to reflect on and improve their woodworking abilities throughout the process. In the process of woodworking, it is difficult to produce a precise Luban Lock in one attempt. Students will develop a sense of quality as they experience the process of iterative modification and optimization.

(1) **Connections to Society.** The purpose of this session is to have students share and exchange information in order to increase their overall social awareness. The woodworking classroom is equipped with an exhibition stand and the teacher has set up individual exhibition cabinets for each class. At the end of the class, students put their semi-finished or finished products into the class cabinet as cultural artifacts for display. However, the scope of this publicity only reaches other students in the woodworking classroom. In order to enhance the students' sense of cultural identity, the teacher will also lead each class in selecting three of the best projects and uploading images of those projects to a public website.
(2) **Conclusive Reflection.** The aim of this session is to combine a variety of evaluation methods to comprehensively evaluate students' learning and achievements. The Luban Lock Project adopts a combination of formative process evaluation and summative evaluation using both self-evaluation and teacher evaluations. During the implementation of the project, students reflect on the two major parts of making the Luban Lock (scribing and sawing) for all four wooden elements (the basic workpiece, general workpiece, and specific workpieces 1 and 2). At last, summative evaluation is carried out on the final cultural products.

3.6 Project Implementation

The C-STEAM project "Woodworking Wizard—The Technical Challenge of the 3-way Luban Lock" was designed for six classes in the 7th grade in Beijing, and the time of the project was four weeks (Fig. 3). The C-STEAM program is an interdisciplinary project-based learning program that integrates elements from the woodworking, geometry, physics, art, science, and architecture curricula [15]. The project inspired students to search for the connotations and merits of mortise and tenon construction, and enhanced students' awareness of the cultural inheritance represented by mortise and tenon construction. After teaching this project four times, we have seen students' labor core literacies develop and improve as a result. Junior high school labor courses pay attention to practice, and teachers mostly use formative evaluation to focus on student learning outcomes. We evaluate the impact of this this project based on students' performance, project completion, and student self-assessment.

(1) **Labor concepts.** Students' self-assessments contain their views on carpentry laborers; the content of the text will be generated to the word cloud map. The word cloud map shows with Fig. 4 shows that students were able to appreciate both the fun and the difficulty, develop a greater respect for labor, appreciate the skill of woodworkers engaged in the process of making, and even express pride in China's mortise and tenon culture.
(2) **Labor ability.** Teachers focused on cultivating the students' woodworking abilities and spatial thinking skills. At the end of the project, 29 students successfully completed one or more 3-way Luban Locks; 78 students were able to complete the

Fig. 3. Participation process and presentation of Luban Lock

production of all three wooden components, but not yet the assembly of the Luban Lock; and the other 6 students have not yet completed the production of the three wooden components. These results show that most of the students were able to correctly and safely master the use of the electric jigsaw and complete their woodworking projects with the drawings. From students' self-assessments of their spatial thinking skills, data shows that students agree that the 3-way Luban Lock project helped them to quickly identify three-view drawings, improved their mental rotation abilities, and helped with their geometric learning (See Fig. 4).

(3) **Labor habits, qualities, and labor spirit.** In the process of making, students can actively cooperate with classmates and appreciate the hard work of quality woodworking. Students understand, recognize, and pass on the spirit craftsmanship and striving for excellence.

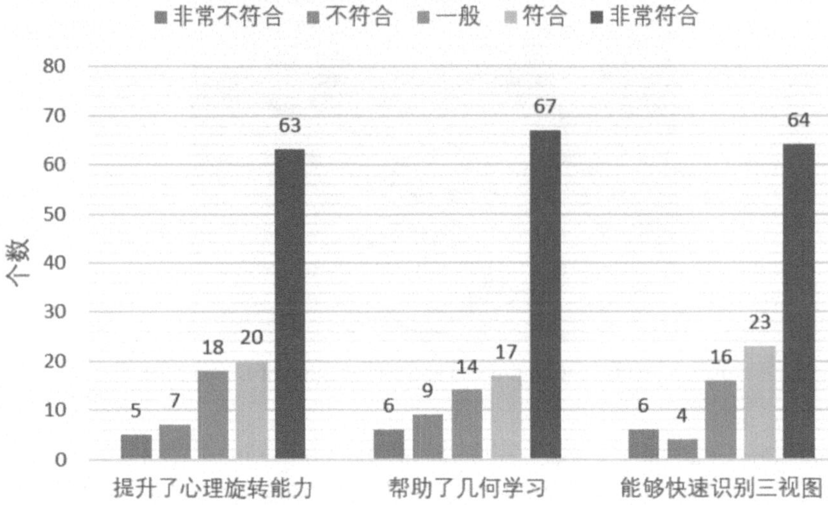

Fig. 4. Data results of cultivating students' laborability

4 Reflections and Recommendations

The Luban Lock Project fully integrates a labor education carpentry curriculum with C-STEAM educational concepts. The content of the project not only reflects the importance of mortise and tenon structures in traditional Chinese culture, but also mobilizes students' personal experience to develop their creative and critical thinking abilities.

(1) **Lead with cultural heritage to stimulate students' interest in learning.** The Luban Lock Project combines the geographical advantages and cultural heritage of Beijing and chooses the Luban Lock as the central task of the program. Through the first three links of the 6C model, we stimulate students' sense of their own cultural identity and inheritance of mortise and tenon construction.

(2) **Learn by doing, reflect on the process, and practice craftsmanship.** Embodied learning emphasizes students' "physical participation" in the learning process [16]. Labor curriculum emphasizes "learning by doing," fostering craftsmanship through an iterative process and encouraging students to experience the geometric beauty of mortise and tenon construction in practice.
(3) **Subdivide the making process to reduce cognitive load on students.** The entire making process is clearly divided into a limited number of steps, so 7th grade students are first exposed to sawing three-dimensional objects before embarking on the full project. Designing 3D resources, learning checklists, and delineation mnemonics can help students clearly understand the assigned production goals.
(4) **Use multiple assessments to promote the development of students' core literacies.** The multiple evaluations of this course include formative process evaluations and summative evaluations. The process evaluation contains the evaluation of students' works-in-progress and students' participation in the course. The summative evaluation is the evaluation of final, optimized project pieces. The program not only involves teachers' assessments, but also students' self-assessments, which provides multiple data points for more comprehensive, high quality, and robust results.
(5) **Reflections on the C-STEAM education model.** When designing the Luban Lock Project under the 6C model of C-STEAM education, the first three sessions are centered on teaching the cultural heritage and background principles of the project. In contrast, the labor curriculum emphasizes hands-on practice, and the length of the first three sessions may be too long. At the same time, the boundaries of the sessions are not clear enough, and the teachers often combine all three cultural background sessions in their teaching. The concept of culture is often abstract. Before the production of the work, students' understanding of cultural identity often remains superficial. In future iterations of this project, a more flexible, spiral approach to the 6C model could be adopted. Students will have a different understanding of culture after the deeper experiences in the last three sessions.

References

1. Yu, S., Hu, X.: STEM education and its model for interdisciplinary integration. Open Educ. Res. **21**(4), 13–22 (2015)
2. Falloon, G., Hatzigianni, M., Bower, M., Forbes, A., Stevenson, M.: Understanding K-12 STEM education: a framework for developing STEM literacy. J. Sci. Educ. Technol. **29**, 369–385 (2020)
3. Wang, S.: The interpretation of '2017 China STEM education white paper.' Modern Educ. **7**, 4–7 (2017)
4. Ministry of Education of the People's Republic of China. The National Compulsory Education Curriculum Standard (2022 Edition) Beijing Normal University Press Beijing (2022)
5. Qian, S., et al.: A comparative analysis of STEM curriculum research at home and abroad in the past decade. J. Compar. Educ. **5**, 133–146 (2023)
6. Zeng, N., Zhang, B., Wang, Q.: A comparative analysis of STEM education research at home and abroad (2008–2017) based on content analysis. Modern Dist. Educ. **5**, 27–38 (2018)
7. Yifei, M.: Policy, dilemma and approach of labor education curriculum reform in the New Era. Curr. Teach. Mater. Method **5**, 20–25 (2023)

8. Zhongwen, D., Tian, D.: Level, competency connotation and evaluation framework construction of compulsory education labor curriculum objective. Curr. Teach. Mater. Method **6**, 152–159 (2023)
9. Lihong, Z., Lijie, L.: From task group to labor project based learning: reflections on school practice of Labor Education. J. Chinese Soc. Educ. **4**, 69–73 (2023)
10. Ring, E.A., et al.: The evolution of teacher conceptions of STEM education throughout an intensive Professional Development Experience. J. Sci. Teach. Educ. **28**(5), 444–467 (2017)
11. Zheng, Y., et al.: K–12 Science, Technology, Engineering, and Math characteristics and recommendations based on analyses of teaching cases in China. Front. Psychol. **13** (2022)
12. Zehui, Z., et al.: Transdisciplinary education for cultural inheritance (C-STEAM): value orientation and classification framework. China Educ. Technol. **3**, 69–76 (2020)
13. Zehui, Z., et al.: Interdisciplinary education for cultural inheritance (C-STEAM): the 6C model and its typical case. Modern Dist. Educ. Res. **2**, 29–38+47 (2020)
14. Chi, S., Qi, Z.: A study on the aesthetic characteristics of Ming-Style furniture. Furniture Interior Des. **2**, 9–11 (2021)
15. Lingyun, G.: Research on the teaching value of traditional art in the context of STEAM: taking the structure of mortise and tenon as an example. Art Educ. Res. **1**, 136–138 (2023)
16. Zhihe, L., et al.: Dilemma and turnaround of deep learning: from disembodied learning to embodied learning on the construction of a deep-embodied learning environment. e-Education Res. **10**, 70–78 (2023)

LIDT Framework and Implementation of Life-Oriented Instructional Design for Preschool Children's Scientific Activities Empowered by Digital Intelligence Technology

Youru Xie(✉), Fangyuan Zheng, and Fanfan Zou

School of Educational Information Technology, South China Normal University, Guangzhou, China
xieyouru@aliyun.com

Abstract. Under the umbrella of education digitalization, the design and implementation of science education activities in kindergarten should approach life and intelligence. This study adopted a theoretical deduction method and evaluation research method to make use of digital intelligence technology to carry out theoretical research and practical exploration centering on life-oriented instructional design and implementation of science activities in kindergarten. Firstly, according to the rules of the scientific activities process as well as the characteristics of preschool children, the general process of preschool children's scientific activities and the connotation of life-oriented preschool children's scientific activities were clarified. Secondly, the content of the life-oriented instructional design of children's scientific activities was sorted out, and the empowering role of digital intelligence technology in the life-oriented instructional design of children's scientific activities was clarified. On this basis, the researchers constructed a LITD framework that emphasizes "Life-Oriented, Instructional Design and Technology-empowered" (LIDT), which is the instructional design for preschool children's scientific activities empowered by digital intelligence technology. Finally, to facilitate practical application, a life-oriented implementation path for preschool children's scientific activities empowered by digital intelligence technology was proposed. An empirical test was conducted through the scientific activity "AR Forest Adventure Tour" for a senior class of kindergarteners, and the effectiveness was analyzed and evaluated using a classroom teaching smart evaluation system. Relevant data indicated that this path could effectively enhance children's scientific literacy, enrich children's learning experiences, and develop children's individual characteristics.

Keywords: digital intelligence technology · preschool children's scientific activities · life-oriented · instructional design

Q. Liu et al. (Eds.): EITT 2024, CCIS 2600, pp. 145–158, 2026.
https://doi.org/10.1007/978-981-95-1525-7_12

1 Introduction

Digitalization in education has endowed early childhood education with new development opportunities, promoting systematic changes in content, structure, and form, enhancing the quality and efficiency, and effectively promoting the high-quality development of preschool education. The Ministry of Education and eighteen other departments jointly issued the "Opinions on Strengthening the Work of Scientific Education in Primary and Secondary Schools in the New Era" [1]. The goal of this policy is to comprehensively improve students' scientific literacy and endeavor to plant the seeds of science in the minds of children [2]. Scientific educational activities are of great importance in satisfying children's curiosity for knowledge, enhancing their scientific inquiry skills. The "Development Guide for Children Aged 3–6" (hereinafter referred to as the "Guide") points out that when conducting scientific education, educators should make full use of nature and real life and guide children to discover, analyze, and solve problems. The characteristics of the physical and mental development of preschool children also determine the life-oriented nature of education. Life-oriented scientific education is the correct method for preschool children. Therefore, this study clarifies the general process of preschool children's scientific activities and the connotation of life-oriented preschool children's scientific activities, constructs an instructional design framework for life-oriented scientific activities for preschool children empowered by digital intelligence technology, and proposes the implementation path of life-oriented scientific activities for preschool children empowered by digital intelligence technology, aiming at providing theoretical guidance and practical references for the implementation.

2 Literature Review

2.1 Research of Preschool Children's Scientific Education

Current research on preschool scientific education mostly focuses on objective attainment, educational content, and activity organization. Yang proposed that science education should focus on natural science content and develop scientific literacy [3]. Wang and Zhang argued that children's scientific activities should be linked to social life and family [4]. Larimore argued that preschool science education must adopt an approach that promotes synergistic development of physical, social-emotional, and cognitive needs [5]. In summary, many scholars have conducted in-depth discussions on the content and implementation of preschool scientific education. However, there are few studies on the design method of preschool scientific activities. Therefore, according to children's age characteristics and their cognitive development level, the design and implementation of scientific activities in a life-oriented way are worthy of further study.

2.2 Research of Life-Oriented Instructional Design for Preschool Children Scientific Activities

Most of the existing research on the life-oriented instructional design of children's science activities focuses on theoretical inquiry, methodological strategies, and value and significance. Chang believed that life-oriented characteristics were fundamental to the kindergarten curriculum [6]. Wan et al. believed that life education could be implemented by setting up life education environments [7]. Hao advocated for using life-oriented materials, such as pots, bowls, and basins, to enrich the teaching forms [8]. García-González and Schenetti explored outdoor spaces and natural environments to provide learning spaces [9]. Su, Zhang and Sun combined the concept of life with curriculum to let children learn through interaction in real life environments [10]. Lei concluded that a kindergarten life curriculum could expand the scope of children's experience, cultivate comprehensive ability, and cultivate their spiritual world [11]. In summary, some scholars have studied the life-oriented characteristics of preschool education, but how to optimize the life-oriented instructional design of preschool scientific activities needs to be further explored.

2.3 Research of Life-Oriented Instructional Design for Preschool Children's Scientific Activities Empowered by Digital Intelligence Technology

The existing research on the life-oriented instructional design of scientific activities empowered by digital intelligence technology mainly focuses on the principles, methods, and effects. Li and Taber analyzed the potential uses of preschool teachers who were learning to understand technologies such as AR, holography, and artificial intelligence [12]. Lee, Jo, Lee, and Kim proposed that humanoid AI robots could enhance children's scientific learning experiences [13]. He and Tang suggested that technology could enrich children's learning choices, extend interaction space, and stimulate creative instincts [14]. In summary, existing research has combined digital intelligence technologies such as AI and AR with preschool science education, but there are few studies on life-oriented instructional design of scientific activities empowered by digital intelligence technology.

3 The Life-Oriented Instructional Design for Preschool Children's Scientific Activities Empowered by Digital Intelligence Technology

3.1 Life-Oriented Preschool Children's Scientific Activities

Preschool Education Guidance Outline (Trial) (hereinafter referred to as the "Outline") stated that health, language, society, science, and art were the major content of preschool education. Preschool science education should focus on stimulating preschool children's interests. According to the rule of the scientific activities process and the characteristics of preschool children, this study argues that the general process of preschool children's scientific activities includes: (1) Create a situation for raising questions, (2) Connect with reality for scientific exploration, (3) Interact and communicate to acquire cognition, and (4) Migrate applications to develop literacy.

Xingzhi Tao proposed: "Life is education, and every day's life is curriculum. Preschool children's physical and mental development is not yet mature, and their cognition of the world is perceptual and intuitive. Therefore, it is necessary to integrate education with life to carry out life-oriented teaching" Some scholars have defined concepts of life-oriented teaching. Xing believed that life-oriented teaching means that teachers should pay attention to the children's life experiences during the teaching process, including life-oriented content and teaching methods [15]. Therefore, this study posits that life-oriented scientific activities for preschool children should emphasize the connection between preschool children's scientific activities and daily life, focused on capturing phenomena from the children's daily lives, selected, and reconstructed scientific content to guide preschool children to engage in practical inquiry activities through teacher-student interaction, group exploration, and shared exchanges. This process was aimed at promoting preschool children to acquire scientific knowledge experience, improve scientific inquiry skills, and enhance scientific emotional attitudes.

3.2 The Content of Life-Oriented Instructional Design of Preschool Children's Scientific Activities

Instructional design serves as a bridge between theory and practice. This study, guided by life education theory, connected cognition theory with the connotation of life-oriented preschool children's scientific activities, sorting out the content of life-oriented instructional design of preschool children's scientific activities. This sorting includes determining teaching objectives, reconstructing teaching content, organizing scientific activities, and conducting teaching evaluations as shown in Fig. 1.

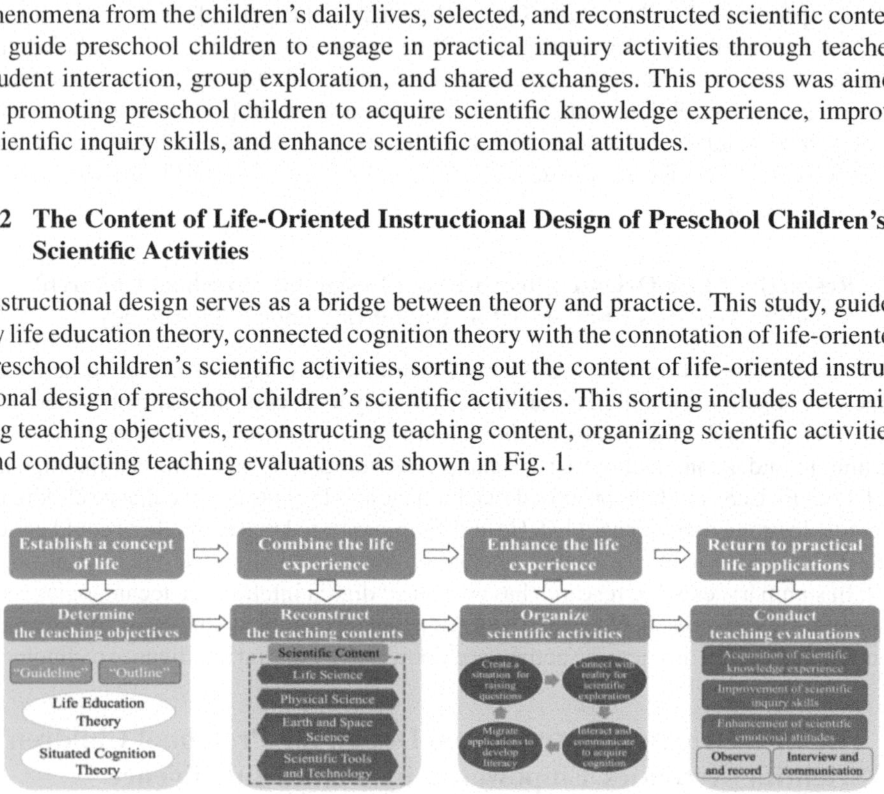

Fig.1. The content of life-oriented instructional design of preschool children's scientific activities.

Establish Concept of Life and Determine the Teaching Objectives. Preschool teachers, guided by life education and situational cognition theories, in accordance with the Guidelines and Outline, should help preschool children to acquire scientific knowledge and experiences, enhance scientific inquiry skills, and foster scientific emotions and attitudes. They should carefully observe phenomena in life, migrate them into scientific knowledge that children can understand, then determine the objectives.

Combine Life Experiences and Reconstruct the Teaching Content. According to the main spirit of the Guideline, the content of preschool science education can be summarized into the following four areas: life science, physical science, earth and space science, and scientific tools and technology. Preschool teachers should carefully observe everyday phenomena that embody scientific principles and combine children's daily life experiences to select and reconstruct the teaching content.

Enhance Life Experience and Organize Scientific Activities. Preschool teachers should pay attention to children's life experiences in the process of inquiry and practice. Firstly, teachers should create a real-life situation and raise scientific questions to stimulate children's curiosity. Secondly, teachers should connect life reality, provide life materials, and guide children to carry out inquiry practice through group cooperation and independent exploration. Thirdly, children should be encouraged to discuss and gain perceptual cognition. Finally, teachers should lead children to migrate applied scientific knowledge and skills to develop scientific literacy.

Return to Practical Life Applications and Conduct Teaching Evaluations. Preschool teachers should lead children to use scientific knowledge to explain life phenomena, carry out scientific inquiry to gain life experiences and cultivate children's scientific emotional attitudes in real-life scenarios. They should use methods of observation and recording as well as interviews to document the learning process and to discuss feelings about science learning. Finally, an evaluation should be conducted on the following aspects: the acquisition of scientific knowledge, the improvement of scientific inquiry skills, and the enhancement of scientific emotional attitudes.

3.3 The Empowering Role of Digital Intelligence Technology in the Life-Oriented Instructional Design for Preschool Children's Scientific Activities

The iterative development of digital intelligence technology has facilitated rapid changes in education. Utilizing digital intelligence technology to empower life-oriented instructional design for preschool children's scientific activities can enhance children's interest in scientific inquiry and practical application abilities. Digital intelligence technology plays an empowering role in the life-oriented instructional design for preschool children's scientific activities in the following aspects: in terms of teaching objectives, it integrates multi-dimensional data to analyze individual differences; in terms of teaching content, it integrates resources from various materials to swiftly reorganize content; in terms of teaching activities, it presents real-life scenarios to provide exploration environments; and in terms of teaching evaluation, it enables senseless automatic observation to evaluate instantly and visually.

Multi-dimensional Data Penetration and Analysis of Personality Differences. Technologies such as artificial intelligence, big data, cloud computing, 5G networks, and natural language processing should be utilized to collect and analyze children's multidimensional performance data, such as learning behaviors, interests, and cognitive characteristics, from classroom activities, area corner activities, and daily life. Therefore, teachers can grasp the needs and characteristics of each child, analyze individual differences, and formulate teaching objectives that are in line with the developmental stage and ability level of each child.

Various Resources Integration and Reconstruct Content Swiftly. Utilizing generative artificial intelligence, scientific experimentation platforms, and digital libraries, preschool teachers should gather scientific educational resources, swiftly reorganize, and agilely reconstruct them into teaching content that aligns with children's cognitive levels and life experiences. Based on preschool children's learning and feedback, teachers should adjust the teaching content and methods to meet children's diverse needs and promote the development of their knowledge, skills, and attitudes.

Realistic Scene Presentation and Provide Exploration Environment. Utilizing advanced technologies such as VR, AR, and somatosensory interactive device, an immersive learning environment should be created for preschool children to stimulate their curiosity and increase their learning interest. Through extensive interaction and scientific inquiry in virtual environments or in the real world, children can intuitively understand and grasp scientific knowledge, which fosters their innovative thinking and problem-solving abilities.

Senseless Automatic Observation and Evaluate Instantly and Visually. Utilizing intelligent sensing devices and data analysis techniques, children should be observed automatically and senselessly, and relevant data about their speech, facial expressions, and body movements should be collected in real-time. Then these data can be presented visually through charts, images, and other graphical forms, assisting teachers in promptly obtaining evaluative data on teaching, dynamically adjusting teaching strategies, and enhancing teaching outcomes.

3.4 LIDT Framework of Life-Oriented Instructional Design for Preschool Children's Scientific Activities Empowered by Digital Intelligence Technology

Based on the content of life-oriented instructional design of preschool children's scientific activities and the empowering role of digital intelligence technology, this study constructed a framework of life-oriented instructional design for preschool children's scientific activities empowered by digital intelligence technology, which emphasizes LIDT, as shown in Fig. 2.

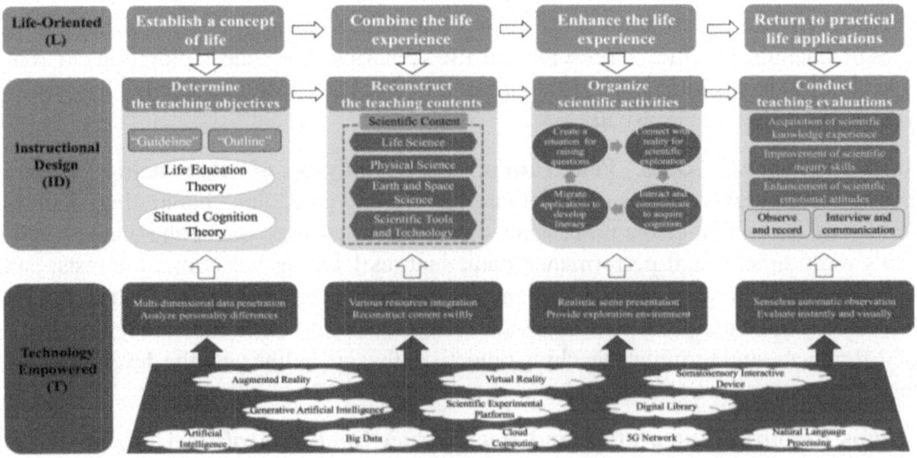

Fig. 2. LIDT framework of life-oriented instructional design for preschool children's scientific activities empowered by digital intelligence technology.

4 The Life-Oriented Implementation for Preschool Children's Scientific Activities Empowered by Digital Intelligence Technology

4.1 The Life-Oriented Implementation Path for Preschool Children's Scientific Activities Empowered by Digital Intelligence Technology

The teaching implementation path is a concretized representation of instructional design and implementation. In accordance with the LIDT framework, this study proposes a life-oriented implementation path for preschool children's scientific activities empowered by digital intelligence technology, which includes transforming teaching concepts, reconstructing teaching contents, innovating instructional design, implementing teaching activities and conducting teaching evaluations, as shown in Fig. 3.

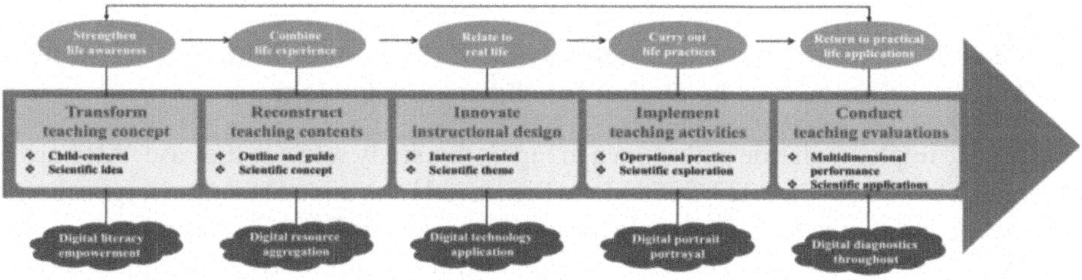

Fig. 3. The life-oriented implementation path for preschool children's scientific activities empowered by digital intelligence technology.

Transform Teaching Concepts. Preschool teachers should strengthen the awareness of life, transform teaching concepts, and establish the concept of child-centered education, guided by the idea of science. They should emphasize the ideas of respecting science and pursuing truth. They should also have digital literacy, which means that preschool teachers need to possess digital awareness, master digital knowledge and skills, understand digital applications, and have a strong sense of social responsibility.

Reconstruct Teaching Content. Preschool teachers should combine children's life experiences with the requirements of the Guidelines and the Outline, reconstructing content that is scientific and enlightening and aligned with the characteristics and process of children's cognitive development, to help children form scientific concepts. Teachers should also make use of digital platforms to select scientific activity content, micro-lessons, and videos, which can serve as materials for implementing scientific activities.

Innovate Instructional Design. Preschool teachers should connect with real-life experiences with the aim of stimulating children's interest in learning, selecting scientific themes that are close to children's lives. With the help of digital tools and equipment, they can reduce the cycle of scientific observation and recording, enhancing the safety of the experimental operation process.

Implement Teaching Activities. Aimed at developing children's scientific literacy, preschool teachers should enhance hands-on operations and practical experiences, guiding children to conduct scientific inquiries through self-exploration and cooperative communication. They should use digital means to depict digital portraits of children, promptly identify issues that arise during the learning process, and provide personalized support and assistance for the children.

Conduct Teaching Evaluations. Preschool teachers should lead children to return to real life, apply scientific knowledge and skills to explain life phenomena and solve life problems. At the same time, teachers should record multi-dimensional performance data in the process of scientific learning and evaluate the effect of children's scientific learning. In addition, teachers should use digital technology to diagnose and improve, record, and evaluate children's learning process and results, then provide diagnosis and feedback on the children's learning.

4.2 The Life-Oriented Implementation Case of Preschool Children's Scientific Activities Empowered by Digital Intelligence Technology

Based on the implementation path shown in Fig. 3, this study selected the case "AR Forest Adventure Tour" from the 2023 Guangdong Provincial Education Dual-Convergence and Entrepreneurship Teachers' Digital Literacy Improvement Practice Activity to implement the LIDT framework. The case comes from the senior class of G kindergarten, with a total of 8 children, benchmarks scientific literacy such as scientific knowledge experience, scientific inquiry skills, and scientific emotional attitudes, and carries out scientific practice activities, the specific implementation of which is shown in Fig. 4.

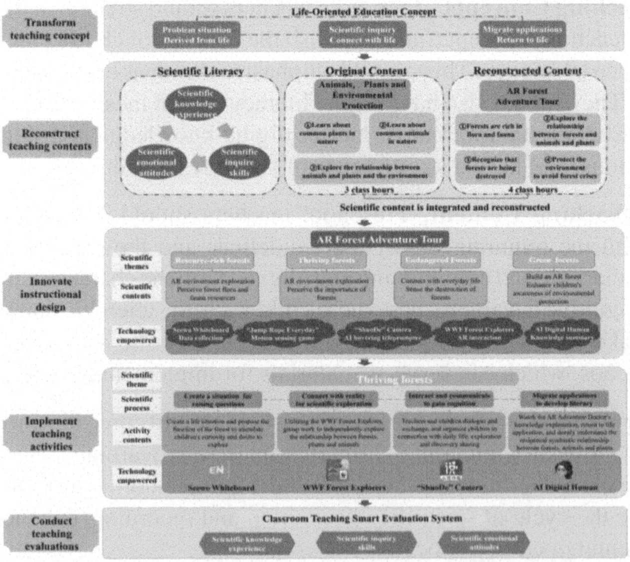

Fig. 4. Implementation of the scientific activity "AR Forest Adventure Tour" in kindergarten.

Transform Teaching Concept. Guided by the concept of life-oriented education, the overall design of the case was in accordance with the idea of "Problem situation, Derived from life; Scientific inquiry, Connect with life; Migrate application, Return to life".

Reconstruct the Teaching Content. This case benchmarked the scientific literacy of preschool children. The original three class hours were integrated and reconstructed into four class hours, as shown in Fig. 4.

Innovate Instructional Design. This case pointed to the scientific literacy of preschool children, and according to the reconstructed life-oriented teaching content, designed the four scientific themes of "Resource-rich Forests", "Thriving Forests", "Endangered Forests", and "Green Forests". It also used digital technologies such as AR, AI digital humans, and other intelligent devices such as Seewo whiteboards and tablets, to innovate the instructional design.

Implement Teaching Activities. This case chose the scientific theme of "Thriving Forests" and relied on intelligent tools and devices such as "WWF Forest Explorer", Dr. AR Explorer digital human, Seewo whiteboard, and tablet computers. The teaching process was organized according to the typical process of preschool children's scientific activities. Through teacher-student interaction, group exploration, and sharing, teachers led children to explore the relationship between forests, plants, and animals. The teaching process was shown in Fig. 5.

Fig. 5. The teaching process of the subject "Thriving forests".

Conduct Teaching Evaluation. This case selected the scientific theme of "Thriving forests" and utilized the classroom teaching smart evaluation system to evaluate the children's scientific literacy.

5 The Effectiveness of the Life-Oriented Implementation for Preschool Children's Scientific Activities Empowered by Digital Intelligence Technology

This study employed a digital evaluation approach by importing video of the science teaching activity "Thriving Forests" into the Classroom Teaching Smart Evaluation System (CSMS), to evaluate the effectiveness of the life-oriented implementation for preschool children's scientific activities empowered by digital intelligence technology. The resulting data were automatically generated and revealed that implementing the LIDT framework could enhance children's scientific literacy, enrich their learning experience, and develop their individual characteristics.

5.1 Enhance Children's Scientific Literacy

The classroom interest index and subject confidence index could comprehensively reflect the scientific literacy of preschool children. CSMS shows that the classroom interest index is calculated based on the effectiveness of children's autonomous learning, effective expression, and cooperative communication. The subject confidence index for children is calculated based on classroom regulation, positive atmosphere, and children's effective expression. These two indicators' maximum scores are 100 points. The interest and subject confidence indexes of this science teaching activity were shown in Fig. 6, where the interest index was 82.9 and the subject confidence index was 88.2, indicating that the children had a high level of scientific literacy.

Indicator	Score
Children's classroom interest index	82.9
Children's subject confidence index	88.2

Fig. 6. Children's classroom interest index and subject confidence index.

The children's interest index was 82.9, with a norm of 63.5, exceeding the norm index by 19.4, indicating that children could express themselves and engage in autonomous learning, and that communication and cooperation among children were also effective. The subject confidence index was 88.2, with a norm of 63.4, exceeding the norm index by 24.8, indicating that teachers could effectively control the activities and foster a positive atmosphere. Overall, this finding suggested that the LIDT framework and its implementation stimulated children's thinking, promoted communication and exchange, improved children's scientific literacy.

5.2 Enrich Children's Learning Experience

The evaluation scores of the learning experience, the comprehensibility of the instructional language, and the difficulty span of the instructional language collectively reflect the learning experience of children. The learning experience is calculated comprehensively based on the number of student participations, the amount of full expression by students, the proportion of positive emotions expressed by students, and the proportion of stable states in student expressions. The learning experience reflects that children participate in all learning stages actively, think about problems carefully, and engage in learning activities wholeheartedly. The comprehensibility of the instructional language refers to the degree to which the instructional language is understandable, and the larger the value, the easier it is to understand. The difficulty span of the instructional language represents the effectiveness of presenting complex concepts in a simple manner, with higher values indicating a greater span and a more pronounced contrast between difficulty levels. As shown in Fig. 7, the comprehensibility of the instructional language was mainly distributed between 0.8 and 1, and the difficulty span was mainly distributed between 0 and 0.2, indicating that the instructional language was highly understandable and that the teaching method conveyed complex concepts in an easy way.

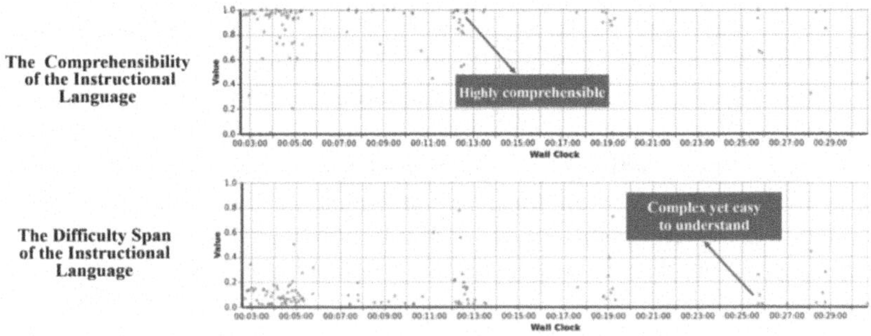

Fig. 7. The comprehensibility and the difficulty span of the instructional language.

The evaluation data showed that the average comprehensibility of the instructional language in this science teaching activity was 0.85. The comprehensibility of the reference vernacular is 0.835. This result indicated that the instructional language was intelligible. The average difficulty span of the instructional language was 0.12. The difficulty span of the reference teaching language is 0.15. This result indicated that the instruction was effectively presented in a way that was complex yet easy to understand. Additionally, the evaluation score for the preschool children's learning experience was 9.8 out of 10, indicating that they could participate in the learning actively, think about problems carefully, and enjoy learning. Overall, this study suggested that the LIDT framework and its implementation could enhance the understandability of preschool children's science activities, convey complex ideas in a simple manner, and improve the preschool children's learning experience.

5.3 Develop Children's Individual Characteristics

Class types and teacher-child interaction curve show the children's participation in activities. Class types reflect the ratio of teacher behavior to children's behavior, while the curve describes the interaction between teacher and children. As shown in Fig. 8, this science teaching activity was practice type, and the teacher-child interaction curve had an inclination angle greater than 45 degrees.

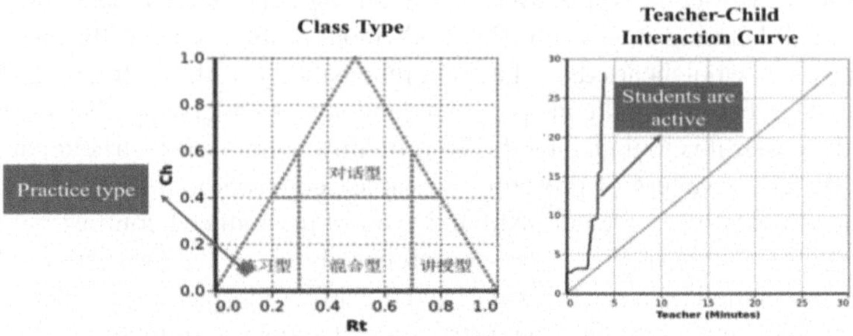

Fig. 8. Class type and teacher-child interaction curve.

Based on the class type, Rt was between 0 and 0.2, indicating a lower frequency of teacher behavior. Ch was also between 0 and 0.2, indicating fewer transitions between teacher's and children's behaviors, with the children's behaviors being more predominant. The teacher-child interaction curve had an inclination angle greater than 45 degrees, suggesting that children's activities were predominant and that the children were actively engaged in the classroom. Overall, this result indicated that the LIDT framework and its implementation provided opportunities for children to engage in autonomous learning and group cooperation, thereby developing their individual characteristics.

6 Conclusion and Discussion

6.1 Conclusion

This study employed theoretical deduction and evaluative research methods to leverage the empowering role of digital intelligence technologies such as AR and AI. This study focused on the design and implementation of life-oriented scientific activities in kindergarten to carry out theoretical research and practical exploration, forming the following conclusions:

Clarified the General Process of Preschool Children's Scientific Activities and the Connotation of Life-oriented Preschool Children's Scientific Activities. This study clarified that there were four steps in the general process of preschool children's scientific activities and pointed out its connotations.

Constructed the LIDT framework of Life-oriented Instructional Design for Preschool Children's Scientific Activities Empowered by Digital Intelligence Technology. This study sorted out the content of life-oriented instructional design for preschool children's scientific activities and clarified the empowering role of digital intelligence technology. On this basis, the LIDT framework was constructed.

Proposed the Life-oriented Implementation Path for Preschool Children's Scientific Activities Empowered by Digital Intelligence Technology and Validated its Effectiveness. This study, based on the LIDT framework, proposed the life-oriented implementation path. This pathway was empirically tested through the scientific activity "AR Forest Adventure Tour" in a senior class of kindergarten. The Classroom Teaching Smart Evaluation System was utilized for evaluation and analysis. The relevant data indicated that this pathway could effectively enhance children's scientific literacy, enrich their learning experience, and develop their characteristics.

6.2 Discussion

During the implementation of life-oriented instructional design, teachers should select different life-oriented teaching materials and resources based on the characteristics of children at various ages. This framework is not only suitable for preschool children but also for students in the lower grades of primary school. Appropriate teaching themes should be chosen, such as using electricity safely or protecting forests. However, due to the generally low level of digital literacy among preschool teachers, the design and implementation of teaching empowered by digital intelligence technology was relatively challenging. Subsequent training for teachers can be conducted to enhance their application and innovation in digital teaching. And returning to practical life applications is a challenging aspect to implement, which requires teachers to identify opportunities for applications. At the same time, parents also need to cooperate with teachers to implement the LIDT framework through collaborative home-school education. Additionally, the practical verification of this study was carried out in some science activity topics for the senior children in kindergarten. In the future, the sample size and scope of implementation can be expanded. It is possible to explore the development and construction of curricula as well as the specific implementation pathways and strategies for life-oriented scientific activities of preschool children. We hope that the LIDT framework can provide theoretical guidance and practical reference for life-oriented teaching and the development of preschool children's scientific activities empowered by digital intelligence technology.

References

1. Ministry of Education of the People's Republic of China: The Opinions of the Ministry of Education and Other Eighteen Departments on Strengthening the Work of Science Education in Primary and Secondary Schools in the New Era. http://www.moe.gov.cn/srcsite/A29/202305/t20230529_1061838.html?eqid=9b0cb5f000037a67000000066476b5b3. Accessed 9 June 2024
2. Ministry of Education of the People's Republic of China: The Person in Charge of the Off-Campus Education and Training Supervision Department of the Ministry of Education Answers Reporters' Questions on the Opinions of the Ministry of Education and Other Eighteen Departments on Strengthening the Work of Science Education in Primary and Secondary Schools in the New Era. http://www.moe.gov.cn/jyb_xwfb/s271/202305/t20230529_1061836.html. Accessed 9 June 2024

3. Yang, Y.: Building a world-class science education system with Chinese characteristics. J. Chin. Soc. Educ. **10**, 1 (2024)
4. Wang, Z., Zhang, H.: Kindergarten Curriculum Implementation Guide Series. Nanjing (1997)
5. Larimore, R.A.: Preschool science education: a vision for the future. Early Childhood Educ. J. **6**, 703–714 (2020)
6. Chang, M.: Construction and implementation of life-oriented curriculum in kindergarten. Stud. Early Childhood Educ. **352**(4), 83–86 (2024)
7. Wan, C., et al.: The practice of life education in kindergarten in Liupanshui City of Guizhou Province in China. Front. Educ. Res. **3**(3), 93–98 (2020)
8. Hao, S.: A Study on the teaching of percussion music with life materials in pre-school music education. In: 2019 International Workshop on Education, Development and Social Sciences (IWEDSS), pp. 1299–1303. Tokyo, Japan (2019)
9. García-González, E., Schenetti, M.: Education in nature and learning science in early childhood: a fertile and sustainable symbiosis. J. Outdoor Environ. Educ. **25**(3), 363–377 (2022)
10. Su, J., Zhang, X., Sun, L.: The innovation of naturalization, life oriented and gamification curriculum of kindergarten in Beijing. Educ. Sci. Res. **326**(5), 56–61 (2022)
11. Lei, C.: The value of life-oriented kindergarten curriculum and its realization. Stud. Early Childhood Educ. **331**(7), 83–86 (2022)
12. Li, X., Taber, K.S.: The Future of Interaction: Augmented Reality, Holography and Artificial Intelligence in Early Childhood Science Education. Singapore (2022)
13. Lee, J., Jo, J., Lee, J.-O., Kim, S.-H.: Incorporating humanoid artificial intelligence (AI) robots into early childhood education. Early Childhood Educ. J. **53**(6), 1–9 (2024)
14. He, K., Tang, Y.: Technology in children's life world: essence, value and limit. J. Shanxi Xueqian Normal Univ. **35**(9), 1–6 (2019)
15. Xing, Y.: Research on the Integration of Daily Life in Collective Mathematics Teaching for Senior Kindergarten Classes. Shandong Normal University (2018)

A Practical Study of AR Technology-Enabled Children's Immersion Learning

Chunmei Duan(✉) and Yuanyuan Qian

Beijing Beidou Star Kindergarten, Beijing, China
59817266@qq.com

Abstract. With the rapid development of information technology, augmented reality (AR) technology, as a new educational tool, is gradually penetrating into the field of preschool, bringing revolutionary changes to the traditional teaching mode. This paper explores how AR technology can help children to realize immersive learning practices and systematically analyzes how AR technology can promote children's cognitive development, emotional participation, and social interaction. This study used a combination of qualitative and quantitative methods, including observation, interviews, and questionnaires, to collect feedback from children and parents. The study found that AR technology can provide rich visual and auditory stimuli, enhancing children's curiosity and exploration and thus improving learning motivation and participation.

Keywords: AR Technology · Children · Immersive Learning

1 Introduction

With the rapid development of the world today, the modernization of education has become the trend of the times. Countries all over the world have taken the modernization of education and subsequent human development as an important measure to promote educational reform and social progress, as well as to enhance the overall strength of their countries. Preschool education is the first education stage in life, and as an important part of basic education, it plays an irreplaceable role in the development of a person's life. China's national program for children's development (2021–2030) clearly states: "We will strengthen media literacy education for children. We will protect children's rights to use and participate in media. We will enrich children's digital life experiences and improve their digital quality" [1].

1.1 The Application of AR Technology in Education

Augmented reality (AR) provides a completely new interactive experience for learners by superimposing digital information on the real world. In the field of education, the application of AR technology is gradually changing the traditional teaching mode, providing a more vivid and intuitive learning environment for learners. Especially in preschool, AR technology has stimulated children's learning interest and creativity with its unique interactivity and immersion.

Q. Liu et al. (Eds.): EITT 2024, CCIS 2600, pp. 159–171, 2026.
https://doi.org/10.1007/978-981-95-1525-7_13

1.2 The Special Needs and Challenges of Preschool

Early childhood is a critical period for the rapid development of individual cognitive, linguistic, and social skills. Therefore, preschool needs to pay special attention to how to create a safe and stimulating environment to support the all-around development of young children. However, traditional teaching methods often do not meet the diverse learning needs of young children, especially in stimulating young children to explore and participate in active limitations.

1.3 The Concept and Importance of Immersive Teaching

Mmersive teaching is a kind of teaching method which can promote learners' deep learning by creating a learning environment with high participation and emotional input. In immersive learning, learners are placed in a context closely related to the learning content, thus enhancing the coherence and meaning of the learning experience. For young children, immersive learning can provide a more natural and intuitive way of learning, helping them to better understand and master knowledge.

1.4 Purpose and Importance of Research

The purpose of this study is to explore the application of AR technology in children's immersive learning, and to analyze its impact on children's learning experiences and teaching effect. Through this research, we hope to provide new perspectives and methods for preschool and to provide empirical support for the further development of AR technology in education.

2 Evidence

2.1 Theory of AR Technology-Enabled Education

At the sixth annual meeting of the 30-member China Education Forum in 2019, Professor Gu Mingyuan said, "In terms of future education, we need to make full use of artificial intelligence, big data, digital and other information technologies to change the model of talent training. The key is to fully understand students' inherent potential." Professor Li Jiang, director of Stanford's program on Artificial Intelligence, Robotics and the Future of Education and the chair of the Global Consortium for Innovative Design, believes that AI utilization is the way everyone needs to think in the future.

AR is a method for adding virtual reality to the real world. AR can be visual, auditory, or even tactile. The main aim is to bring the real world and the virtual world together in the senses. There are three key points of the AR process. The first point is 3D environmental understanding. It mainly depends on object/scene recognition and location technology. The second point is research on AR technology. These technologies are divided into video perspectives and optical perspectives, as well as other representatives of light field technology (mainly named for Magic Leap) and holographic projection. The third point is human-computer interaction, which is used to make the virtual information interact with the human-computer superposition, for example, voice, gesture, posture, face, and so on. Virtual reality and augmented reality (VR/AR) technology is the result of the development of modern artificial intelligence. As a new modern educational technology, AR technology has the characteristics of interest, immersion, interaction, and visibility.

2.2 Current Status of AR Applications in Preschool

In recent years, the rapid development of AR technology has been increasingly applied in the field of education. Researchers have explored the potential of AR technology at different stages of education, especially in preschool settings, where it is believed to provide a more intuitive and interactive learning experience. The literature shows that AR technology can improve children's learning motivation and participation through visual and auditory enhancement.

2.3 Theoretical Basis of Immersive Learning

The theory of immersive learning holds that the learning effect is better when the learner learns in a highly intellectually and emotionally engaged environment. This kind of learning style emphasizes the interaction between the learner and the learning content, as well as the emotional experience in the learning process. The application of the theory of immersive learning in preschool can help to enhance children's interest and participation in learning and thus promote their all-round development.

3 Research Content

3.1 AR Technology Enables the Selection of Samples of Literary Works

AR technology uses artificial intelligence technologies such as AR as a medium, including multimedia, 3D modeling, real-time video display and control, multi-sensor fusion, real-time tracking and registration, scene fusion, and other new technologies and means. The design of this research was based on artificial intelligence and visual elements of the AR interaction to enhance the physical effect of poetry selected for literary appreciation.

This AR research project was submitted for approval to the HKSAR Innovation and Technology Commission technology voucher scheme. The research project created an experimental group of poetry samples to provide student subjects with an AR experience as shown in Fig. 1.

The poetry sample selected was the Putonghua version of a famous ancient Chinese poem, Bai Juyi's "Spring Trip to Qiantang Lake".

3.2 AR Technology Enables Young Children to Learn

Young children can interact with AR virtual objects, using play as learning to achieve immersive experiences or to actively to explore knowledge, thus enhancing the motivation of young children to learn. When children use AR to learn, they tend to have higher satisfaction, experience more pleasure, and are willing to repeat the experience.

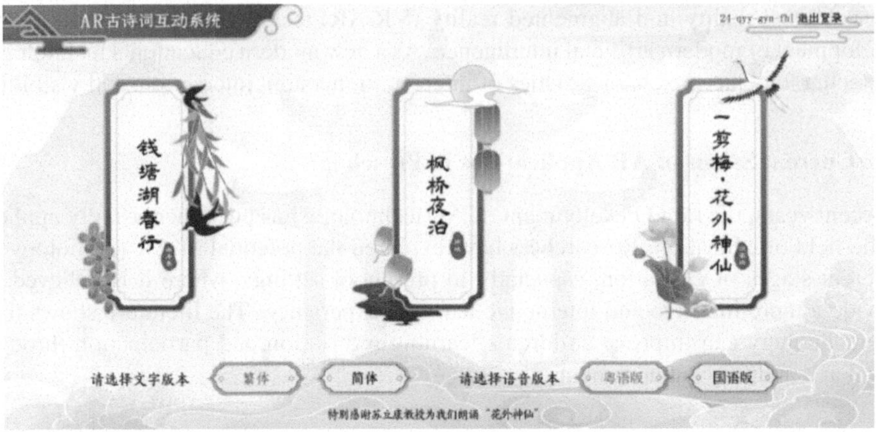

Fig. 1. AR sample interface.

(1) The Selective Construction and Interactive Presentation of Information Guides the Attention Input. The virtual education information with good form and content presented by AR is designed to attract the attention of individuals, so as to guide and keep the attention of children. In terms of content, AR is designed to stimulate children's interest in learning with interesting interactions to make them pay attention continuously. In the actual learning process, AR learning content is generated in the interaction between children and learning objects, which children cannot only see but also touch and manipulate. During this interaction, children's attention is easily attracted and maintained for a period of time.

(2) 3D Virtual Space is Created to Enhance Behavior Participation and Individual Experiences. While children's attention is guided, AR's vivid virtual information can make children produce strong near body space effects, can promote the participation of body posture and behavior, and enrich individual experience.

(3) The Stimulation of Multimodal Information Enriches the Experience of the Object. The stimulation of multiple senses by multi-modal information, as well as the mutual complement, co-induction and guidance of various information, can help to arouse positive emotions and enhance the individual's situational feeling, and thus enrich the individual's knowledge of and experience about the object.

4 Research Methods

4.1 Study Design

This study adopts a mixed-method research design, combining qualitative and quantitative research methods to comprehensively evaluate the application of AR technology in children's immersive learning. The study was divided into two phases: the first phase was a qualitative study. Through observations and interviews, we could better understand children's behavior and experiences in the AR-immersive learning environment.

The second phase was a quantitative study. During this phase we collected children's and parents' acceptance of and satisfaction with AR teaching through a questionnaire.

4.2 Participants Chosen

The subjects were young children between the ages of 3 and 6 years, as well as their parents. Participants were selected taking for different genders, ages, and backgrounds to ensure that the findings were representative and universal.

4.3 Data Collection Methods

(1) Observation: The researchers observed the children's behavior reactions, interaction frequency, and learning participation in AR teaching activities.

(2) Interviews: Semi-structured interviews were conducted with children and their parents to explore their feelings about, views of, and suggestions for AR teaching.

(3) Questionnaire: A questionnaire was designed to collect quantitative data about the children's learning outcomes and the parents' satisfaction.

4.4 Methods of Data Analysis

(1) Qualitative data were encoded and subjects were analyzed through content analysis to identify key patterns and topics.

(2) Quantitative data were analyzed using statistical software for descriptive and inferential statistics, such as t-tests and analysis of variance, to test hypotheses and assess relationships between variables.

5 AR-Immersive Teaching Practice Case

5.1 AR Technology Contributes to the Appreciation and Application Experience of Literary Works

At the beginning of 2024, the principal of the Beijing Beidou Star Kindergarten, Duan Chunmei, presided over the work. There were 13 groups of 39 teachers participating in COR8·AR teaching design of literature appreciation at their own expense. The teachers not only actively experienced the opportunities brought by innovation and technology to teaching practice, but also took part in the evaluation of the teaching design of COR8 · AR. The following figures show Middle 10 class Qian Yuanyuan teachers in their first attempt to introduce AR technology into a series of children's games (Fig. 2).

Fig. 2. Qian Yuanyuan introduces AR-immersive technology into classroom practice.

5.2 Authors and Affiliations

(1) Active Object: In this study, eight 5-year-old children from a kindergarten in the Beijing Military Region were selected as subjects. The research methods included classroom observations, case interviews and analysis of the children's work, and a questionnaire.

(2) Activity Materials: AR samples were provided of "Spring Trip to Qiantang Lake" as described above, as well as of the ancient poem "Spring Dawn" and pieces of Guzheng music. An Android phone loaded with all-in-one multimedia, a multi-angle map, and African drums were also provided.

(3) Activity Process:

a) *The front link:* Teachers introduced the conversation and lead the topic.

b) *The middle link:* Teachers used an Android phone to show the children the AR sample of "Spring Trip to Qiantang Lake". The children were guided into preliminary appreciation of ancient poetry in terms of implied meaning, and the beauty of the ancient poetry. With AR video, the teachers encouraged the children to use their own language to boldly express their feelings of the spring scenery so as to experience the taste of poetry. Showing the children an AR picture, the teachers sang the ancient poem "Spring Dawn" in cadence to express the joy of the spring scenery. The teachers introduced the author of "Spring Dawn" to expand the children's awareness. The children then stood in front of an AR screen background using a multi-angle map. The children were encouraged to combine the poem "Spring Dawn" while playing rhythms on African drums and singing the poem.
c) *The rear link:* The teachers put AR samples of "Spring Trip to Qiantang Lake" in the art area of the classroom to encourage children to paint, craft, and clay sculpture the beautiful spring scenery that inspired the ancient poetry, so as to help children understand the artistic conception of poetry more deeply.
d) *Link extension:* The teachers encouraged the children to read the scenery carefully and taste the language. The teachers asked the questions: What are your favorite AR images of spring? What features do you find in these images? The children were then directed to expand on the poetry. They opened the phone to the AR samples and described the contents of the picture. They then added pictures and engaged in group discussions. Finally, to arouse the children's environmental awareness, the children were encouraged to design an environmental protection poster. Teachers recorded the results of one-on-one interviews, and then told the children to strive to be "Excellent environmental protection propagandists". After completion of the exercise, the teachers reviewed the children's performance.

5.3 An Empirical Study of AR Technology in Literary Appreciation

(1) Case interviews: Why's he in there? Case interview transcription (see Fig. 3): "I found that Song Qingping was very cute. I saw the screen above and felt that he was eating grass. I wanted to go in and have a look. It was different from usual games. I felt very happy. I liked the birds chirping inside. I like AR-experience games. Next time I want to see, I also want to explore.

(2) Questionnaire Statistical Analysis: Eight 5-year-old children from Middle 10 classe of Beijing Beidou Star Kindergarten participated in the AR experience. Eight other children and parents who had not participated in AR were invited to participate in the online questionnaire. The online questionnaire and the automatic statistical analysis were published at https://coldXCKC.COM/COR8. The basic data are as follows (Fig. 4 and Fig. 5).

The figures are extracts of the questionnaire content. These results demonstrate that the children's and parents' feedback is concrete; the effect is clear at a glance.

Fig. 3. The girl's self-discovery response.

6 Results

6.1 The Learning Experience of Young Children

The results show that the application of AR technology in the teaching of "Spring Trip to Qiantang Lake" greatly enriches children's learning experience. Through AR samples, young children can directly see the spring scene described in ancient poetry. This visual stimulus enhanced their understanding and memory of the poetry content. Children showed a high degree of participation and interest in these activities. They actively expressed their feelings about spring through language and action, showing the positive impact of an AR-immersive learning environment.

6.2 Evaluation of Teaching Effect

The data collected by questionnaire, teaching observations and case interviews show that AR-immersion teaching has a significant effect on improving children's language ability, creativity, and emotional expression. Children's ability to interact and express themselves in an AR environment has been significantly improved. They are more confident and fluent in reciting and performing ancient poems. In addition, the use of AR technology also promotes children's awareness and appreciation of traditional culture.

6.3 Research Analysis

In the above research, AR technology was used to create a sensory simulation of the virtual environment, with the help of mobile phones, computers, and other devices to

The proportion of the members in the experiment of the game experience class in AR-enabled kindergarten

	To participate	Non-participants
No. Of persons/one	8	8

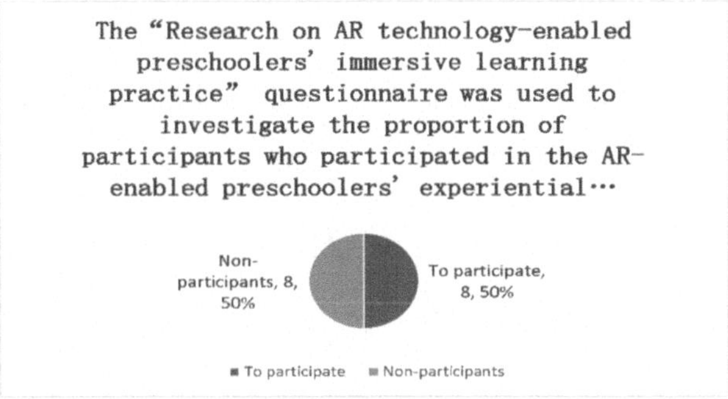

The ratio between parents and children

	children	parents
No. Of persons/one	16	16

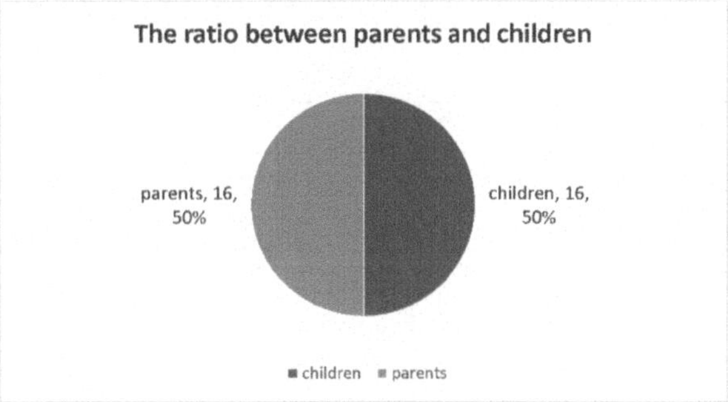

In the questionnaire survey, the proportion of after-class reaction of the members who had participated in AR experience class.

	Like it very much	Like it	Didn't mention it
Number	8	0	0

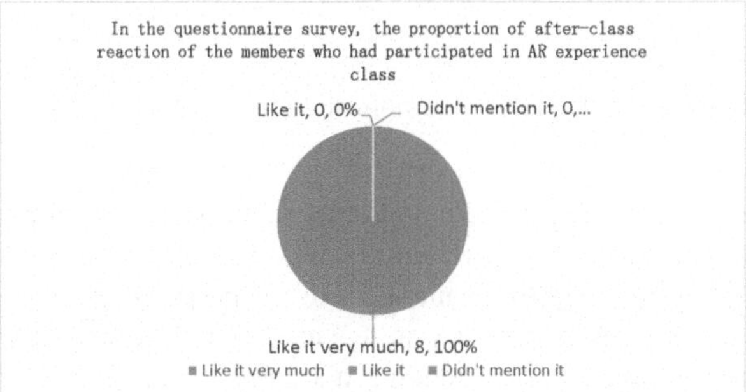

Fig. 4. Data statistical analysis of the online questionnaire regarding participation in the AR experience

Topic: Please describe the biggest change of children's AR experience after participating in the "Spring trip to Qiantang Lake".

(1) Children

1	I like ancient poetry, I like to recite ancient poetry every night.
2	I like this game, I will play it again.
3	I like teachers, like ancient poetry.
4	I like Tang poetry.
5	I feel very interesting, also want to play this AR game lesson, I also like to explore, also like ancient poetry.
6	I fell in love with Tang poetry, as well as ancient poetry.
7	I like ancient poetry. I learn one every night before I go to bed.

(2) Parents

1	My children like it very much, very happy, but also like reading ancient poetry.

Topic: Please State Your latest findings for the children attending the experience class?

(1) Children

1	I found that there are rivers, fish, mountains, small pavilion, as well as fish, ponies and will move.
2	There are boats, fish, and clouds, than usual game activities fun, usually do not have this feeling, that day I was very happy to see if I can bite the grass.
3	My favorite is the boat, mountains, and clouds, but also change, and Lotus leaves, water and water of those flowers, trees are full of birds call.
4	I especially like, feel very good, better than other games, also want to last time, especially fun.
5	I found Qingping Song very cute. When I saw the screen above, I felt that he was eating grass. I wanted to go in and have a look. It was different from usual games. I felt very happy. I liked the birdsong inside. I liked AR experience games, next time I want to see it, I want to explore it.

(2) Parents

1	Ordinary games can not be projected onto the screen, this can be projected with a mobile phone, which boats, water, stones, flowers and ordinary different, you can touch.
2	It's different from the games we usually play, and it's made by cell phone.
3	My child found a mountain, and small fish, and a boat, but also found a mountain.

Topic: What are your suggestions for this pilot exploration?

(1) Children

1	I like this game very much and want to play it again next time.
2	I hope to play this game again next time.
3	I feel very fun, also want to play this AR game lesson.
4	I want to explore AR next time.

(2) Parents

1	I'm looking forward to it.
2	I love it.
3	I hope we have this game next time.
4	I feel very good about that game.

Fig. 5. Summary of the online questionnaire for regarding the AR experience

achieve virtual and real-world interaction of objects, allowing children to experience the AR environment. The focus of this research was for children to use AR to appreciate the ancient poem, "Spring Trip to Qiantang Lake" as part of their spring activities. The purpose of appreciating ancient poetry is to enhance the ability of thinking through appreciation, analysis, and creative reading activities. Therefore, with the aid of the AR information technology that provided an AR sample of "Spring Trip to Qiantang Lake", a lively AR-interactive teaching context was enabled. Through the construction of AR animation-mediated activities, the three implementation paths of cooperative inquiry mode, continuous and dynamic extension, and exchange evaluation allowed the children to visualize abstract concepts, diversified teaching time, and promoted the internalization of children's habit transfer.

The right hemisphere is concerned with comprehensive and holistic listening, including imagination, perception, and intuition, while the left hemisphere is concerned with language and language-related functions. Therefore, the key to the realization of the artistic conception lies in activating the right hemisphere so that the children enjoy the taste of poetic language in the AR environment using a combination of means such as sounds and pictures. The goal was to make use of the children's life experiences related to poetry, to arouse the children's affinity, and to shorten the psychological distance between poetry and the children, so that children entered a kind of hazy lyrical mood in which they felt the poetry. Therefore, the activities mainly relied on AR information technology to lead the children through research-based learning. The activity successfully trained the children's creative thinking ability, problem discovery and solution, and exploration of new ideas in a multi-sensory, multi-angle, multi-element way so that every child could start a small universe—they could create a masterpiece full of intellectual spark.

(1) The Interaction of the Individual with the Environment: The age characteristic of young children is that they rely on touch, smell, sight and hearing to explore and play, which not only satisfies their curiosity but also greatly boosts their cognitive progress. When AR is introduced into preschool education, it allows children to interact with their environment through multiple sensory channels, such as vision, hearing, touch, and movement. Thus, AR meets the basic learning needs of all young children and allows for adaptation to different learning styles. Traditional teaching is often limited to vision and hearing, and it is difficult for children to become fully involved. However, by using the immersive teaching mode of AR technology, children can participate in games and learning through the comprehensive use of the senses to maximize their learning effect and growth potential. The application of AR can fully stimulate children's multi-channel participation, providing rich and varied interactive experiences, thus enhancing the fun of learning and encouraging children to learn more actively.

(2) The Combination of Virtual and Real Constructs: The application of AR technology in preschool enables children to perceive the information dimensions of the real world and virtual worlds at the same time in their daily learning environment. As a key element of this educational context, the virtual environment not only expands the physical boundary, but also seamlessly connects the digital information field with the physical world to create a multi-dimensional spatial experience. Instead of relying on two-dimensional images, young children explore and understand abstract concepts and objects in three-dimensional virtual scenes, which can improve cognitive efficiency and help shape their spatial understanding. At the same time, through immersive media such as AR, children can experience the interaction between virtual and real, further enhancing their subjective cognition and experience of both.

(3) The Cooperation Between Each Other: In traditional childcare settings, various factors often lead to insufficient consideration of the individual needs and thinking of each child, and the communication between teachers and children only stays at the basic level of conversation. To some extent, this weakens the depth of teacher-child interaction, which in turn affects the quality of play activities and learning experiences, further limiting the healthy development of children. In contrast, the application of virtual reality technology presents a different scene. Instead of isolating users from reality, AR technology allows teachers to interact and collaborate with young children. With the aid of AR technology, the teaching activities not only encourage cooperative learning between teachers and children, but also promote joint participation in the exploration, proposal, and solution of problems. It can effectively encourage children to acquire new knowledge, construct individual cognitive frameworks, and enhance social communication ability.

7 Conclusion and Outlook

This study reveals the potential and challenges of AR technology in children's immersive learning. AR technology provides an attractive and interactive learning environment for young children, which helps to promote their learning motivation, cognitive ability, and creativity. However, to realize these potentials, AR technologies need to be further optimized and integrated with the learning needs and educational goals of young children. Although this study provides valuable insights, there are some limitations. For example, the relatively small sample size of the study may limit the generalizability of the results. In addition, the study focused on short-term learning effects, and the effect of AR on children's long-term learning and development is not clear. Future studies should expand the sample size and conduct long-term follow-up studies to more comprehensively assess the effectiveness of AR in preschool. Future research should also focus on the long-term effects of AR and how to maximize its use in preschool through teachers' professional development and teaching strategies.

In the future, there is still much to explore, such as determining the effects of AR-immersive learning on children's long-term cognitive and emotional development, and how to improve AR technology in order to better adapt to the learning needs of young children and protect their health. It is believed that in the near future, with the continuous progress of technology and in-depth research, AR technology will play a greater role in preschool.

AR-immersive technology is an important engine for building educational power. In the face of the new wave of scientific and technological revolution, education should actively take the initiative for change.

References

1. Outline for Children's Development in China (2021–2030). State Department of The People's Republic of China (2021)
2. Zhang, H., Li, M.: The application of augmented reality in preschool. Mod. Educ. Technol. **32**(2), 45–52 (2022)
3. Wang, L., Zhao, Q.: The influence of immersive learning environment on children's cognitive development. Educ. Res. **40**(4), 123–130 (2021)

4. Lin, X.P., Xie, W., Zeng, J., Wang, S.: AR technology and the application of literary appreciation. Chinese Scientific Papers Online, Beijing (2020). http://www.paper.edu.cn/releasepaper/content/202002-9
5. Lin, H., Wang, W.: Research on interactive learning model of children based on AR technology. Educ. Inf. Forum **29**(6), 105–112 (2021)
6. Yangli, C., Xiangxue, Z., Wenbing, C.: An analysis of the mechanism of AR-enabled learning and its explanation-based on a comparative experiment of reading AR and paper picture books. Mod. Educ. Technol. **32**(9), 77–81 (2022)
7. Ting, Y., Xiaoping, Y.: Application of augmented reality in kindergarten teaching: value logic and action strategy. J. Shaanxi Pre-Sch. Norm. Univ. **32**(9), 37–39 (2019)

Research on the Construction and Application of a Multi-agent Collaborative Digital Volunteer Education Model

Shiyuan Lin[1](✉), Xiuting Li[2], Luoying Huang[3], Tianqi Duan[2], Chong Wang[4], and Jialuo He[4]

[1] Institute of Education for Lifelong Learning and Sustainable Development, Beijing Research Institute of Education Sciences, Beijing, China
linsy1981@126.com
[2] Lab of Learning Design and Learning Analysis, Beijing Normal University, Beijing, China
[3] College of Education, Hebei University, Baoding, China
[4] Digital Volunteer Teaching Department, Tencent Sustainable Social Value Organization, Beijing, China

Abstract. Supporting education in rural areas is the key to supporting and helping educationally undeveloped areas to achieve high-quality development. Because it can effectively avoid the problems of low economic efficiency and weak continuity in traditional teaching, a digital form of teaching that takes advantage of information technology has gradually become a new type of teaching. How to enhance resource delivery to empower educational and teaching innovation has become a key and contentious point in the study of digital volunteer education. Based on the primary goals of promoting balanced allocation of educational resources, innovating curriculum teaching, and helping the construction of the teaching team in disadvantaged areas, this study constructs a multi-agent collaborative digital volunteer education model that is effective. This model uses collaborative participation of multiple types of relevant subjects, supported by online and offline integrated learning spaces, with the teaching community as the key, the dual-teacher live interactive teaching model as the core, and driven by evidence-based teaching and learning support services. The application of this model found that students achieved three-dimensional collaborative development in knowledge, ability, and attitude, and that the professional quality and level of local teachers were also improved. The multi-agent collaborative digital volunteer education model constructed in this study provides a reference for the innovative development of digital volunteer education.

Keywords: digital volunteer education · multi-agent collaboration · integrated learning space · live teaching by two teachers · evidence-based teaching · teaching community

Q. Liu et al. (Eds.): EITT 2024, CCIS 2600, pp. 172–181, 2026.
https://doi.org/10.1007/978-981-95-1525-7_14

1 Research Background

Teaching in rural areas is an important activity to support education in disadvantaged areas. It follows the compensatory principle and promotes the realization of educational equity. The traditional teaching model of sending personnel to teach in these areas, especially short-term teaching, has achieved certain results. But there are also practical problems such as increasing the economic burden on local governments and schools, planning for the personal safety of, responsibilities for, and risks of teacher training, and recognizing how personnel changes affect the continuity of teaching [1]. With the application of modern information technology in the field of education, digital volunteer education through online teaching platforms, and the participation in learning through multimedia devices have the advantages of providing efficient and convenient delivery of rich and high-quality educational resources; digital volunteer education has also received widespread attention and expectation. However, the existing model of digital volunteer education is still considered an innovation in teaching and learning through technology application, and it still mainly utilizes lectures and demonstrations, with insufficient interactivity [2, 3], which limits its effectiveness and scope. The reason for these limitations is that the systematic design and implementation of digital volunteer education are not well recognized, and the means to fully connect multiple subjects and elements is ignored. To fully unleash the potential of digital volunteer education in promoting changes in teaching methods and improving teaching quality, and to promote the coordinated development of local teachers, students, volunteer teachers, and other types of subjects, it is urgently necessary to explore new digital volunteer education models.

Based on this foundation, this study aims to construct a new model of a multi-agent collaborative digital volunteer education model to solve the problems of low effectiveness, insufficient economy, and difficulty in ensuring safety in traditional teaching. At the same time, it aims to break through the bottlenecks of existing digital volunteer education, such as outdated teaching methods and difficulty in sustaining teacher team building, to achieve the multiple goals of high-quality resource delivery, teaching model innovation, and local teacher team building, and to further enhance the self-generating function of underdeveloped areas and weak schools.

2 Literature Review

2.1 Research on the Value of Support Teaching in Rural Areas

The value pursuit of "why teach" is the key to defining the content and methods of teaching. In addition to promoting student development, helping teachers in weak schools to improve their teaching abilities has become an important value appeal of digital volunteer education [4]. "Famous Teacher Classroom" emphasizes the use of online training communities and the demonstration of outstanding teachers to promote the professional development of teachers in undeveloped areas and in weak schools [5], which is similar to the digital volunteer education model. Therefore, digital volunteer education has the dual value appeal of promoting student development and teacher development.

2.2 Current Status of Research on Digital Volunteer Education

In determining how to best carry out digital volunteer education, some studies focus on the spatial and temporal reconstruction of teaching and pay attention to forming a hybrid teaching model [6]; some studies focus on the establishment of teaching communities [7], building teaching communities from the perspective of volunteers and local teachers, and carrying out hybrid teaching that combines online and offline space integration. Some studies focus on training communities while promoting the professional development of local teachers. They also explore teaching methods and teaching reform paths suitable for local schools. Some studies are based on the perspective of data governance, emphasizing the use of learning situation analysis data to carry out precise teaching evaluation and support services [8].

In summary, digital volunteer education has gone beyond the simple goal of resource delivery and has focused more on improving teaching quality through innovation in teaching models and achieving the value of both promoting student development and local teachers' professional development. In line with this value pursuit, digital volunteer education has followed a trend of focusing on online and offline space integration, the establishment of teaching communities, the construction of training communities, and precise learning support in its promotion measures.

3 Multi-agent Collaborative Digital Volunteer Education Model

3.1 Goal-Oriented

The goal of the multi-agent collaborative digital volunteer education model is to build a new teaching method that is different from traditional on-site teaching, and at the same time is superior to low-level digital volunteer education. This new teaching model has three goals: the first is to promote the balanced allocation of educational resources, not only pursuing the shared allocation of high-quality teachers, but also pursuing the sharing of high-quality courses and their resources; the second is to innovate curriculum teaching through digital methods and build new learning spaces [9], thereby providing environmental support for the reform of educational and teaching methods, innovating teaching methods [10], improving the quality of education and teaching with more flexible and appropriate course content and teaching methods, and promoting student development; the third is to promote the construction of the teaching staff in underdeveloped areas and schools and to promote the professional development of local teachers through various methods such as joint teaching and collaborative teaching and research to cultivate a high-quality, professional and innovative teaching staff for the sustainable development of underdeveloped areas and schools.

3.2 Model Construction

Guided by these three goals, combined with the results of relevant research and practice on digital volunteer education, and giving full play to the important advantages of information technology and digitization, this study has constructed a digital volunteer education model. This model is shaped by the collaborative participation of multiple subjects, which in turn is formed by the collaborative participation of multiple types of relevant subjects and based on online and offline integrated learning spaces. The teaching community is the key and the dual-teacher live interactive teaching model is the core. This model is driven by evidence-based teaching and learning support services, as shown in Fig. 1, to guide and promote digital volunteer education from the low-level of delivering teaching resources to the high-level of empowering educational and teaching innovation.

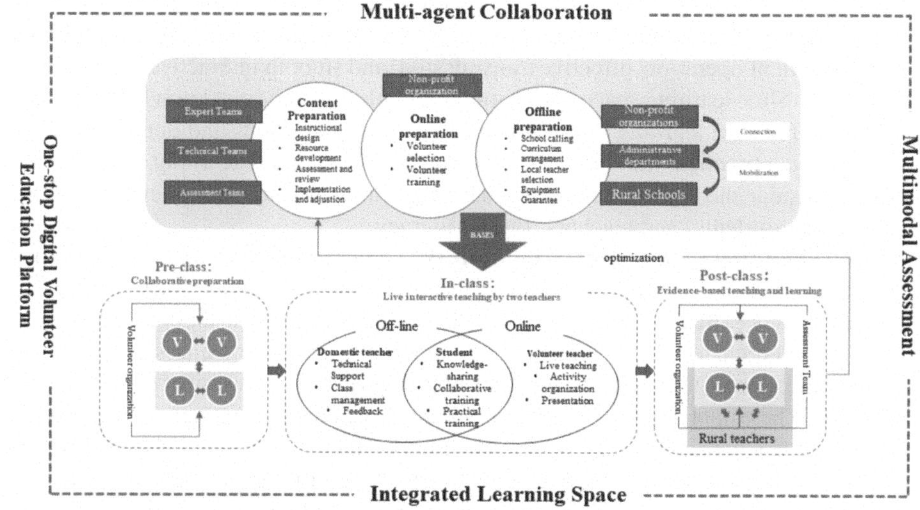

Fig. 1. Multi-agent collaborative digital volunteer education model.

(1) **Multi-agent Collaborative Participation.** The multi-agent collaborative digital volunteer education model integrates various institutions such as teaching institutions, public welfare organizations, education administrative departments in the areas receiving teaching, schools receiving teaching, third-party evaluation institutions, as well as many individuals such as volunteers, local teachers, leaders of education administrative departments, school principals, and evaluation experts. Different institutions and individuals work together to promote the development of teaching. Teaching institutions are mainly responsible for curriculum design and development, digital volunteer education platform construction, and digital volunteer education hardware. Public welfare organizations are responsible for volunteer selection, organization, training, and communication with education administrative departments and schools in the areas receiving teaching. Education administrative departments are responsible for teaching curriculum reviews and school management. Schools are responsible for the application of teaching curricula, local teacher selection, and other objectives. Third-party evaluation institutions are responsible for evaluating teaching and making recommendations for improving curriculum and teaching.

(2) **Online and Offline Integrated Learning Space.** The learning space is the key place where teaching and learning activities take place, and it is closely related to teaching and learning activities and learning outcomes [11]. Digital volunteer education has a natural technical bent that relies on information technology and takes advantage of the benefits of information technology. The learning space that integrates information technology is the key component of the construction of the digital volunteer education model. The online and offline integrated learning space is an important element of the multi-agent collaborative digital volunteer education model. It is the key to effectively connect online and offline teaching activities and enhance students' learning experience. Among them, the online learning space is mainly based on convenient operation, effective transmission, and smooth interactive communication; the offline learning space mainly plays the role of promoting knowledge sharing and collaborative learning [11], supporting group discussions and demonstrations, and offering practical training [12]. It is worth noting that the learning space should fully consider the appropriateness of hardware and software equipment conditions as regards students' and teachers' digital literacy.

(3) **Digital Volunteer Education Community.** The establishment of a collaborative teaching relationship between volunteer teachers and local teachers during the teaching process is extremely important [10]. This collaboration can support and achieve the improvement of teaching and learning efficiency and effectiveness [13], and it is also of great value to the professional growth and ability improvement of local teachers [14]. The establishment of a digital volunteer teaching community composed of volunteer teachers and local teachers, and the clarification of the collaborative mode between the two types of teachers is the key to the multi-agent collaborative digital volunteer teaching model. This study designed the collaborative teaching mode of volunteer teachers and local teachers using course stages. In the pre-class stage, the volunteer teachers and local teachers jointly carried out teaching design in the interactive community of the volunteer teaching platform, designed teaching resources and methods, and jointly formed a teaching plan. In the mid-class stage, the volunteer teachers taught online and interacted with local teachers and students in real-time through a variety of interactive tools. The local teachers organized students to discuss in groups offline, offered hands-on practice, and provided students with learning incentives and guidance.

(4) **Double-teacher Live Interactive Teaching Method.** To effectively support the collaborative teaching of the digital volunteer education community, the specific teaching organization and implementation methods need to be further clarified. This study proposed a collaborative teaching method with the dual-teacher live interactive teaching method as the core, that is, with the support of modern information technology, to achieve multi-dimensional and multi-category real-time interaction between multiple pairs of subjects (e.g., volunteer teachers and local teachers, volunteer teachers and students, local teachers and students, and students and students). Its essence is ability of the teaching community members to cooperate and coordinate with each other in the process of teaching implementation to complete the teaching and learning activities. The dual-teacher live interactive teaching method is based on the close cooperation between online teaching teachers and on-site local teachers. Online teachers mainly teach live and organize Q&A, while on-site teachers

mainly provide technical support, classroom management, and problem feedback. The development of the dual-teacher live interactive teaching process and the use of teaching methods vary depending on the nature of the course, teaching objectives, and student characteristics. The specific design and implementation of teaching activities need to be adaptively adjusted according to actual conditions.

(5) **Evidence-based Teaching and Learning Support Services.** The design, organization, and implementation of teaching activities are all based on the digital learning space that integrates online and offline learning and is supported by modern information technology, so that the process and results of digital volunteer education can be recorded and stored in an objective and rich data form, thus laying the foundation for the evidence-based analysis, diagnosis, and improvement of digital volunteer education. This study specifically proposed two evidence-based teaching and learning support services to promote the effectiveness of digital volunteer education: first, in the course and resource construction link, "design evidence-based" is promoted, and questionnaires, interviews, on-site observations, data analyses, and other channels and means are used to achieve the goals of the national curriculum while respecting the actual situation of regions, schools, teachers, and students; second, in the teaching process, "teaching evidence-based" is achieved through digital volunteer education platforms and digital equipment in physical classrooms build data collection, analysis, feedback, improvement mechanisms and processes based on the normalized teaching and learning process. These platforms help volunteer teachers and local teachers find problems and adjust teaching strategies and methods in a timely manner.

4 Application Effectiveness of Multi-agent Collaborative Digital Volunteer Education Model

4.1 Quasi-experimental Research Design on the Effectiveness of the Digital Volunteer Education Model

(1) **Research Purpose.** The application practice of the multi-agent collaborative digital volunteer education model aims to comprehensively evaluate and effectively verify the effectiveness of the model, which is specifically implemented in two aspects: promoting student development and promoting teacher professional development.

(2) **Application Scenario.** The multi-agent collaborative digital volunteer education model constructed in this study was applied to the digital volunteer education project initiated by Tencent SSV Digital Volunteer Education Laboratory. The four courses of calligraphy, science, career enlightenment, and information technology in primary and secondary schools in Gansu and Chongqing were used as examples to evaluate the effect of this model in teaching practice. The teaching practice cycle is one year, with calligraphy, science and information technology courses offered once a week and career enlightenment courses offered once every two weeks.

(3) **Research Design and Sampling.** During the implementation process, stratified sampling was conducted among regions and school types, and classes participating in digital volunteer education were selected as experimental classes. Parallel classes in the same school that did not participate in teaching were selected as control

classes. The experimental class adopted a multi-agent collaborative digital volunteer education model to carry out course learning, while the control class did not carry out teaching courses. The effectiveness was verified by a longitudinal comparison method of pre-test, process monitoring, and post-test of the experimental class, as well as a horizontal comparison between the experimental class and the control class. Specifically, before teaching began, a pre-test was conducted on the subject literacy level of students in the experimental class and the control class. During the experiment, observation records and interviews were used for follow-up evaluation. After teaching was completed, a post-test of students' subject literacy was conducted, and interviews with local teachers and volunteers were conducted for summary and evaluation.

(4) **Methodology.** Three types of tools were developed for the evaluation of student development and teacher development effectiveness. The first type of tool is a questionnaire survey tool, which includes seven questionnaires, such as the rural basic information questionnaire, the student subject literacy questionnaire, the local teacher development effectiveness questionnaire, and the supporting teacher development effectiveness survey. The second type of tool is the classroom observation record form for the interactive teaching of two teachers during a live broadcast. The third type of tool is the interview outline for multiple subjects such as students, local teachers, supporting teachers, principals of the supported schools, and public welfare organizations.

(5) **Data Collection and Analysis.** The evaluation of the effect integrated multiple tools (e.g., questionnaire surveys, interviews, and non-intrusive observations), collected multimodal and multidimensional data, and focused on the mutual verification and complementarity of quantitative data and qualitative data. 2,765 students, 30 teachers and 76 volunteers were surveyed, and a total of 8,652 questionnaires were collected, and the collected data were analyzed using software such as SPSS, descriptive statistical analysis, paired sample t-tests, and other methods. Online interviews were conducted with the help of Tencent Conference. Thirty-four students, 16 teachers, 17 volunteers, six principals, and three members of voluntary agencies, a total of 76 people, were interviewed. The average interview time per person was approximately 30 min. The data were recorded through conference screen recording and back-to-back coding was performed using Nvivo; classroom observations collected data by combining live listening in the conference room with analysis of screen recording playback. A total of 62 lessons in 23 classes in 22 schools were observed and analyzed by back-to-back coding.

4.2 Analysis of Application Effectiveness

(1) **Effectiveness of Student Development.** Firstly, four questionnaires related to the development level of students' subject literacy were analyzed for validity and internal consistency. Among them, the total explanation of the scientific literacy questionnaire was 53.97%, the factor loading of each item was between 0.520 and 0.728, and the total Cronbach's coefficient was 0.92. The total explanation of the calligraphy literacy questionnaire was 57.65%, the factor loading of each item was between 0.517 and 0.793, and the total Cronbach's coefficient was 0.90. The total explanation of

the information literacy questionnaire was 65.67%, the factor loading of each item was between 0.526 and 0.868, and the total Cronbach's coefficient was 0.89. The total explanation of the professional literacy questionnaire was 57.52%, the factor loading of each item was between 0.505 and 0.840, and the total Cronbach's coefficient was 0.87. From the above analysis, it can be seen that all four questionnaires have good structural validity and internal consistency.

Secondly, the paired sample t-test method was used to comprehensively analyze the student subject literacy questionnaire data of four subjects including calligraphy and science. The results are shown in Fig. 2. As can be seen, students who adopt the multi-agent collaborative digital volunteer education model have significantly improved their subject knowledge, abilities, and attitudes. Analysis of the interview results found that in terms of knowledge, the volunteer teaching courses not only effectively improved students' depth of understanding of subject knowledge, but also expanded the breadth of knowledge. In terms of abilities, the volunteer teaching courses trained students' thinking patterns, improved their study habits, and improved their hands-on skills. In terms of attitude, students' interest and love for learning were significantly enhanced, and their learning initiative was significantly improved. This demonstrates that the multi-agent collaborative digital volunteer education model has a three-dimensional collaborative development trend of "knowledge-ability-attitude" in promoting students' development effectiveness.

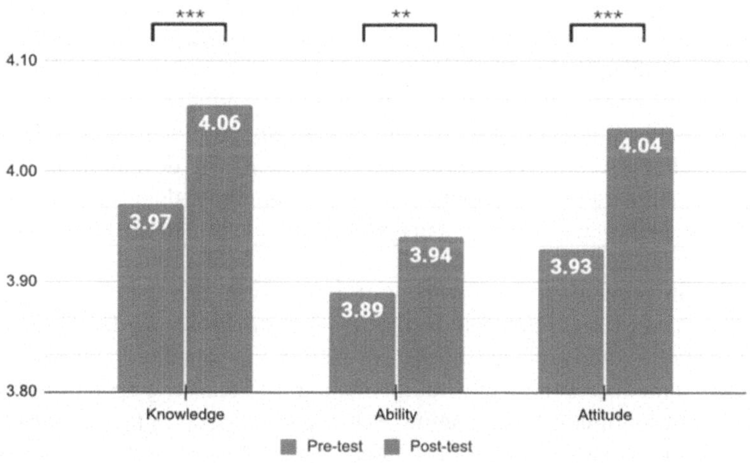

** indicates $p < 0.01$, *** indicates $p < 0.001$

Fig. 2. Effect of digital volunteer education on promoting students' subject literacy development.

(2) **Effectiveness of Teacher Development.** The multi-agent collaborative digital volunteer education model has been demonstrated to promote the overall improvement of the quality of the teaching staff. This teaching model provides support for local schools so that they can carry out teaching and research activities. The school has promoted the circulation of high-quality teaching resources during the teaching project

by organizing teachers to attend classes and rotate classes, by organizing teachers of the same subject to exchange information, and by expanding the radiation gain of the teaching project on the construction of the local teacher team. In the interviews, a principal mentioned that the school's organization of local teachers to attend and observe volunteer teachers' classes was helpful to the teachers at the school; 40% of local teachers shared their information in various ways about teaching content, teaching methods, listening experience, and so forth. Digital volunteer education courses have also promoted the multi-dimensional improvement of teachers' knowledge, abilities, concepts, and attitudes. 60% of local teachers said that while assisting volunteer teachers, they improved their understanding of subject knowledge; 100% of local teachers said that they learned the volunteer teachers' course activity design strategies, learning motivation stimulation strategies, teaching language organization strategies, and more. The value of online teaching with two teachers stimulated a stronger sense of responsibility, mission, and enthusiasm for teaching.

5 Conclusion and Suggestions

The multi-agent collaborative digital volunteer education model is based on the three goals of delivering high-quality resources, innovating teaching models, and promoting the development of local teachers. It pursues the dual value of student development and teacher professional development, integrates the collaborative participation of multiple institutions, supports the innovation of teaching and learning with online and offline integrated learning spaces, and forms a digital volunteer education community composed of volunteers and local teachers. It takes dual-teacher live broadcast teaching as the main method and provides support services for teaching and learning based on multimodal data. This digital volunteer education model not only promotes the overall development of students, but also effectively promotes the professional development of local teachers through the creation of a robust teaching community, dual-teacher live interactive classroom, collaborative teaching and research, joint lesson preparation, and cooperative teaching.

The improvement and development of digital volunteer education requires continuous systematic design, not only in the relatively microscopic learning environment, course resources, and teaching mode, but also in the relatively macroscopic teaching organization mode, resource integration method, and teaching operation mechanism so as to carry out continuous optimization. At the same time, the appropriateness of teaching courses, the information literacy of students and local teachers, the technical conditions of teaching areas and schools, and the coordination between volunteer teachers and local teachers, require attention.

References

1. Caiping, X., Qimin, F.: Urban-rural integration of teacher resource allocation under the information environment: theory and conception. Audio Vis. Educ. Res. **4**, 11–13+49 (2007)
2. Huichen, Z., Wenhui, L., Kele, Y.: Problems, causes and countermeasures of online teaching for college students supported by cloud classrooms. China Educ. Inf. **2**, 13–18 (2020)

3. Yufang, W.: The current dilemma and improvement strategies of 'Internet +' art teaching in Northwest ethnic minority areas. Audio Vis. Educ. Res. **2**, pp. 86–91 (2023)
4. Caiping, X., Xiangyang, H., Ruihua, W.: On the revolutionary impact of information technology on educational development. Educ. Res. **6**, 22–29 (2012)
5. Liguo, Z., et al.: 'Famous teacher classroom' in the perspective of symbiosis: essence, real dilemma and solution. J. Audio Vis. Educ. Res. **44**, 44–50+59 (2023)
6. Jian, Z., Shaoqing, G.: Research on the quality assurance mechanism of 'Internet +' teaching support services in normal colleges. China Audio Vis. Educ. **10**, 98–104 (2021)
7. Junru, S., Shaoqing, G., Xiangchun, H., Dong, W., Jiayang, W.: Research on the construction and application of the 'relay' teaching model of normal school students under the conditions of 'Internet +'. China Audio Vis. Educ. **2**, 112–121 (2021)
8. Jianwei, L.: Solving rural education problems with digital empowerment, January (2024)
9. Salas-Pilco, S.Z.: The impact of COVID-19 on Latin American STEM higher education: a systematic review. In: 2022 IEEE World Engineering Education Conference (EDUNINE), Santos, Brazil, pp. 1–6. IEEE (2022)
10. Zhang, W., Zhao, J., Kou, C.: Research on online supporting education based on the model of urban and rural interactive classrooms. In: 2022 10th International Conference on Information and Education Technology (ICIET), Matsue, Japan, pp. 204–208. IEEE (2022)
11. Kocak, O.: Adapting the flipped classroom model to a design course in online learning environments: a case study. Int. J. Art Des. Educ., 12481 (2023). jade
12. Mariano, J.S., Cordova, S.G.V.: E-sports online learning module amidst COVID-19 pandemic. Interact. Learn. Environ., 1–10 (2022)
13. Duncan-Howell, J.: Teachers making connections: online communities as a source of professional learning. Br. J. Educ. Technol. **41**, 324–340 (2010)
14. Chung, T.-Y., Chen, Y.-L.: Exchanging social support on online teacher groups: Relation to teacher self-efficacy. Telemat. Inform. **35**, 1542–1552 (2018)

Can Information Technology Bridge Cultural Differences? Design and Effectiveness Analysis of COIL Based on Intercultural Communication

Baojun Xia[1(✉)], Deniz Ünsal[2], Shuye Zhuang[1], and Xingya Chen[1]

[1] School of Information Technology in Education, South China Normal University, Guangzhou, China
xiaobaojun@m.scnu.edu.cn

[2] School of Communication and Culture, Royal Roads University, Victoria, BC, Canada

Abstract. As an emerging pedagogical model based on information technology, Collaborative Online International Learning (COIL) involves cross-cultural teaching and collaboration between students and teachers from different countries. In light of the current demand for developing students' intercultural communicative competence in the context of globalization and the trend of digital transformation in education, COIL has gained widespread attention as an alternative to traditional international education programs that require physical travel. This study focuses on the Sino-Canadian COIL cooperative program in Interpersonal Communication and Intercultural Communication, aiming to explore and construct a new type of course teaching mode by introducing the COIL model and combining it with the theory of intercultural communication. The quantitative data tests conducted with Chinese students show that the model can significantly enhance students' intercultural communication skills, promote intercultural communication in diverse ways, improve students' intercultural competence in an all-rounded and multidimensional way, and truly embody the integration of technology and culture in educational practice.

Keywords: Intercultural communication competence · Information technology · Collaborative Online International Learning (COIL)

1 Introduction

Internationalization in higher education advocates for preparing students for an interconnected world, aware of cultural sensitivities and differences and competent to navigate and adapt to different intercultural relations and contexts. It also encourages higher education institutions to develop the tools, policies, and skills to become adaptive to working internationally and interculturally. Last but not least, it invites instructors to reflect on teaching and learning and develop new pedagogical models responsive to the changes in higher education in global contexts.

In a world rapidly globalizing with the movement of goods, ideas and people, Collaborative Online International Learning (COIL) has been developing as a pedagogical

Q. Liu et al. (Eds.): EITT 2024, CCIS 2600, pp. 182–192, 2026.
https://doi.org/10.1007/978-981-95-1525-7_15

methodology that promotes intercultural learning experiences in digital settings in higher education. The creation of "globally networked learning environments" through experiences such as COIL contributes to rethinking and reevaluating learning methodologies and practices.

As an emerging teaching mode rooted in information technology, COIL not only fosters mutual understanding among students from diverse cultural backgrounds but also serves as a critical bridge to connect different cultures and facilitate international exchange. However, can COIL truly facilitate cross-cultural communication through the creation of online teaching contexts and the design of teaching activities? Moreover, how can the most effective COIL teaching approaches be designed? And what impact do such experiences have on students' intercultural communication skills? These are the concerns addressed in this study.

Given the background, problems, and challenges outlined above, this study aims to build a COIL teaching model grounded in cross-cultural communication using the COIL concept and model as its foundation. It seeks to explore the integration of COIL into the course teaching model, employ effective teaching strategies, foster a blend of synchronous and asynchronous instruction in the classroom, and assess the impact of this teaching design on fostering collaborative online international learning. The objective is to improve learners' communication and teamwork skills within multicultural settings, offering a theoretical and practical framework for higher education in an era of intelligentization and informationization.

2 Literature Review

The Center for Collaborative Online International Learning (the COIL Center) at the State University of New York (SUNY) was established in 2006. It has become "an international point of reference for practitioners interested in this approach to curricular internationalization" [1]. SUNY COIL offers workshops and resources to instructors and administrators to help them create and maintain interactive and intercultural digital learning environments. SUNY COIL Global Network serves as a community of practice with events, networking opportunities, and resources [2]. COIL is based upon developing team-taught learning environments where teachers from two cultures work together to develop a shared syllabus, emphasizing experiential and collaborative student learning [1]. In this framework, the educator acts as a facilitator for students who are working on a collaborative task, while the students serve as experts for one another and learn through interacting with each other, constructing and building on knowledge [3].

COIL falls under the social-constructivist educational approach of collaborative learning and places a focus on learning through social interaction, which is a keystone to developing intercultural competence. COIL becomes part of an internationalized curriculum, providing students with a significant cross-cultural experience during the learning process. COIL integrates internet-based learning, project-based learning, and group collaboration, emphasizing multi-level interactions and cooperation among teachers and students. Its specific elements include establishing contact between teachers, negotiating a common syllabus and learning objectives, determining the duration of interactions, designing project activities, selecting collaborative methodologies and technological tools, and monitoring student work and learning progress. Students from both faculties communicate online under the guidance of the common syllabus, working together to build effective international and intercultural teams, discussing course assignments and content, completing project-based activities, and reflecting on the academic content of the course and the intercultural exchanges that occur.

COIL is a relatively new approach, and literature and research on COIL is still emerging. As a result, there is a dearth of experimental research on the effectiveness of COIL. Research into COIL is a growing field but the literature available on COIL mostly consists of descriptions of best practices, case-studies, recommendations on implementation or evaluations [3]. The earliest empirical studies on COIL can be traced back to 2014, with a continuous increase in publications after 2019. Research into COIL has burgeoned in recent years, focusing on areas such as nursing, language learning, academic culture, intercultural competence, and entrepreneurship education. The breadth of its applications underscores the versatility of COIL as an effective tool in any discipline, serving various learning goals. Moreover, current research primarily concentrates on the exchange of COIL programs among countries such as Mexico, the United States, South Africa, and Denmark. However, domestic educational research on COIL in China is still in its nascent stage, and there is a paucity of studies focusing on the feedback of the COIL model within the learning process of Chinese students. Furthermore, the adoption of this teaching model in China has not been widely introduced, leading to a lack of relatively mature teaching models and practices in cross-cultural virtual cooperation.

In general, COIL courses are designed around three main elements. Both studies reinforce the importance of the three main COIL components: pedagogy, technology, and cross-cultural learning. COIL projects implement the shared syllabus approach through pairing faculty members from differing international institutions with the purpose of co-creating a curriculum for their respective students. In a classroom environment, students are placed into groups to learn effective communication methods that aid other group members' success. Students learn from one another by understanding and listening to different perspectives. This educational approach creates positive interpersonal relationships between students and faculty, advocates critical thinking, promotes self-esteem, and stimulates positive independence [4]. Moreover, students in this study enjoyed working collaboratively with their international colleagues to build individual intercultural and organizational communication theory and practice [5]. Students noted that this course increased their global perspective and how they view other cultures [5]. As to IT, all activities relating to this project were conducted online using Microsoft Teams™ and Zoom™ [6]. Also, recent COIL research has incorporated virtual reality (VR) into COIL programs [7], creating cross-cultural VR activities to encourage cultural exchange.

Although COIL provides rich learning experiences and opportunities for cross-cultural communication, implementing it in the classroom presents many challenges. The main ones are the challenge of working with technology [8], different languages [9], different cultural practices [10], different time zones, different study levels [8], and students' general lack of commitment and co-operation [4].

3 COIL Instructional Design and Implementation

3.1 Our Experience/Process

We established a partnership between a Canadian and a Chinese university based on two undergraduate courses in the field of Communication: an Interpersonal Communication Across Cultures course offered through the BA Program in Professional Communication at Royal Roads University, and an Interpersonal International Communications course offered through the BA Program at South China Normal University (SCNU) in Guangzhou. In this first collaboration, we limited student numbers to ten at Royal Roads University and twenty at SCNU. This collaborative learning experience focused on mutual learning of cultural values and practices that impact interpersonal communication in different cultural contexts.

Our COIL preparation started with a reflection on the learning outcomes of our existing courses and of how COIL modules would enhance and enrich the intercultural learning for the students and the instructors. An analysis of the needs and expectations of students helped us in planning the outcomes and our approach to course activities and resources.

We participated in two workshops offered by SUNY COIL to be able to design and plan our courses and COIL modules. In these workshops, discussions of the COIL model, the intercultural aspects of learning, and the experiences of some workshop participants who had implemented COIL in their courses gave us insights to design and plan our course learning outcomes, material, activities, and assignment.

We integrated the COIL modules into our undergraduate courses that are offered in person and on campus. The total course duration was 11 weeks at Royal Roads and 16 weeks at SCNU. The COIL modules were integrated into the final 6 weeks of each course, and so before the start of the COIL modules, we were able to prepare the students academically and communication-wise for the approaching international online collaboration.

The COIL Collaboration took place over a 6-week period, from May 6 to June 16, 2024, and focused on online collaborative learning in English, utilizing both synchronous and asynchronous formats. This initiative connected students from SCNU in China with those from Royal Roads University in Canada. Ahead of the collaborative course commencement, students from each institution were independently divided into groups. Chinese students were organized into 3 groups of 7 members each, while Canadian students formed 3 groups of 3 to 4 members each. This initial interaction helped students from both nationalities get to know one another better and fostered a stronger foundation for effective collaboration within the program. During the COIL experience, each Chinese group was paired with a corresponding Canadian group to establish a one-to-one cooperative relationship. During COIL time, they worked in teams in their home university and collaborated online with teams from the partner university to complete an assignment (i.e., they engaged in team-to-team collaboration).

This six-week online collaboration included shared readings, recorded lectures, a single live Zoom orientation session featuring breakout rooms for icebreakers and introductions, and online forum communication among students, culminating in a joint team assignment. Students worked within their home university teams and collaborated with teams from the partner university to complete this assignment.

Since our courses were also delivered in person on campus, we were able to provide our students with a blended learning model. This model combined an online international component, allowing students to connect with their international peers, and an in-person regular learning environment where they could reflect, debrief, and collaborate with their domestic classmates.

3.2 Our Aims and Objectives

Our COIL collaboration focused on mutual learning of cultural values and practices that impact interpersonal communication in different cultural contexts. Students explored theories of cultural adaptation by focusing on experiences of international students at host university campuses and cities. Through readings, videos, discussion, and reflection they were invited to:

- Explain the formation and maintenance of social networks in Canadian and Chinese cultural contexts.
- Recognize the impact of different perceptions of self and priorities in relationship to communication practices in Canadian and Chinese cultures.
- Reflect on theories of cultural adaptation from an interpersonal communication perspective.
- Discuss the process of cultural adaptation for international students.
- Develop communication strategies to support the cultural adaptation of international students.

3.3 Information Technology Platform

This model included synchronization with Zoom for real-time collaboration and asynchronous features for Moodle-based online learning. Zoom, as a synchronous online learning platform, provided an easy-to-use video conferencing tool for educational purposes.

This online collaboration took place on the Moodle Innovate platform hosted at Royal Roads University. All students and instructors were able to access this online course platform, and the integrated discussion forum allowed students to communicate with each other. Zoom was used for the orientation session.

In these modules, students in Victoria from Royal Roads University who were taking a course on Interpersonal Communication Across Cultures collaborated with students in Guangzhou from SCNU who were taking a course on Interpersonal International Communication.

3.4 Six-Week COIL Cooperative Instructional Design

The classroom content was divided into two main phases. In the first phase (spanning one week), both courses were taught mainly through normal lectures by the course instructors. In the second phase, COIL (Cooperative International Learning) cooperation was carried out.

Prior to the official start of the collaborative program, students were asked to create a self-introduction video, which was then posted on the online platform. Everyone could greet and comment on each other. The objectives and timeline of the program were

Table 1. Course assignments and communication patterns

DATE	Class Schedule	Learning Task	Platform
Week1	Live Orientation Meeting between two universities 7:30–8:30am (May 8)	Post a personal introductory video in the Introductions forum on the Innovate Now website Presentation videos of campus life in small groups in the "Intercultural Teamwork" forum	synchronization (Zoom)
Week2	Lecture and Discussion: Acculturation and cultural adaptation theory and practice In class activity: Students work as a team on developing questions to send to partner university teams	As a team, post questions on the Intercultural Teamwork Forum in response to an introductory video of the other team member's campus life	synchronous (Innovate Now)
Week3	Lecture and Discussion: Communication in Chinese/Canadian culture: Recorded lecture by teachers In class activity: Students work as a team on their responses to questions sent by partner university teams	Post responses to questions in the Intercultural Teamwork Forum	synchronous (Innovate Now)
Week4	Round-table discussion: Teams share their research reviews for their report	As a team, present the first draft of the Welcome Strategy Report for International Students at the Intercultural Teamwork Forum	synchronous (Innovate Now)
Week5	Presentations: Teams present their reports to each other for feedback In class activity: Students work on their feedback for the partner university teams	After carefully reading the other team's report, post feedback and revisions to the draft report in the Intercultural Teamwork Forum	synchronous (Innovate Now)
Week6	Round-table and discussion on final reports End of module wrap-up	Submit the final report to your own mentor and also post the report on the Intercultural Team Collaboration Forum so that the collaborating team can see the revised final version of the report	synchronous (Innovate Now)

presented in the first week of the course. Additionally, the course study documentation, referred to as "Course Material," was also posted in advance on the Moodle platform.

Table 1 below shows the group learning tasks assigned during the second phase of the course's COIL collaboration (spanning five weeks). During this phase, students on both sides engaged in synchronous or asynchronous discussions about the weekly learning tasks via the Innovate Now website.

For synchronized collaborative learning, the Zoom platform was utilized for a synchronous online collaborative learning session during Week 1 of the COIL Collaborative course. Students from both schools participated in the classroom together and viewed each other through the Zoom interface. Both sets of students accessed the same Zoom room during class time and participated in the session with the lead teacher from each school.

The design of the asynchronous exchange follows a sequence of tasks, comprising three phases: information exchange, comparison of cultural practices, and collaborative outputs.

First, students were asked to post a personal introduction video on the introduction forum, introducing their campus life to one another as a group. Subsequently, after watching the campus life videos of each other's schools, the Chinese and Canadian students asked questions about culture, life, and interpersonal communication on the Innovate Now website forum and answered each other's questions. Following this task, after gaining insights into each other's culture and campus life, each group collaborated to complete the international student welcome strategy report for their respective school. They carefully reviewed the report of the other country's group to make necessary adjustments and finalize the international student welcome strategy report.

A questionnaire survey was conducted to gather feedback from students in China regarding their perceptions and experiences with the COIL program. The results of this survey are discussed later in this paper.

The schedule of modules and the joint assignments required several steps to be completed as noted in Table 1. Each step necessitated collaboration within home teams and with partner university teams, following the weekly schedule to accomplish each task.

4 The Effect of Applying the Teaching Model Based on COIL Intercultural Communication

Using the hands-on project for the Intercultural International Collaboration course, the authors conducted both a pre-test and a post-test to track the development of students' intercultural competence. To supplement the quantitative data and to explore students' deeper insights into their experiences, the team utilized the focus group method, a qualitative research approach guided by the researcher's specific questions, and the authors also collected data through group discussions.

Based on the objectives of the program to cultivate intercultural communication competence, the questionnaire scale used in this study was designed based on the multidimensional model of intercultural competence proposed by Byram with reference to the Self-Assessment of Intercultural Competence questionnaire developed by Fantini, and in conjunction with the actual situation of Chinese university students' intercultural competence level. The scale assesses four dimensions: intercultural attitude, intercultural knowledge, intercultural skills, and intercultural awareness. This categorization aligns with the views on intercultural competence articulated by scholars [11]. These dimensions are measured using a Likert scale. To assess intercultural awareness, specific items were included in the questionnaire, which was also designed with a Likert scale in mind. Before and after the program's implementation, pre-test and post-test surveys were conducted in the experimental class to track changes in the same group of research subjects. These surveys revealed significant improvements in the four major evaluation indices. By synthesizing the above survey information, we find that COIL significantly improves students' intercultural communication skills.

4.1 Notable Enhancement in Intercultural Communication Skills

After distributing and statistically analyzing the online questionnaire with SPSS, the alpha reliability coefficient was found to be 0.836, indicating good reliability. The results of the questionnaire are presented in Table 2. As observed in Table 2, the mean difference in pre-test and post-test scores for overall intercultural competence among students following the COIL model was −8.722. The significance level (Sig. = 0.000) was less than 0.05, demonstrating a statistically significant difference. This indicates that the teaching model can substantially improve students' intercultural competence.

Table 2. Paired samples T-test for pre-test and post-test of intercultural competence before and after COIL

		(Math.) Pairing Difference					T	df	Sig. (Two-tailed)
		Average Value	Standard De-via-tion	Stand-ard Error Mean	Difference 95% Confidence Interval				
					Lower Limit	Limit			
Pair 1	Pre-test Post-test	−8.722	5.200	1.226	−11.308	−6.137	−7.117	17	0.000

Based on the objectives of the program to cultivate intercultural communication competence, the questionnaire establishes the four dimensions of intercultural attitude, intercultural knowledge, intercultural skills, and intercultural awareness, referencing the Likert scale. The dimensions of intercultural awareness were specifically set, and the questionnaire was compiled accordingly. The dimensions of intercultural competence were further analyzed, and the results of the analysis are presented in Table 3. In Table 3, "A" represents intercultural attitude, "B" represents intercultural knowledge, "C" represents intercultural skills, and "D" represents intercultural awareness, The pre-test and post-test scores for each dimension are labeled as "1" and "2," respectively.

As observed in Table 3, the mean difference in scores between the pre-test and post-test for the four major dimensions is less than 0, and the significant value is less than

0.05. These values indicate that the implementation of this COIL teaching model has a significant effect on the intercultural communication competence of students taking the Interpersonal International Communication course, particularly in terms of intercultural attitude, intercultural knowledge, intercultural awareness, and intercultural communication skills. They also demonstrate that the COIL teaching model significantly enhances the students' intercultural attitude, intercultural knowledge, and intercultural awareness within their intercultural communication competence in the Interpersonal International Communication course.

Table 3. Paired samples T-test for pre-test and post-test for each dimension of thinking skills before and after COIL

		(Math.) Pairing Difference					T	df	Sig. (Two-tailed)
		Average Value	Standard De-via-tion	Stand-ard Error Mean	Difference 95% Confidence Interval				
					Lower Limit	Limit			
Pair 1	A1-A2	−.778	0.808	0.191	−1.180	.376	−4.082	17	0.001
Pair 2	B1-B2	−2.333	1.534	0.362	−3.096	−.1571	−6.454	17	0.000
Pair 3	C1-C2	−2.056	2.127	0.501	−3.114	−.998	−4.099	17	0.001
Pair 4	D1-D2	−3.566	2.572	0.606	−4.835	−2.277	−5.865	17	0.000

4.2 Analysis of Intercultural Communication Attitudes and Abilities

The post-test scale questions revealed that participants harbored a generally positive attitude toward engaging with individuals from diverse cultural backgrounds, as indicated by a mean score of 4.18. Moreover, the majority were able to communicate in a relaxed and comfortable fashion. Regarding their comprehension of cultural disparities, participants exhibited a robust understanding of their own cultural norms but a comparatively shallow grasp of those from other cultures, suggesting a need for enhancement in this area. Concerning linguistic proficiency and cross-cultural communication tactics, the participants fared well overall, demonstrating confidence in modifying their verbal and non-verbal conduct as well as employing effective cross-cultural communication strategies.

4.3 Analysis of COIL Project Experience and Perception

Among the major barriers encountered in the COIL program, language barriers were the most significant, accounting for 65% of the total challenges faced (Fig. 1). The time difference and the variety of communication software used were also common issues. Participants generally agreed that the COIL project successfully enhanced domestic internationalization, opened up interdisciplinary perspectives, stirred curiosity, and taught the ability to connect knowledge to real-world problems, earning a positive reception and confirmation from the participants.

In summary, while most participants demonstrated a positive attitude and proficiency in intercultural communication, there is scope for improvement in their understanding of other cultural norms and styles of interaction. The COIL program has proven effective in fostering internationalization, interdisciplinary collaboration, and stimulating interest in learning, yet it remains essential to address and improve upon the challenges posed by language barriers and jet lag.

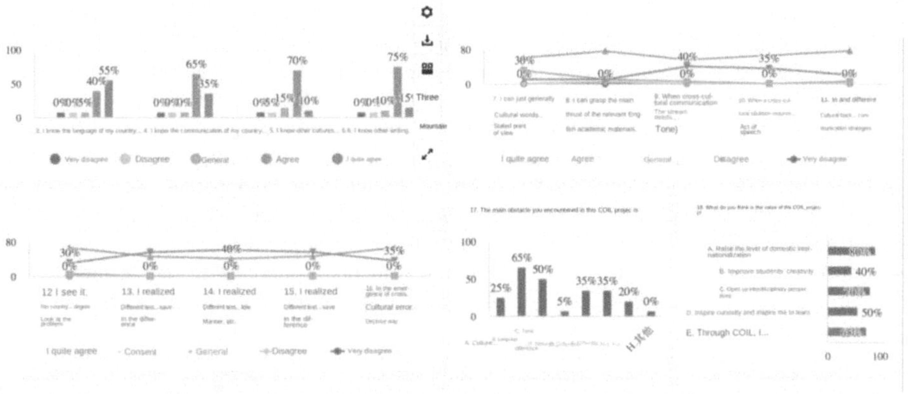

Fig.1. Analysis of COIL project experience and perceptions

5 Conclusions and Discussion

The COIL model creates authentic intercultural communication practice scenarios for students by integrating Internet technology, project learning, and group collaboration, which compensates for the shortcomings of the traditional teaching model in terms of intercultural practice experience. The model not only significantly enhances students' intercultural skills, awareness, and attitudes but also stimulates their interest and desire for intercultural communication. Through the implementation of the COIL-based teaching model, the course achieved significant educational outcomes and effectively enhanced students' intercultural communication skills.

Quantitative data analysis shows that students' scores on the four dimensions of intercultural communication—intercultural attitude, intercultural knowledge, intercultural skills, and intercultural awareness—have significantly improved. All four dimensions of attitude, knowledge, skills, and awareness were significantly improved, with a mean score difference of −8.722, indicating the effectiveness of the COIL model in cultivating intercultural competence. In the qualitative study, participants highly evaluated the COIL program experience. The COIL program promotes intercultural communication through diverse means and enhances students' multidimensional intercultural communicative competence.

Although the COIL program has achieved positive results, it has also revealed some shortcomings, including the compactness of exchange time, which limits in-depth communication, and the scarcity of exchange opportunities, which affects the frequency of cross-cultural interactions. For this reason, future programs may consider improving the curriculum by increasing the exchange time and adding contextualized activities to promote deeper cultural communication. These improvements can further enhance the effectiveness of the COIL model, provide students with a richer cross-cultural learning experience, and help them become internationally competitive media talents in the era of globalization.

References

1. Rubin, J.: Embedding collaborative online international learning (COIL) at higher education institutions: an evolutionary overview with exemplars. Int. High. Educ. **2**, 34 (2017)
2. Skagen, D., McCollum, B., Morsch, L., Shokoples, B.: Developing communication confidence and professional identity in chemistry through international online collaborative learning. Chem. Educ. Res. Pract. (2018). https://doi.org/10.1039/C7RP00220C
3. Deardorff, D.K.: Identification and assessment of intercultural competence as a student outcome of internationalization. J. Stud. Int. Educ. **3**, 241–266 (2006). https://doi.org/10.1177/1028315306287002
4. Vahed, A.: Factors enabling and constraining students' collaborative online international learning experiences. Learn. Environ. Res. **25** (2021). https://doi.org/10.1007/s10984-021-09390-x
5. West, H., Goto, K., Borja, S.A.N., Trechter, S., Klobodu, S.: Evaluation of a collaborative online international learning (COIL): a food product analysis and development project. Food Cult. Soc. **27**(1), 152–173 (2022). https://doi.org/10.1080/15528014.2022.20694
6. Ganassin, S., Satar, M., Pham, T.N., Regan, A.: Designing a collaborative and virtual exchange-embedded virtual summer school. J. Virtual Exch. **6**, 1–14 (2023). https://doi.org/10.21827/jve.6.39812
7. Liu, Y., Shirley, T.: Without crossing a border: exploring the impact of shifting study abroad online on students' learning and intercultural competence development during the COVID-19 pandemic. https://online.suny.edu/introtocoil/suny-coil-what-is/. Accessed 11 June 2025
8. Naicker, A., Singh, E., van Genugten, T.: Collaborative online international learning (COIL): preparedness and experiences of South African students. Innov. Educ. Teach. Int. (2021). https://doi.org/10.1080/14703297.2021.1895867
9. Hackett, S., Janssen, J., Beach, P., Perreault, M., Beelen, J., van Tartwijk, J.: The effectiveness of collaborative online international learning (COIL) on intercultural competence development in higher education. Int. J. Educ. Technol. High. Educ. **20**(1), 5 (2023). https://doi.org/10.1186/s41239-022-00373-3
10. Ramírez, K.C.: Influences of academic culture in collaborative online international learning (COIL): differences in Mexican and U.S. students' reported experiences. Foreign Lang. Ann. **53**, 438–457 (2020). https://doi.org/10.1111/flan.12485
11. Deardorff, D.K.: Intercultural competence: mapping the future research agenda. Int. J. Intercult. Relat. **48**, 3–5 (2015). https://doi.org/10.1016/j.ijintrel.2015.03.002

Design, Implementation, and Application of a 3-D Virtual Simulation Visualization Experiment System for Camera Calibration

Yuxin Tang[1], Huigang Qu[2], Taotao Long[1], Yuqin Yang[1], Di Chen[1], and Jia Chen[1(✉)]

[1] Central China Normal University, Wuhan, China
jc@ccnu.edu.cn
[2] Texas A&M University, College Station, USA

Abstract. Three-dimensional (3-D) virtual simulation experiment systems simulate 3-D visualizations and human-computer interactions. In the Camera Calibration (CC) experiment class, it will often expend a huge amount of budget to equip each group of students with a precision calibration tool which is expensive and easily damaged. For this problem, this paper proposes to design an online 3-D virtual simulation experiment system for the CC experiment by using 3-D virtual reality (VR) visualization technology. According to the specific process of the CC experiment, a 3-D virtual simulation experiment software system for the CC experiment based on unity3D is designed and implemented. This experiment system has been integrated into the virtual simulation experiment teaching management information system of Central China Normal University, so that students can overcome the limitations of traditional experiments in time and space, and log into the online platform at any time to do the CC experiment. In the practical application of experimental teaching, teachers and students have a high degree of acceptance of this virtual experimental system, which proves that the system is a successful application of virtual reality visualization technology to support "VR + Education."

Keywords: 3-D visualization · visualization system · virtual simulation experiment · visualization in education

1 Introduction

Three-dimensional (3-D) virtual simulation experiment systems simulate 3-D visualizations and human-computer interactions. Experimental teaching is an important part of pedagogic activities. It can provide practical verification of theoretical course content, improve students' interest in learning, deepen their understanding of theoretical knowledge, and cultivate innovative thinking and scientific reasoning. Although universities dedicate significant financial resources to the purchase and update of experimental apparatuses every year, the rapid expansion of the number of students and the continuous emergence of new technologies leaves many universities struggling with the lack of experimental apparatuses and the damage and maintenance of precision instruments

Q. Liu et al. (Eds.): EITT 2024, CCIS 2600, pp. 193–203, 2026.
https://doi.org/10.1007/978-981-95-1525-7_16

[1]. The development of 3-D virtual reality (VR) visualization technology provides an effective way to solve this problem [2–4].

In the Camera Calibration (CC) experiment of the course Computer Vision Foundation, each group of students needs to be equipped with a precise calibration tool (such as cube calibration block), but the precision calibration tool is expensive and easy to damage. In a traditional CC experiment class, it is expensive to equip each group of students with costly and fragile precision calibration tools. The class is further limited by experiments that can only be operated during class time. In this CC experiment, we develop a 3-D virtual simulation experiment software system for the CC experiment based on unity3D is designed and implemented. This virtual simulation experiment system enables students to overcome the limitations of traditional experiments in time and space, and access the platform remotely for the CC virtual simulation experiment.

2 Camera Calibration Experiment

2.1 Definition of Camera Calibration

A mathematical model of camera geometric imaging must be established to determine the relationship between the 3-D world coordinates of a point on the surface of an object and its corresponding 2-D pixel coordinates on the image plane. The parameters in the image formation model need to be estimated by experiments. The process of establishing these parameters is called camera calibration [5, 6].

2.2 Theoretical Basis of Camera Calibration

As shown in Fig. 1 and (1), we can use a 3×4 transformation matrix to express the relationship between the 3-D world coordinates (x_w, y_w, z_w) of a point on an object and its corresponding 2-D pixel coordinates (u, v) on the image plane.

To determine this transformation matrix, a special precision calibration tool is placed in front of the camera (cube calibration block, as shown in Fig. 2).

Taking the linear camera imaging model shown in (1) as an example, we can select n corresponding point pairs $[(u, v) - (x_w, y_w, z_w)]$ and form linear equations to calculate the internal and external parameters of the model in advance. Using that information, we can bi-directionally solve the 3-D world coordinates of any point in space or its corresponding 2-D pixel coordinates on the image plane with each other through the imaging model with known parameters where f is the focal length; dx, dy represent the pixel width and the pixel height respectively; $R_{3\times3}$ represents rotation matrix (extrinsic camera parameter); $t_{3\times1}$ represents translation vector (extrinsic camera parameter) [7].

$$z_c \begin{bmatrix} u \\ v \\ 1 \end{bmatrix} = \begin{bmatrix} \frac{1}{dy} & 0 & u_0 \\ 0 & \frac{1}{dy} & v_0 \\ 0 & 0 & 1 \end{bmatrix} \begin{bmatrix} f & 0 & 0 & 0 \\ 0 & f & 0 & 0 \\ 0 & 0 & 1 & 0 \end{bmatrix} \begin{bmatrix} R & t \\ 0^T & 1 \end{bmatrix} \begin{bmatrix} x_w \\ y_w \\ z_w \\ 1 \end{bmatrix}. \quad (1)$$

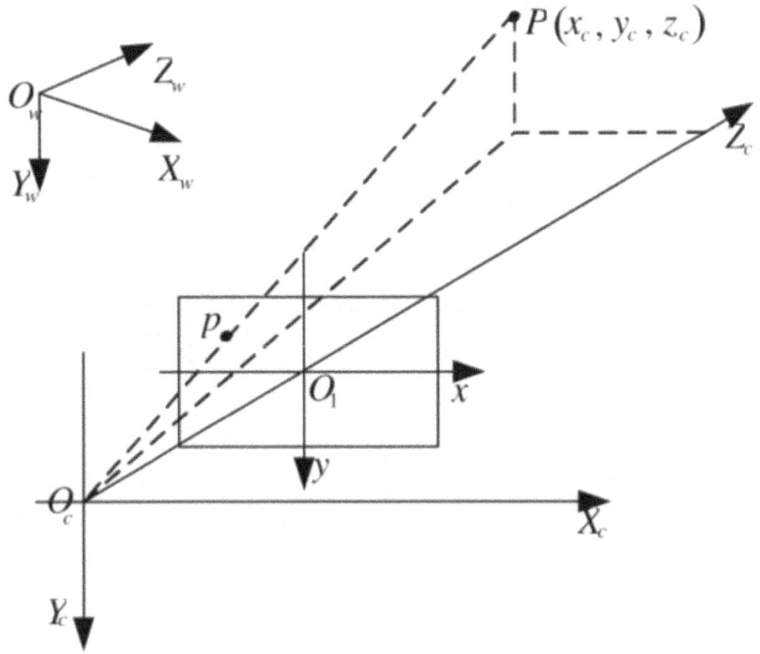

Fig. 1. Different coordinate systems [6].

2.3 Apparatuses Needed for the Experiment

As shown in Table 1, some apparatuses are needed in this experiment and they are expensive and easy to damage. So, we propose to design an online 3-D virtual simulation experiment system for the CC experiment by using 3-D virtual reality (VR) visualization technology.

2.4 Experiment Steps

The traditional camera calibration experiment steps are as follows [6].

1) The camera is placed according to the actual needs;
2) The cube calibration block is placed so that it appears in the image plane of the camera;
3) Select a corner point of the cube calibration block, and record its 3-D world coordinates (x_w, y_w, z_w) and its corresponding 2-D pixel coordinates (u, v) on the image plane.
4) Repeat the previous step to record dozens of corresponding point pairs [(u, v) – (x_w, y_w, z_w)];
5) According to the least square method, an overdetermined equation is solved to obtain the projection matrix M;
6) The internal and external parameters of the camera are further solved by decomposing M.

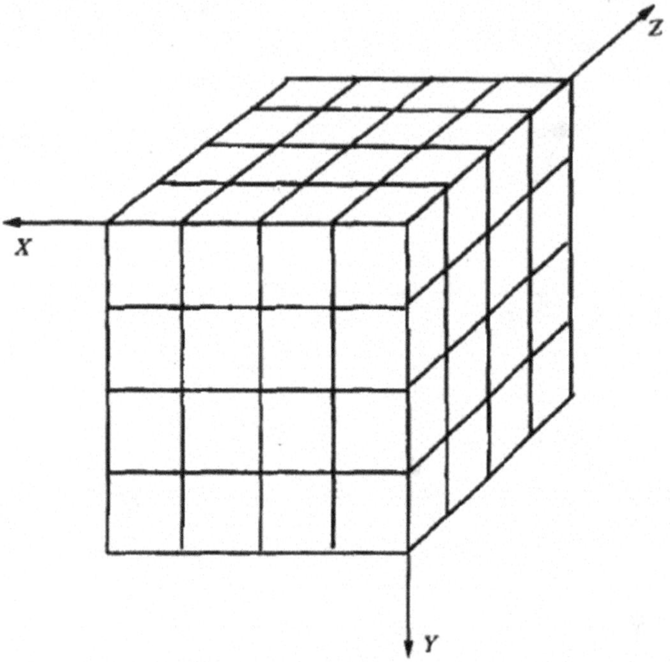

Fig. 2. Cube calibration block [6].

Table 1. Equipment needed for the experiment

Name	Amount (piece)	Unit price (RMB yuan)
Cube calibration block	1	6800
Canon 60D	1	4600
Camera tripod	1	378

3 Design and Implementation of 3-D Virtual Simulation Visualization Experiment System

Professional teachers need to be familiar with disciplinary knowledge and demonstrate it to students as vividly as possible. Whether it is theoretical teaching or experimental teaching, teachers need to carefully design and prepare teaching resources before class. But the workload and complexity of designing and implementing the virtual simulation experiment system far exceed the traditional teaching resources construction [9]. Especially for 3-D virtual simulation experiment system, we need to use 3-D visualization technology.

3.1 Design of Virtual 3-D Visualization Scene

As seen in Fig. 2 and the operation steps, this experiment requires users to select some corner points of the cube calibration block, and record 3-D world coordinates (x_w, y_w, z_w) and corresponding 2-D pixel coordinates (u, v) on the image plane for each point, so we select the camera's view as the screen output, and the 3-D visualization scene displayed on the screen is the content captured by the camera. The 3-D model of the cube calibration block is shown in Fig. 3.

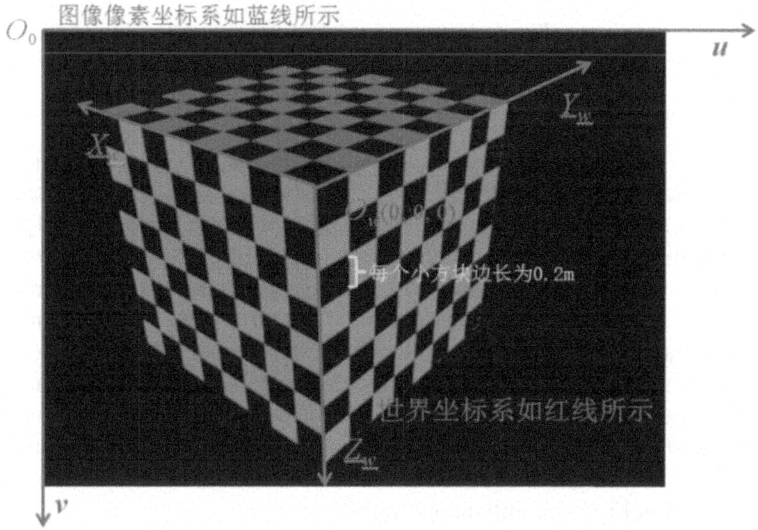

Fig. 3. 3-D model of the cube calibration block.

3.2 Design of Function Modules

The virtual simulation experiment needs to simulate the real experiment operation, so we designed the platform that can provide feedback and operation similar to real experiment. The main function modules are shown in Fig. 4, including:

1) The orientation and position of the virtual camera are initialized randomly for each experiment to avoid the repetition of experimental data;
2) Users can adjust the orientation and position of the virtual camera when an experiment begins;
3) Users can select and mark the points on the calibration block many times;
4) The virtual reality visualization system can output the 3-D world coordinates (x_w, y_w, z_w) and the corresponding 2-D pixel coordinates (u, v) on the image plane of the selected point.

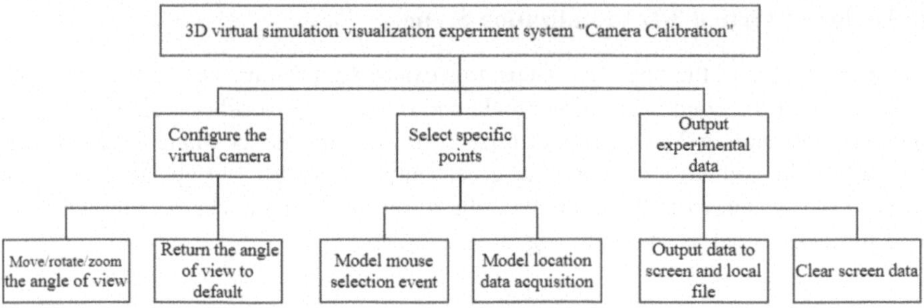

Fig. 4. Function module diagram

3.3 Design of Software

Unity3D is a cross-platform game development engine, which can create high-quality 3-D virtual simulation visualization system [12]. 3-D virtual display, virtual roaming and virtual simulation based on unity3D have become hot research topics [13, 14]. Our 3-D virtual simulation experiment system adopts C# to develop and design interactive scripts. According to the function module diagram of the experimental system, we designed the software structure diagram, which is divided into four parts: camera control, information display, status management and button interaction. The camera control module and the button interaction module can directly respond to the users' operation, and the information display module can display the necessary feedback information in the window on the screen. The status management module can save and manage software status information and data, and users cannot interact with it directly. The system software structure diagram is shown in Fig. 5, mainly including:

Fig. 5. Software structure diagram.

(1) **Configure the virtual camera:** After it has been started, the system will enter the camera configuration mode by default, and users can change the orientation and position of the virtual camera by operating the mouse on the screen. We can detect mouse events in real time through the update() function in the script bound to the virtual camera, and change the orientation and position of the virtual camera according to the mouse movement, so as to realize the functions of dragging the

left mouse button to pan, dragging the right mouse button to rotate and operating the virtual camera with the wheel. When users align the camera window to the middle of the cube calibration block, the multi-angle shooting centered with the cube calibration block can be completed.

(2) **Select specific points:** Implement the function of selecting specific points. The use case is: click the mouse on the screen, and if you click the cube calibration block, you can obtain the image pixel coordinates of the clicked point on the screen and the corresponding 3-D world coordinates of the clicked point on the cube, and record the obtained information.

(3) **Output data:** The output of data is divided into three parts: First, the information window on the left side of the software interface displays the selected points information in real time (as shown in Fig. 6); Second, after clicking the "Output data" button, users can output all the selected points to a txt file, which is convenient for later data analysis and processing; Third, after clicking the "Clear window" button, users can clear the previously selected data and information.

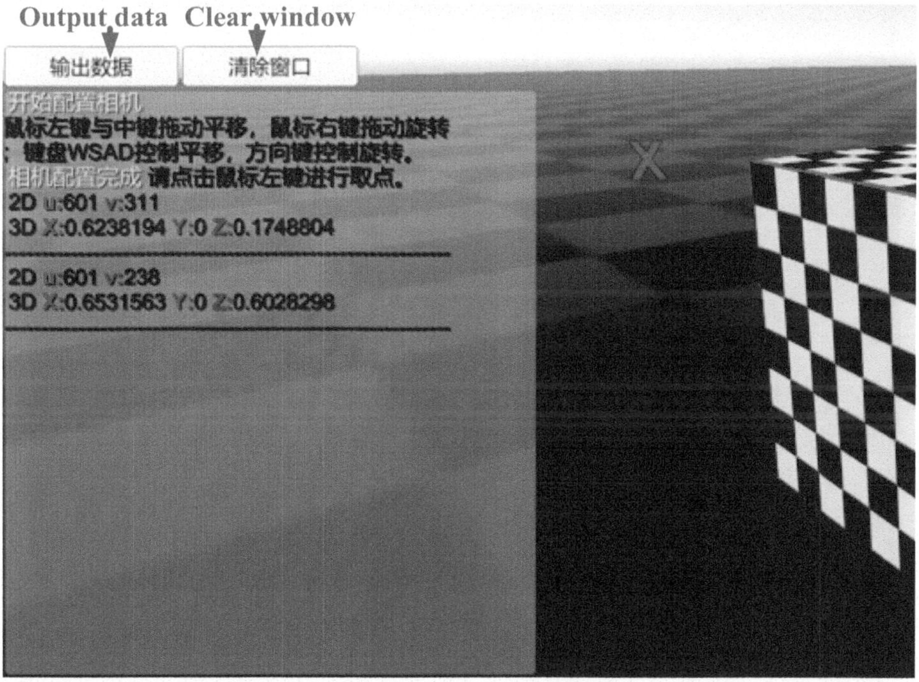

Fig. 6. Information output interface.

4 Application of the 3-D Virtual Simulation Visualization Experiment System

4.1 System Description

Based on the SDK interface of the Unity platform, the system is divided into PC Version (as shown in Fig. 7) and Web Version (as shown in Fig. 8). The system is simple and easy to operate. After obtaining the experimental data, the experimental results can be obtained by using the Octave software, which greatly reduces the time cost, overcomes the limitation of time and space, and enables users to conduct this online experiment at any time when necessary. In addition, the 3-D virtual simulation visualization experiment system has been applied to the CC experiment of the course Computer Vision Foundation in Central China Normal University. Both students and teachers can log into the virtual simulation experiment teaching management information system of CCNU at any time.

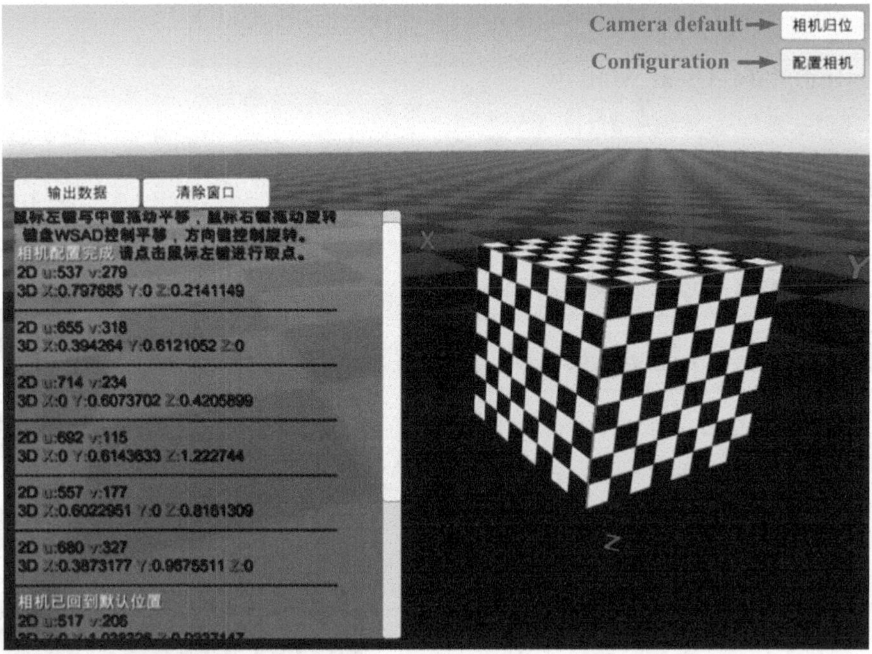

Fig. 7. Screenshot of PC version experimental system.

4.2 System Operation Steps

Virtual simulation visualization technology meets the interaction needs between users and objects in the digital environment. When necessary, it needs to use specific devices to simulate the audio, visual and tactile environment of the real world as much as possible, so that users can have a better experience. The operation steps of the virtual experiment

Fig. 8. Screenshot of web version experiment system.

system mainly include preparation, data acquisition, calculation and analysis, result output, and so on. The specific experimental steps are as follows:

1) Enter the system (PC version or web version), choose the resolution and image quality of the system, and then start the experiment;
2) Enter the start interface, use the mouse or keyboard to make the virtual cube calibration block in the state that three faces (at least two faces) can be displayed in the acquisition window, and click the button "configuration" in the upper right corner to complete the configuration;
3) According to the prompt window, use the left mouse button to select at least six points. During the point selecting process, users can see the image pixel coordinates and the corresponding 3-D world coordinates of the selected points in the prompt window. It is important to note that in order to ensure the accuracy of the experimental results, it is recommended that users try to select more than 10 pairs of points on three surfaces (at least two faces) of the cube calibration block;
4) After the previous operation is completed, click the "output data" button to see that the user's data are output to the preset position for subsequent use;
5) Repeated experiments can enhance students' familiarity with the experimental process, so the system enables users to start the next experiment without quitting. But before the next experiment, users can click the "camera default" button to return the camera to its original orientation and position (return the angle of view to default) and clear the data in the information window;
6) After the experiment, users can shut down the system (PC version)/log out (WEB version).

5 Conclusion

Significant budgetary resources are needed to equip each group of students with a precision calibration tool that is easily damaged in the CC experiment of the course Computer Vision Foundation. For this problem, this paper proposes to design a 3-D virtual simulation visualization experiment system for the CC experiment. This paper presents the design, implementation and application of this 3-D visualization system for the CC experiment. According to the traditional experiment process, the virtual experiment system simplifies the tedious operation in the traditional experiment, improves the teaching efficiency while making the students understand the experiment content, saves time and money for purchasing experimental apparatuses in the university laboratory, and enables the experiment to overcome the limitation of physical condition and be carried out anytime and anywhere.

Acknowledgments. We thank all anonymous reviewers for their help. This work is supported by National Natural Science Foundation of China (No. 62477018), Joint Fund of the Ministry of Education for Equipment Pre-research (No. 8091B02072406), Hubei Provincial Natural Science Foundation (No. 2021CFB539), the Fundamental Research Funds for the Central Universities (No. CCNU24JCPT005 and No. CCNU22QN011), and Open Fund of Hubei Key Laboratory of Digital Education (No. F2024E06).

References

1. Wang, L., Chen, J., Huang, J., Wang, S.: Reform and practice of experimental curriculum system for electronic information specialty. Res. Explor. Lab. **37**(3), 147–149 (2018)
2. Chen, J.: Design and application of virtual simulation experiment based on VR narrative, M.S. Thesis, Department of Educational Technology, Central China Normal University, Wuhan, China (2018)
3. He, S., Zhou, L.: Software technology of transportation virtual simulation experiment teaching center. J. Syst. Simul. **28**(2), 322–328 (2016)
4. Jia, Y., Zhao, X., Zhao, J., Kong, L., Huang, K.: Research on virtual reality simulation system for crew cabin of special vehicle. J. Syst. Simul. **30**(1), 132–138 (2018)
5. Szeliski, R.: Computer Vision: Algorithms and Applications, 2nd edn. Springer, New York (2018)
6. Ma, S., Zhang, Z.: Computer Vision. Science Press, Beijing (1998)
7. Chen, J.: Research on key technologies of multi-view video based 3D human body shape and motion tracking system, Ph.D. dissertation, SMEA, Harbin Institute of Technology, Harbin, China (2012)
8. Liu, Y., Su, L., Wu, Y., Yu, L.: Principles and standards of constructing virtual simulation teaching resources. Exp. Technol. Manag. **34**(5), 8–10 (2017)
9. Yan, M., Zhu, L., Lou, J., Wang, G.: Construction of virtual simulation experimental teaching center for traffic information and control. J. Electr. Electron. Educ. **39**(2), 131–135 (2017)
10. He, K.: The deep integration of information technology and subject teaching. Educ. Res. **38**(10), 88–92 (2017)
11. Zhu, Z.: Educational information: top of educational technology. China Educ. Technol. **2**, 5–8 (2001)
12. Liao, H., Yang, Y.: Design of somatosensory interaction and application in three-dimensional virtual experiment. Distance Educ. J. **31**(1), 54–59 (2013)
13. Yao, P., Chen, Z., Tong, J., Qian, L.: Virtual simulation system of cutter suction dredger based on Unity3D. J. Syst. Simul. **28**(9), 2069–2075 (2016)
14. Chen, J., Hu, P., Huang, Z.: Design and realization of 3D virtual campus roaming system based on Unity3D. In: Liu, Q., Mısır, M., Wang, X., Liu, W. (eds.) CENet2018 2018. AISC, vol. 905, pp. 485–492. Springer, Cham (2020). https://doi.org/10.1007/978-3-030-14680-1_53

15. Wang, J., Wei, X.: Hot status and cold thinking of virtual experiment. China Educ. Technol. **4**, 126–129 (2011)
16. Wei, S., Yang, X., Fu, J.: The research of cognitive friction in virtual experimental operating system. Mod. Educ. Technol. **19**(9), 121–123 (2009)

AI-Driven Learning Analytics and Assessment

Optimizing the Assessment System of Integrated Course Design Through Causal Inference

WenXia Feng(✉), Bin Duan, and Yi Kuang

School of Automation and Electronic Information, Xiangtan University, Xiangtan, China
3071225205@qq.com

Abstract. This paper focuses on the optimization of the assessment system of comprehensive course design. It uses the switched-mode power supply (SMPS) design competition as an example to discuss using the causal inference method to establish reasonable scoring criteria. This paper reveals the limitations of traditional probability statistics in dealing with complex causal relationships by analyzing the competitors' performance in different experiments and the "data generating process" behind them and it adopts the causal inference model to assess the competitors' abilities more accurately. In the research, the competitors' ability to decompose higher-order problems and the difficulty of experiments are considered potential variables, and the complex interactions between these factors are clarified through constructing causal diagrams, especially in identifying and controlling confounders and mediating variables. On this basis, this paper innovatively proposes an assessment strategy, the Enhanced Causal-Weighted Scoring Method, which aims to account for differences in experimental difficulty while more accurately reflecting competitors' abilities in higher-order problem decomposition. The advantages of the method in scoring are demonstrated by comparing it to traditional percentage scoring. This method solves the scoring problem and provides a scientific and reasonable educational evaluation strategy for comprehensive course design and related competition fields. In addition, given the current rapid development of artificial intelligence (AI) where AI still lacks causal thinking as well as data generation processes, the results of this research provide the theoretical support and practical orientation containing causal logic for the development of an AI scoring system, which promotes the growth of intelligence and depth of the educational evaluation system.

Keywords: Optimization of the Assessment System · Data Generating Process · Causal Inference · Higher-Order Problem Decomposition Ability · Enhanced Causal Weighting Scoring Method

1 Introduction

Recently, certain universities in the United Kingdom are re-examining students' grades, with some students who were originally admitted based on arithmetic averages (e.g., 88/100) receiving notices of withdrawal of their offers. In contrast, others are asked to provide weighted averages (e.g., 3.5/4) to re-assess their eligibility. This incident reveals that subtle differences in grading criteria may lead to very different results. Therefore, it is crucial to establish scientifically sound evaluation guidelines in education.

Q. Liu et al. (Eds.): EITT 2024, CCIS 2600, pp. 207–219, 2026.
https://doi.org/10.1007/978-981-95-1525-7_17

Exploring the nature of the problem reveals the limitations of traditional probabilistic statistical methods in dealing with complex cause-and-effect relationships, especially in assessing students' performance in integrated course design, where scores on a superficial level alone do not reflect the actual abilities of students when solving complex problems. To overcome this problem, this paper draws on the approach of Robert Ness in his latest book, *Causal AI* [1], which aims to shed light on the nature of the problem, namely that differences in the data generating process (DGP) lead to different data outcomes. DGP refers to the underlying stochastic mechanisms behind the generation of observed data. Specifically, it covers the interactions between all the underlying variables that produce a particular data stream.

In this paper, we focus on optimizing a comprehensive course design assessment system and use a switched-mode power supply (SMPS) design competition as an example of causal reasoning to establish reasonable scoring criteria. In the study, we consider contestants' performances in different experiments and the data generation process behind them and employ a causal inference model to assess their abilities more accurately. In particular, we considered contestants' ability to solve complex problems with experimental difficulty as potential variables. We clarified the complex interactions between these factors by constructing causal diagrams, especially identifying and controlling confounders and mediating variables.

Since existing scoring methods (e.g., Percentage Scoring, Cumulative Scoring) usually rely on traditional probability statistical techniques, these techniques are very effective in dealing with simple stochastic processes. Still, they are inadequate when confronted with complex causal relationships. Traditional methods ignore the effects of potential variables such as experimental difficulty and contestants' ability to decompose higher-order problems, resulting in less accurate and fair scoring [1]. Therefore, based on this foundation, this paper innovatively proposes the Enhanced Causal-Weighted Scoring Method (ECWSM), which aims to more accurately reflect contestants' abilities in advanced problem decomposition and other aspects while taking into account the differences in experimental difficulty, and demonstrates the robustness of the method when dealing with outliers and accuracy. ECWSM not only solves the scoring problem but also provides a scientifically sound strategy for educational assessment in the field of integrated course design and related competitions.

In addition, considering that the current state of artificial intelligence (AI), although developing rapidly, is still insufficient in causal reasoning and understanding of the data generation process, the results of this study provide theoretical support and practical guidance for the development of AI scoring systems with causal logic, which will promote the development of intelligent and in-depth educational assessment systems.

2 Theory and Methodology

Causal inference, as one of the scientific methods, has been developing rapidly in recent years and is widely used in biomedicine [3], economics [4], and other disciplines. Most of the world's data are observations rather than results of experiments. When randomized experiments are not feasible due to infeasibility, cost, or ethical reasons, causal inference enables data scientists to simulate the experiments and estimate the causal effects from the observed data. Judea Pearl proposed that causal models incorporate the concepts of science, knowledge, and data into a specific contextual framework [5]. The complexity of causality can be sorted out and demonstrated by causal diagrams, which are graphs consisting of dots and arrows, where the dots represent the variables. The arrows represent the causal directions, as shown in Fig. 1. There are three basic patterns of causal diagrams: linking, fork-joining, and collision-joining. Combining these three patterns can represent the complex structure of causal relationships. In this paper, we effectively describe the joint probability one-to-many relationship by constructing a causal diagram of the data generation process and the potential variables hidden therein from both linking ensemble and fork-joining.

This paper focuses on assessing performance data, which are nodes in a larger network of interpersonal activities and interpersonal cognitive structures [6], thus making it all the more important to express the data generation process data in the network explicitly. Literature [1] describes that the data generation process consists of observed, latent, and intermediate variables. An observed variable is a factor that can be directly observed. A latent variable is a factor that cannot be directly observed in the process but affects the observed data. Latent variables are often used to explain the complex relationship or structure between the observed data in statistical modelling and causal inference. An intermediate variable is a factor that is not directly observed but has an effect on the observed data. Intermediate variables are not directly observed, and strictly speaking they are not latent variables in the traditional sense because they are not fixed individual attributes; instead, they are dynamically calculated based on behavioral outcomes [1].

3 Comprehensive Course Design Background Setting

One study explores assessment approaches within the unique context of Extended Reality (XR) learning environments, which offer immersive and contextualized experiences. Four unique cases of XR-based assessment are presented, exploring the context of each case to contribute to a richer understanding of assessment for learning in XR environments and providing direction for future assessment design in this emerging field [7].

As it is evident that context has an undeniable role in assessment, this section will construct a data-generating context for these data to evaluate the integrated design better.

3.1 Background Data

To refine the context of the data generation process for the nature of the problem, we assume that there is a comprehensive SMPS design competition consisting of two experimental tests. In Experiment 1, if contestant A answered 1 incorrectly and answered 19 correctly, and if contestant B answered 3 incorrectly and answered 37 correctly, then contestant A has an error rate of 5% (1/20), and contestant B has an error rate of 7.5% (3/40). Because 5% < 7.5%, contestant A wins. In Experiment 2 contestant A answered 12 incorrectly and 28 correctly, and contestant B answered 8 incorrectly and 12 correctly. Therefore, contestant A has a 30% (12/40) error rate and contestant B has a 40% (8/20) error rate. Because 30% < 40%, and contestant A wins. Suppose the totals of the two experiments are calculated. In that case, contestant A has a total of 13 wrong answers and 47 right answers, and contestant B has a total of 11 wrong answers and 49 right answers, and therefore contestant A has an error rate of 22%. Because 22% > 18%, contestant B wins. So which contestant has actually won?

The problems described above were processed as shown in Table 1. Table 1 data can indicate that contestants participate in a total of two experimental designs; each experiment contains several test points, with each observation point using judging criteria for the successful completion of the record as "Correct" and unsuccessful completion or completion of the wrong record as "Error". To eliminate the interference of human scoring, the system can judge right and wrong and repeat the count; the total number of right and wrong can be capped at 40. A more detailed setup of the SMPS design competition is needed to complete the narrative background.

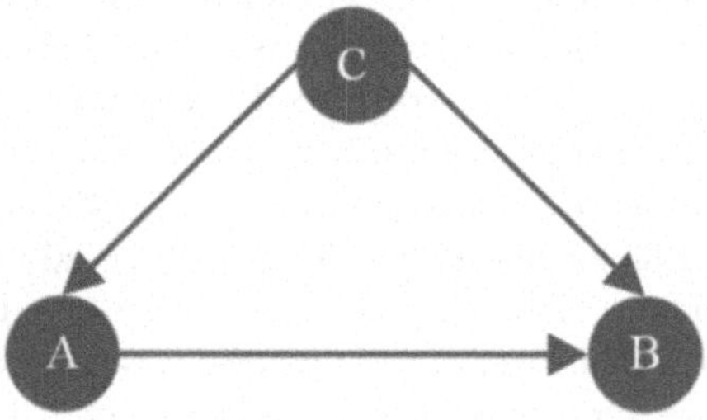

Fig. 1. Example of a cause inference diagram.

Table 1. SMPS Design Contest Contestant Scoring.

SMPS Design Competition	Contestant A		Contestant B	
	Error	Correct	Error	Correct
Experiment 1	1	19	3	37
Experiment 2	12	28	8	12
Total	13	47	11	49

3.2 Settings

Placing the comprehensive course design in the middle of a specific SMPS design competition makes it easier to rationalize the development of our structure-based scoring criteria and rubrics.

There are two experimental designs included in the SMPS design competition, as shown in Table 2.

Overall, the critical points assessed in the two experiments in the competition are as follows:

(1) **Experimental Technical Challenge.** Contestants are first required to complete a series of SMPS design tasks based on practical operation and technical realisation, which focus on testing the contestants' understanding of basic principles, hands-on practical ability, and mastery of engineering specifications.

Table 2. SMPS design competition experiment setup table.

Competition Experiments	Competition Experiments Observation Points
Experiment 1 (Analyze the effect of operating conditions and operating modes on the power loss and efficiency of a boost regulator)	1. Analyze the effect of operating conditions on the power loss of a boost regulator. Contestants need to understand the effect of different operating conditions (e.g., input voltage, output voltage, load) on the power loss of a boost regulator and be able to quantify this effect
	2. Analyze the effect of operating conditions on the efficiency of a boost regulator. Contestants need to understand the impact of different operating conditions on the efficiency of a boost regulator and be able to quantify this effect
	3. Understand the basic conversion process. Contestants need to understand the basic conversion process of a boost regulator, including the principles and mechanisms by which an input voltage is converted to an output voltage
Experiment 2 (Analyze the effects of operating conditions, peak current control loop and voltage feedback compensation on the load transient response of a boost converter)	1. Analyze the effects of operating conditions on the load transient response of a boost converter. Contestants are required to analyze the effects of different operating conditions (e.g., input voltage, output voltage, load changes) on the load transient response of a boost converter and be able to quantify such effects
	2. Analyze the effect of peak current control slope compensation on the transient response of a boost converter load. Contestants need to understand the impact of peak current control slope compensation on the transient response of a boost converter load and be able to analyze its effect on system performance

(*continued*)

Table 2. (*continued*)

Competition Experiments	Competition Experiments Observation Points
	3. Analyze the effect of voltage feedback compensation on the transient response of the boost converter load. Contestants need to understand the role of voltage feedback compensation on the transient response of the boost converter load and be able to analyze its effect on system performance
	4. Understand the effect of control system parameters on converter stability. Contestants need to understand the impact of control system parameters (e.g., peak current control loop, voltage feedback compensation) on the stability of a boost converter in the face of rapidly changing conditions and be able to analyze how to improve system stability by adjusting these parameters

(2) **Preliminary System Design.** As the difficulty gradually increases, the second session may involve a more complex or comprehensive SMPS system design, which requires contestants not only to be able to complete the design and debugging of the unit modules independently but also to consider the integration of the overall system functions and performance optimization.
(3) **Advanced Problem Decomposition and Innovative Design.** Compared with Experiment 1, Experiment 2 focuses more on advanced problem analysis and solving ability, as well as the depth of knowledge of cutting-edge issues in the field of SMPS design. The challenges in this phase are architectural innovation of SMPS systems, key technology breakthroughs, and decomposition and reconstruction of complex engineering problems.

This paper will analyze the overall data generation process in the context of the event setup of the SMPS design competition described above.

4 Latent Variables in the Data-Generating Process

In this section, we will analyze the overall data generation process regarding the competencies assessed by the tournament and extract potential variables to study the relationship between the data joint probability one-to-many.

4.1 Confounder Variable

To solve the misinterpretation of the intrinsic laws of the data or the bias on model fitting, it is necessary to construct a causal inference model of the data generation process.

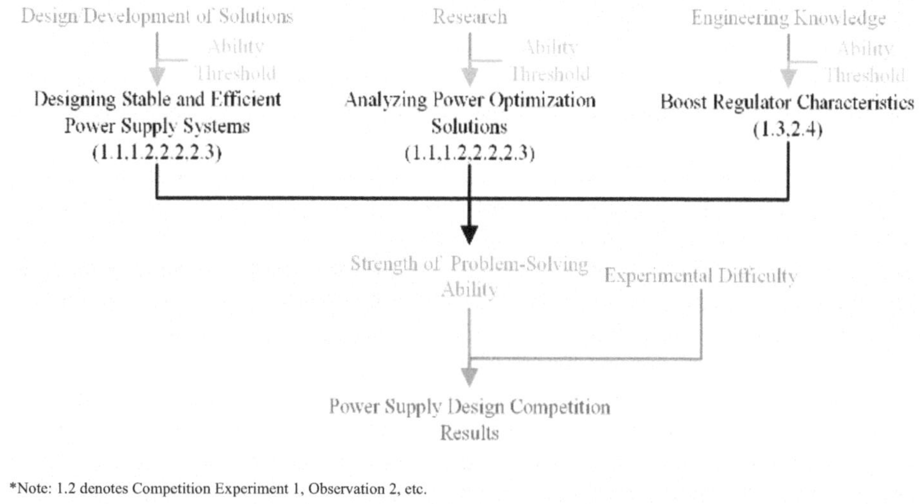

Fig. 2. DGP from competition observation 1.

Starting from the comprehensive ability of the contestants assessed in the SMPS design competition, the relevant ability assessed in the competition and the experimental difficulty of the competition are taken as potential variables, the systematic reconstruction of the observation points assessed in the competition is taken as an observational variable, and the problem decomposition ability of the contestants is taken as an intermediate variable, to construct a causal map of the DGP of the observation points of the competition (Fig. 2).

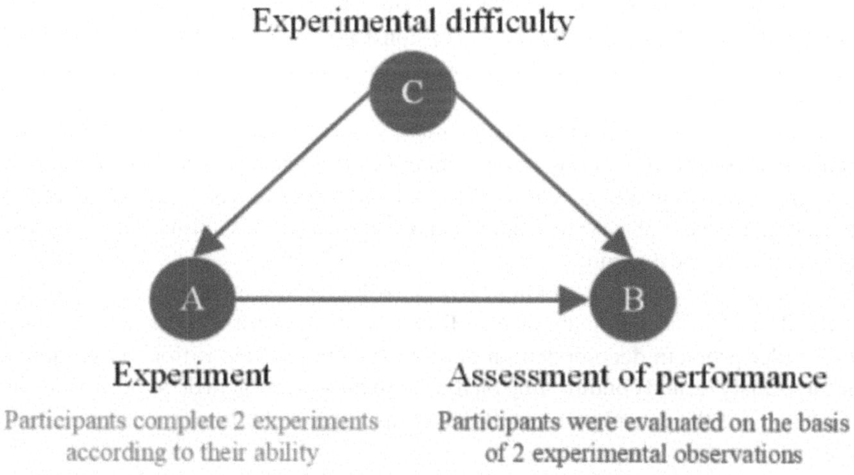

Fig. 3. Causal diagram for experimental difficulty as a confounder variable.

From Fig. 2, it can be seen that only when the contestants' ability threshold reaches a certain level can they complete the observation points of the competition experiment, and only when the contestants' problem-solving ability intensity is greater than or equal to the difficulty of the competition experiment can they have a considerable competition result.

In the science of causality, the process of fork-joining A ← C → B can be seen in Fig. 3. The confounding variable C will make A and B statistically correlated, and this false correlation can be eliminated by "take the variable as a condition". In terms of the points of the competition experiment and the process of generating data from the competition observation points, the difficulty of the competition experiment affects the results, and it is evident that experiment 2 is more difficult, so "difficulty of the experiment" is used as a confounding variable.

To objectively estimate the effect of contestant level on performance, it is important to control for confounding variables. It is, therefore, more reasonable to stratify the data by looking at the data from Experiment 1 and Experiment 2 separately and then averaging them. Taking the arithmetic mean, contestant A had an error rate of 17.5% (average of 5% and 30%) and contestant B had an error rate of 23.75% (average of 7.5% and 40%). Because 17.5% < 23.75%, contestant A won the competition.

4.2 Mediating Variable

Retracing the content of the competition experiment, i.e., the main points of the assessment, the two experiments were compared and analyzed at the level of cognitive ability.

From the analysis presented in Table 3, it can be seen that Experiment 2 improves the stability of the converter in the face of rapidly changing conditions by controlling the system parameters and reflecting higher-order cognitive needs, so the contestants also have to have a certain advanced problem decomposition ability to complete the contest assessment points. Figure 2 is the basis for constructing a causal diagram of the data generation process for the competition observation point containing advanced problem decomposition ability.

From Fig. 4, it can be seen that the design of a stable and efficient SMPS system and the analysis of the SMPS optimization scheme need contestants to have higher-order problem decomposition ability when the higher-order problem decomposition ability is insufficient to pass the observation point of Experiment 2. This ability indirectly affects the contestants' contest results.

In the science of causality, linking A → N → B can be seen in Fig. 5; control B will block the flow of information about A to B or B to A. According to the above analysis, "Higher-order problem decomposition ability problem decomposition" is suitable as a mediator to analyze, and conditioning on the competitor's problem decomposition ability will make one of the causal paths ineffective, resulting in the level of experimentation. This causal path does not work.

To objectively estimate the effect of contestant level on the results, the data stratification method by looking at the data from Experiment 1 and Experiment 2 individually and then averaging them could not be used, i.e., the way the results were assessed in A was no longer applicable, and the data aggregation process was more reasonable. For the

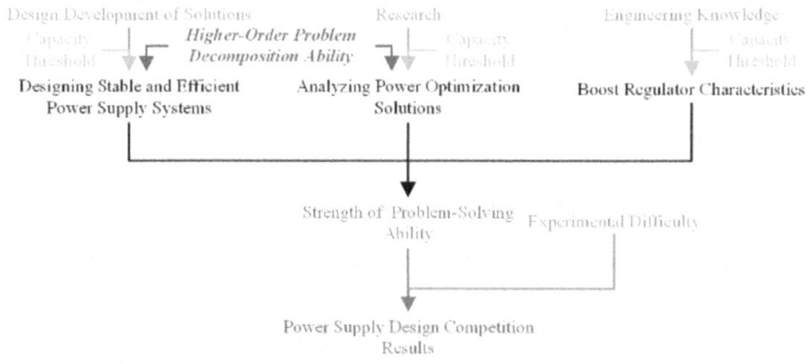

*Note: 1.2 denotes Competition Experiment 1, Observation 2, etc.

Fig. 4. DGP from competition observation 2.

overall results, aggregating the two samples of experimental performance, contestant A had a 22% (13/60) error rate and contestant B had an 18% (11/60) error rate. 22% > 18%; therefore, contestant B won the competition.

Table 3. Competition Experiment Comparison Table.

Attributes	Experiment 1 (Power Losses and Efficiency)	Experiment 2 (Load Transient Response)
Problem Area	Loss and Efficiency Optimization	Control System Dynamic Response
Cognitive Level	Intermediate to Advanced	Advanced
Problem Decomposition	Advanced Performance Optimization	Advanced Design of Control Systems

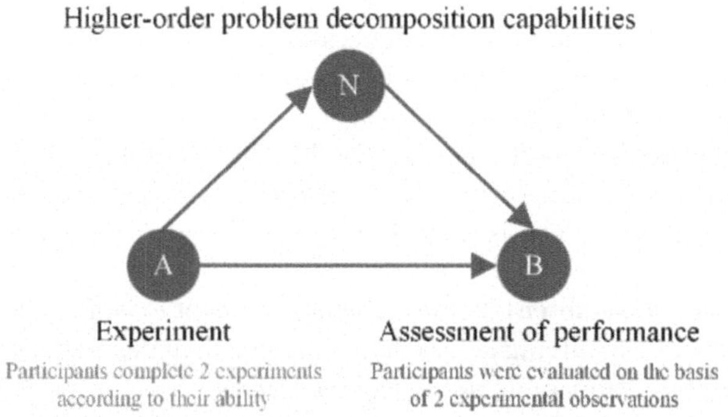

Fig. 5. Causal diagram for experimental difficulty as a moderating variable.

The ECWSM is a scoring system that combines the principles of causal reasoning with adjustments for experimental difficulty to create a fair and comprehensive scoring system that accounts for contestant differences in higher-order thinking and problem-solving abilities while taking into account the difficulty of different experiments.

4.3 ECWSM Key Steps

The main steps of ECWSM include the following:

(1) **Define Latent and Observed Variables.** They determine which variables can be measured directly and which need to be inferred indirectly from other variables.
(2) **Construct the Data to Generate a Causal Diagram.** Clearly expresses cause and effect relationships between variables to help understand the data generation process.
(3) **Confound the Variable Control.** Identify confounders that may affect the fairness of scoring and stratify the data to ensure that comparisons at different difficulty levels are fair.
(4) **Mediate Variable Analysis.** Identify mediating variables and assess their role in the relationship between performance and outcomes, distinguishing between direct skill effects and indirect strategy effects.
(5) **Score Calculations:**
  1) *Base scores.* Calculate raw scores for each experiment, e.g., error rate.
  2) *Difficulty adjustment.* Adjust the base score based on the difficulty of the experiment. Correct answers to more difficult experiments may be given higher weights. For example, the difficulty coefficient for Experiment 2 was set at 1.2, reflecting its experimental difficulty and cognitive demands.
  3) *Competency weighting.* Assign competency weights based on the performance of problem-solving skills (especially higher-order problem decomposition ability). For example, contestants with higher-order solid problem decomposition ability may have an additional percentage added to their score in Experiment 2.
  4) *Final Score.* The adjusted scores of the two experiments are weighted and summed in a specific proportion (e.g., 50% each) to arrive at the final total score.

4.4 Specific Implementation Steps

(1) **Conduct Data Collection and Preprocessing.** Collect experimental data from all contestants, including the number of errors and corrections and each experiment's specific content and difficulty level.
(2) **Determine Difficulty Coefficient.** Set the difficulty coefficient for each experiment through expert evaluation or historical data analysis. For example, the difficulty coefficient for Experiment 1 is set to 1, and the difficulty coefficient for Experiment 2 is set to 1.2.
(3) **Assign Ability Assessment Values.** Design or adapt existing scales to assess the contestants' problem-solving and higher-order problem decomposition abilities and convert them into comparable numerical indicators.
(4) **Calculate Adjusted Scores:** Experiment 1 Score = Error Rate * Experiment 1 Difficulty Coefficient * 100.

Experiment 2 Score = Error Rate * Experiment 2 Difficulty Coefficient * (1 + Higher-order Problem Decomposition Ability Weighted Value) * 100.

The higher-order problem decomposition ability weighting is determined based on the competency assessment results and is reflected in the additional competency embodied in Experiment 2.

(5) **Aggregate and Calculate the Total Score.** The adjusted scores of the two experiments are weighted and summed by a certain percentage (e.g., 50% Experiment 1, 50% Experiment 2) to arrive at the final total score to obtain Table 4.

Table 4. ECWSM Adjusted Scores Table.

SMPS Design Competition	Experiment 1 (DC = 1)[a]		Experiment 2 (DC = 1.2,HD = 0.2)[b]		Total Error Score
	Error Rate	Score	Error Rate	Score	
Contestant A	5%	5	30%	43.2	24.1
Contestant B	7.5%	7.5	40%	57.6	32.55

[a]DC (Difficulty Coefficient).
[b]HD (Higher-order Problem Decomposition Ability Weighted Value).
[c]Assuming that both experiments are equally important, i.e., each is weighted 50%.
[d]Exception data: 1 attempt made, 40 errors.

Table 5. Repeat experimental dataset.

SMPS Design Competition	Contestant A		Contestant B	
	Error	Correct	Error	Correct
Experiment 1	2	38	3	37
Experiment 2	12 + 40d	28	16	24
Total	54	66	19	61

As seen from the above tables, after fully considering the experiments' difficulty and the contestants' ability to decompose the higher-order problems, B's total error score was significantly higher than A's, and the final contestant, A, won the competition.

4.5 Repeat Experiment

To verify the validity of the ECWSM, this study used a repeated experiment to compare the ECWSM with two scoring methods, percentage scoring and an outlier in contestant A's Experiment 2, to demonstrate that the ECWSM demonstrated robustness and accuracy in dealing with outliers. To visually see how the percentage scoring compares to both ECWSM, the percentage of contestants who experimented correctly was used to calculate the score whereas the replicated experimental dataset in Table 5 is derived from real student experimental data.

Table 6. Percentage Scoring vs. ECWSM

SMPS Design Competition	Percentage Scoring		ECWSM	
	A	B	A	B
Experiment 1 Score (ECWSM: DC = 1)	95	92.5	95	92.5
Experiment 2 Score (ECWSM: DC = 1.2, HD = 0.2)	0e	60	100.8	86.4
Average Score	47.5	76.25	97.9	89.45
Total Score	95	152.5	195.8	178.9

Since the number of errors is greater than the number of corrects, A's Experiment 2 score is 0.

As can be seen in Table 6. The comparison between percentage scoring and the ECWSM highlights the importance of considering the impact of exceptional circumstances when assessing contestant performance. The traditional percentage scoring method fails to consider these unusual circumstances, leading to unfair assessments, particularly for contestant A, whose total score dropped significantly due to an extreme event.

In contrast, the ECWSM method effectively addresses these issues by adjusting scores based on error rate, which more accurately reflects each contestant's ability.

- For contestant A, in the presence of extreme anomalous data, the ECWSM avoided the negative impact of the outliers on the overall score by adjusting his score, increasing his score from 95 to 195.8.
- For contestant B, in the absence of anomalous data, the ECWSM similarly provided a more reasonable score reflecting its stable performance, with contestant B's score increasing from 152.5 to 178.9 on the traditional scoring method.

As a result, the ECWSM method is more effective in ensuring the accuracy and fairness of the scoring and, thus, more accurately reflects the true abilities of the two contestants, compared to the traditional percentage scoring method, both in the context of the presence of special anomalies and in the regular performance assessment.

This result validates the robustness and accuracy of ECWSM in dealing with outliers, further proving its potential application value of ECWSM in the educational assessment system and highlighting its advantages over traditional percentile scoring methods.

5 Conclusions and Future Work

This paper introduces the Enhanced Causal Weighted Scoring Method (ECWSM) for evaluating contestants' performance in SMPS design competitions. ECWSM employs carefully selected latent variables to provide a comprehensive assessment, transforming complex probability distributions into intuitive causal levels. This method addresses underlying data structures and mechanisms, enhancing the scientific rigor of the evaluation process.

Results indicate that stratifying experimental difficulty as a confounding variable offers a novel perspective on evaluation outcomes. Conversely, they incorporate problem decomposition ability as a mediating variable, which yields distinct conclusions upon data aggregation. These findings underscore the importance of accurately identifying and managing potential variables to ensure a fair and precise evaluation system.

To validate the effectiveness of ECWSM and further improve the strategy, future experiments will be conducted in the following areas:

- Cross-domain application: apply ECWSM to other course design competitions or assessment scenarios, such as mechanical design or software development, to observe the adaptability and effectiveness of the method in different environments.
- User feedback: collect feedback from students and teachers to understand the problems they encountered using ECWSM and make suggestions for improvement to optimize the scoring mechanism continuously.

In addition, with the development of AI technology, we can consider combining ECWSM with AI technology to develop an automated intelligent scoring system that will establish a feedback mechanism to optimize the assessment process and enhance user experience continuously.

Acknowledgments. This study was funded by Teaching Reform Research Project for General Undergraduate Colleges and Universities in Hunan Province (202401000570); Research and Practice Program of New Engineering Science in Hunan Province (202012).

References

1. Ness, R.: Causal AI. Manning (2024)
2. Mislevy, R.J., Behrens, J.T., Dicerbo, K.E., Levy, R.: Design and discovery in educational assessment: evidence-centered design, psychometrics, and educational data mining. Educ. Data Min. **4**(1), 11–48 (2012)
3. Emmons, S., et al.: Causal inference approaches reveal both positive and negative unintended effects of agricultural and urban management practices on instream biological condition. J. Environ. Manag. **361**, 121234 (2024)
4. Xing, W., Chen, C., Xue, L.: Deep learning-based causal inference architecture and algorithm between stock closing price and relevant factors. Electronics **13**(11) (2024)
5. Pearl, J., Mackenzie, D.: The Book of Why: The New Science of Cause and Effect. Penguin, New York (2018)
6. Mislevy, R.J.: Sociocognitive Foundations of Educational Measurement. Routledge, London (2018)
7. Chen, X., Zhang, Y., Wu, W., Luo, Y., Shi, Y.: Exploring assessment approaches of STEM learning in extended reality environments. Int. J. Technol. Teach. Learn. **18**(2), 120–136 (2022)

Evaluation and Continuous Improvement of Graduation Design Based on Causal Inference

Yi Wen(✉), Bin Duan, and Yi Kuang

School of Automation and Electronic Information, Xiangtan University, Xiangtan, China
420730602@qq.com

Abstract. Professional accreditation of engineering education has become a crucial tool for ensuring and enhancing the quality of engineering education. How to conduct scientific and effective evaluation and continuous improvement has become a prevailing topic. This study focuses on integrating the causal inference method into a graduation design evaluation system, adopting the causal diagram and the structural causal model as the core analytical tools, and selecting a case study of a university to reconstruct the teaching improvement process and assess the causal effect of the learning process on the quality of students' graduation theses by gradually introducing the process evaluation. Meanwhile, through multivariate data analysis, the supervising instructor was identified as a confounding variable and this variable was deconfounded to reveal the true causal effects of each link. Continuous improvement measures were then proposed. The results demonstrated that the causal reconstruction process confirmed the instructional improvement measures from the data perspective, and the deconfounding method could effectively shield the effects of confounders. Through the application of the causal inference method, this study offers a multi-perspective and multi-level evaluation approach, overcoming the limitations of the traditional single evaluation method, and provides an empirical basis and strategic direction for the improvement of engineering education quality.

Keywords: Engineering accreditation · Causal inference · Continuous improvement · Curriculum quality assessment

1 Introduction

The rapid development of artificial intelligence technology offers innovative opportunities to the field of education [1]. The "Innovation Action Plan for Artificial Intelligence in Higher Education" emphasizes the need to accelerate the creative application of AI in education and adhere to the integration of science and education.

Professional certification in engineering education has become a vital tool for ensuring and enhancing the quality of engineering education. The concept of continuous improvement is not only the core principle of engineering certification but also its most critical component. The establishment of an evaluation mechanism for the attainment of graduation requirements is the primary goal in the context of continuous improvement.

Q. Liu et al. (Eds.): EITT 2024, CCIS 2600, pp. 220–233, 2026.
https://doi.org/10.1007/978-981-95-1525-7_18

The evaluation of students' annual learning outcomes is a crucial source of information for the continuous improvement of teaching. In the digital age, the limitations of traditional single-method education evaluation methods are increasingly obvious. To meet the requirements of engineering certification, it is necessary to adopt a multi-perspective and multi-faceted evaluation methodology that aligns with the "outcome-based education theory". Most course evaluation methods, which rely on the calculation of scores, lack a description of the data generation process.

The evolution of evaluation methodology has witnessed a transition from a focus on singular evaluations to a more diversified approach, emphasizing multi-perspective and multi-faceted methodologies. Similarly, the scientific formulation of public policies is also undergoing a transition from reliance on subjective experience to an objective evidence-guided model. Compared to traditional evaluation methods, causal inference provides a novel perspective by utilizing reliable causal information between variables. Structural Causal Models (SCMS) facilitate the transformation of structural knowledge about the data generation process, thereby assisting educators and policymakers in analyzing and resolving the complexities inherent in educational practice and optimizing evaluation outcomes [2, 3].

This study conducts a comparative analysis of graduation projects based on evaluation data, taking the graduation design of an electrical engineering program at a specific university as a case study. The reconstruction of the evaluation improvement process utilizes causal science methods. Multivariate data are analyzed to identify confounding variables, and debiasing is conducted. The results of the data analysis have led to the formulation of improvement strategies.

2 Methodology

2.1 Causal Diagram

The causal diagram is typically represented using a Directed Acyclic Graph (DAG) [4]. A simple causal diagram is illustrated in Fig. 1.

The diagram consists of nodes representing variables and directed edges representing causal relationships. For example, an arrow from X to Y indicates that X directly affects Y. These diagrams have distinct features: the "directed" aspect shows the direction of the causal relationship, and the "acyclic" nature ensures no closed loops, maintaining consistency and non-circularity in the causal relationships.

The graphical representation of causal diagrams makes complex causal relationships intuitive. Compared to handling symbolic or numerical parameters of individual systems, this graphical language is more understandable and acceptable [5] (Fig. 2).

2.2 SCMs

Formally, an SCM is an ordered triple <U, V, F>, where:

U is a set of exogenous variables, whose values are determined by factors external to the model;

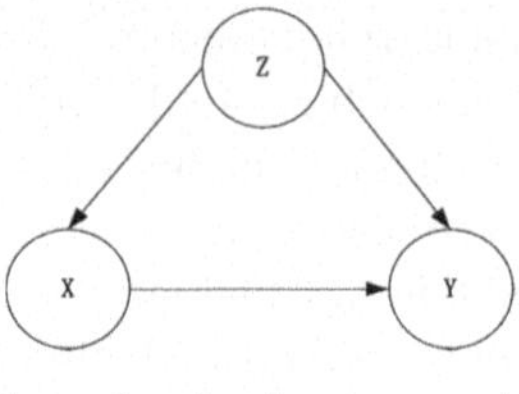

Fig. 1. Example of a simple causal diagram.

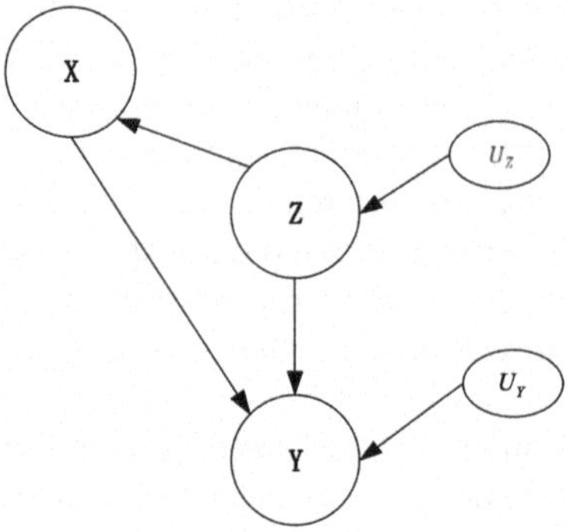

Fig. 2. SCM diagram.

V is a set of endogenous variables, whose values are determined by factors internal to the model;
F is a set of structural equations that express the value of each endogenous variable as a function of the values of other variables in U and V.

Referring to the above diagram, we establish an SCM:

$$U = \{X, U_Z, U_Y\},$$

$$V = \{Z, Y\},$$

$$F = \{f_Z(X, U_Z), f_Y(X, Z, U_Y)\},$$

in the model, X, U_Z, U_Y are called exogenous variables. U_Z, U_Y are called unobserved variables. Z, Y are endogenous variables.

The nature of the SCM framework allows researchers to go beyond simple correlation analysis and delve deeper into causal mechanisms [6]. It also allows for a critical assessment of the potential impact of data sources and collection methods on conclusions. This not only improves the accuracy and reliability of inferences but also provides a strong theoretical basis for policy formulation, experimental design, and understanding of complex systems.

2.3 Intervention Analysis with Confounding Variables

(1) Confounding Bias. The classic case of Simpson's paradox illustrates that when studying the correlation between two variables, there may be two sets of data where the tendency to discuss them separately or together leads to opposite conclusions [7]. The relationship between the variables is distorted by a third variable, leading to a spurious correlation. This third variable is called the confounding variable. In Fig. 1, the central variable T in the forked structure is a confounding variable for X and Y. The spurious correlation caused by confounding variables is known as confounding bias. When studying causal relationships using observational data, it is important to avoid confounding bias to infer the correct causal relationships.

(2) Mathematical Expression of Intervention. The second step in the causal ladder is the intervention, through which the causal effect of the treatment variable on the outcome variable can be obtained.

In the causal field, the do-operator is introduced to define intervention. Specifically, 'intervention X = x' is expressed as 'do(X = x)'. The average causal effect between variables is expressed as:

$$ACE = E(Y = 1|do(X = 1)) - E(Y = 0|do(X = 0)). \tag{1}$$

(3) Adjusting Expression to Calculate Causal Effects. When intervening on variables, it is necessary to regenerate the sample dataset. The gold standard for avoiding confounding is a randomized controlled trial. However, due to the limitations of randomized controlled trials, another approach is to eliminate confounding bias and identify causal effects between variables using observational research data. The following is the derivation of the formulae based on the graphical model analysis:

- Intervening in graphical models corresponds to breaking all edges pointing to node X.

$$P(Y = y|do(X = x)) = P_m(Y = y|X = x), \tag{2}$$

- intervening in graphical models corresponds to breaking all edges pointing to node X.

$$P(Z = z) = P_m(Z = z), \tag{3}$$

- before and after the intervention, the conditional probability P(Y = y |Z = z, X = x) is unchanged.

$$P(Y = y|Z = z, X = x) = P_m(Y = y|Z = z, X = x), \tag{4}$$

- after the intervention, variables Z and X are independent of each other.

$$P_m(Z = z|X = x) = P_m(Z = z) = P(Z = z), \tag{5}$$

thus, based on the probabilistic relationship before and after the intervention, the causal effect expression is converted to the pre-intervention expression:

$$P(Y = y|do(X = x)) = \sum_z P(Y = y|Z = z, X = x)P(Z = z), \tag{6}$$

the causal effect of the intervening variable on the outcome variable can be achieved through observational data only, with adjustments made to the variable Z to enable the calculation of the intervening causal effect, and thus the control of confounding bias [8].

2.4 Research Process

The research process encompasses the following steps (Fig. 3):

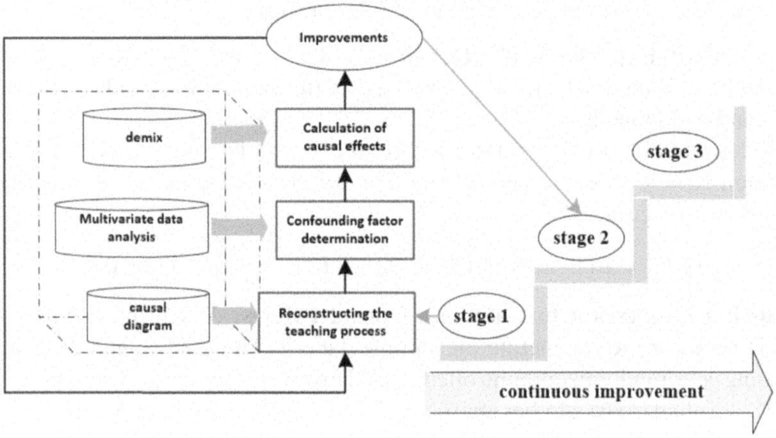

Fig. 3. Flowchart of the study.

1) Construct a causal graph for course evaluation and reconstruct the teaching improvement process. Reconstruct the teaching evaluation improvement measures during data processing and analysis.
2) Utilize graphical model analysis methods to adjust for confounding based on observational data, obtain the true causal effects, identify deficiencies in the evaluation process, and analyze issues.
3) Propose improvement measures. Based on the analysis results, propose targeted intervention measures to enhance the teaching process.

The whole process of continuous improvement based on evaluation data should be:

Stage 1: Establish a causal graph for course evaluation, calculate causal effects, and analyze existing problems.
Stage 2: Based on the results of the analysis, propose and implement improvement measures.
Stage 3: After implementation, evaluate the improvement results, propose improvement measures for the next stage, and form a closed loop of "implementation - evaluation - feedback - improvement".

3 Data Processing and Analysis

3.1 Data Source and Processing

In this study, the raw data were the "Graduation Design" course achievement calculation and total assessment score assessment tables (including the scores of different assessment sections) from the Faculty of Mechanical and Electrical Engineering of a specific university for the years 2017–2019. After removing those with missing assessment data, a sample size of 113 students was obtained.

The data were standardized and processed to be dimensionless due to the inconsistency of the total scores of different learning sections.

3.2 Reconstruction of Teaching Process with Causal Inference

Our understanding of the world is based on causal assumptions. By basing candidate features on implicit causal models and describing the effects between variables, our understanding of systems can be improved. Using the DAG diagram approach, researchers can infer causal hypotheses and make design and statistical analysis decisions. In exploring the impact of study chapters on the students' theses, we used causal diagrams and causal modeling to reconstruct the instructional improvement process. The causal hypotheses were visualized by constructing DAG to model the data generation process.

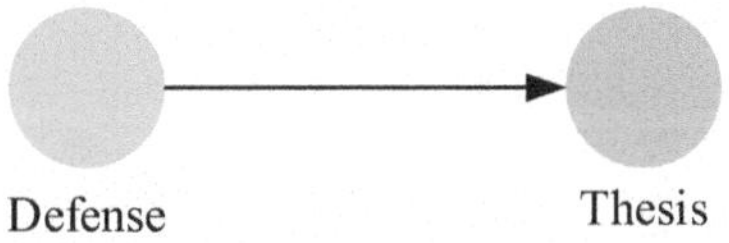

Fig. 4. Causal diagram 2017.

Referring to the school's 2017 syllabus on "Design for Graduation", the defense score was used as the basis for the dissertation assessment. The causal hypothesis was established as follows: the defense reflects the quality of the thesis. The constructed causal diagram is shown in Fig. 4.

The SCM 2017 was stated as follows:

$$Graduation\ Thesis = F(Defense),$$

using linear regression in R to validate the model fit, including tests for goodness of fit and significance of explanatory variables. The results are as follows (Table 1).

Table 1. 2017 Mode result

R2	Estimate	Pr(>|t|)
0.3294	1.4446	<2.2e−16

The R-squared value of the model is 0.3294, indicating weak explanatory power [9]. However, the p-value is less than 0.05, indicating that the defense has a significant effect on the final thesis. This finding suggests that it is not sufficient to evaluate the quality of a thesis based on the defense score alone. A model that evaluates the thesis score based on a single assessment phase cannot fully reflect the entire research process and the quality of the thesis.

Therefore, in the 2017 self-evaluation report, the Faculty proposed the following improvement actions: to break down the thesis evaluation from the original thesis defense into three stages of the entire thesis design process—opening session, mid-term review, and defense session—for process evaluation to monitor the quality of the entire process. Based on this report, the causal hypothesis was proposed: implementing process evaluation will help improve thesis quality. The constructed causal diagram is shown in Fig. 5.

The SCM2018 is as follows:

$$\begin{cases} Thesis = F_1(Opening, Mid - term\ inspection, Defense) \\ Mid - term\ inspection = F_2(Opening) \end{cases},$$

data for 2018 were analyzed using R to produce the following statistical indicators (Table 2).

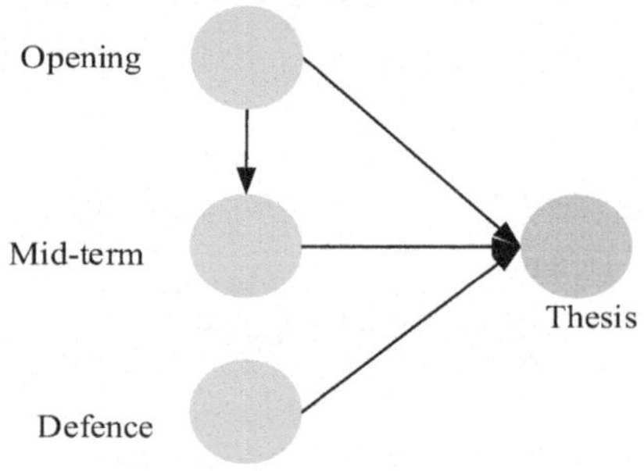

Fig. 5. Causal diagram 2018.

Table 2. 2018 Mode results

Fitness Index	Model Results	General Standards
CFI	0.837	>0.7
GFI	0.855	>0.7
AGFI	0.550	>0.7
SRMR	0.634	<0.1
RESEX	0.162	<0.1

The CFI and GFI both meet the standard of 0.7 or above, but the XGFI, SRMR, and RESEX values do not meet the standard. However, the Goodness of Fit Index (GFI) is

Table 3. 2018 Casual effect

Stage	Total Effect	Direct Effect	Indirect Effect
Opening	0.186	0.062	0.124
Mid-term	0.283	0.283	0
Defense	0.806	0.806	0

0.855, which is greater than the 0.3294 of the 2017 model, indicating an improvement in the model and proving the effectiveness of the improvement measures, yet the fit is not fully up to standard.

As shown in Table 3, the defense session has the greatest impact on the graduation thesis, while the mid-term review, particularly the opening session, has a negligible effect on the outcome. This analysis reveals that the majority of undergraduate engineering design works focus on experimentation, numerical simulation, and system design, with a strong emphasis on hardware and software design. This aspect was not reflected in the 2018 model, which led to an emphasis on the defense session in the analysis. Consequently, in the next assessment, a section for the design work was added, with the causal hypothesis that adding the design work section can promote the improvement of graduation design quality. The causal diagram is shown in Fig. 6.

The SCM2019 is as follows:

$$\begin{cases} Thesis = F_3(Opening, Mid - term, Design\,work, Defense) \\ Design\,work = F_4(Mid - term) \\ Mid - term = F_5(Opening) \end{cases}.$$

data for 2019 were analyzed using R to produce statistical indicators (Table 4).

The fitness indices all meet the standards of structural equation modeling. The GFI is 0.962, indicating that the model has a good fit. The causal effects of each learning section on the graduation thesis were also calculated, as shown in Table 5.

Adding the design work section has increased the indirect effects of the opening session and mid-term review. Additionally, the effect of the design work on the graduation thesis is 0.751, demonstrating the effectiveness of adding the design work section.

Through the analysis of three years of data and the reconstruction of causal models, it is shown from the data perspective that the introduction of a process evaluation system is effective, especially in the 2019 causal model, where the addition of the design work section significantly enhances the model's explanatory power. This finding indicates the importance of multidimensional evaluation for comprehensively assessing the achievements of undergraduate engineering design works.

In traditional teaching evaluation systems, improvements often rely on teachers' experience and personal judgment, lacking systematic and data-driven support. This makes it challenging to quantify the effectiveness of such improvements, especially in complex educational settings. In contrast, structural causal modeling captures and guides these intuitions, providing a framework for evaluations, problem analysis based on assessment outcomes, and continuous improvements. This approach aligns

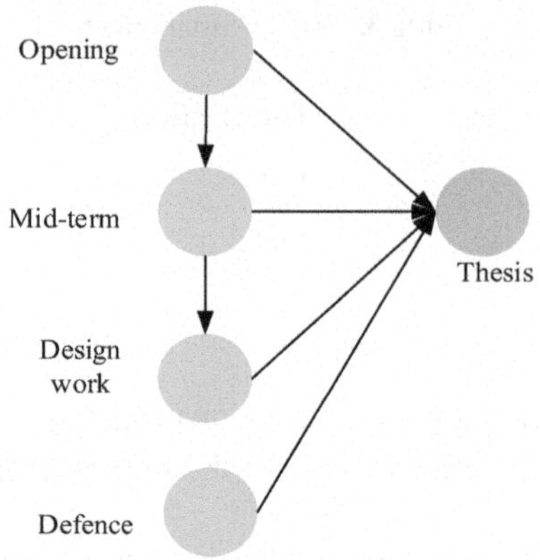

Fig. 6. Causal diagram 2019.

Table 4. 2019 Mode results

Fitness Index	Model Results	General Standards
CFI	0.901	>0.7
GFI	0.962	>0.7
AGFI	0.860	>0.7
SRMR	0.077	<0.1
RESEX	0.086	<0.1

Table 5. 2019 Casual effect

Stages	Total Effect	Direct Effect	Indirect Effect
Opening	0.289	0.080	0.209
Mid-term	0.465	0.126	0.339
Design work	0.751	0.754	0
Defense	0.427	0.427	0

with outcome-based education theory, offering greater explanatory power and scientific validity compared to traditional methods. Despite efforts, many institutions using conventional techniques have seen limited progress in recent years.

Previous research demonstrated that reconstructing the teaching process through causal modeling not only validated teaching improvement measures from a data-driven perspective but also confirmed the effectiveness of causal approaches in practical teaching scenarios. For comprehensive evaluation efforts, the key lies in achieving continuous improvement to enhance teaching quality. This leads to the question: How can we utilize causal methods to propose effective improvement measures for the upcoming academic year?

4 Continuous Improvement

4.1 Determining Confounding Variables

The causal model reflects problems in the teaching process. Looking at the final causal effects table, there is still a phenomenon of low direct effects in the initial and intermediate stages. Combining the opinions of the expert evaluation group and empirical data from the "Graduation Thesis Section Impact Questionnaire Survey", a detailed analysis of the graduation design process was conducted.

The survey was addressed to final-year students of the School of Information Engineering at a specific university. A total of 150 questionnaires were distributed, of which 142 were returned, of which 133 were valid, giving an effective rate of 93.7%.

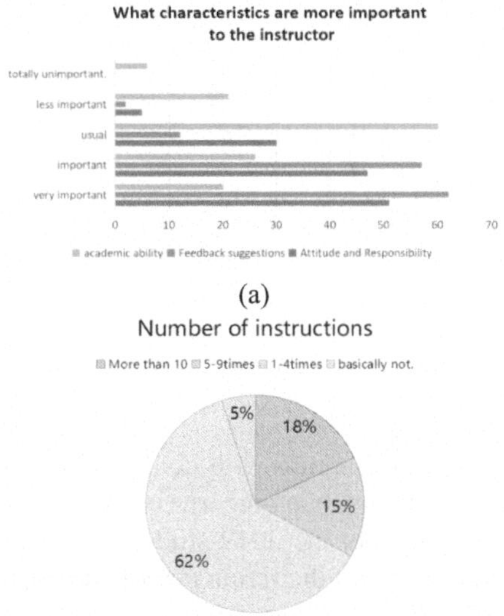

(b)

Fig. 7. Presentation of selected data from the questionnaire.

The statistical analysis of the survey results shows that the students believe that the attitude and sense of responsibility of the guidance teachers are more important than the academic level of the teachers themselves (Fig. 7a). There is also a significant difference in the frequency of guidance provided by teachers (Fig. 7b). This finding suggests that the guidance mode of the guiding teachers may be an important moderating factor affecting early assessment and final grades.

The expert evaluation group found that the trends in students' grades under different guiding teachers were different. Some teachers' students had consistently high initial grades but with differences in the final grades. Based on the comprehensive feedback from the questionnaire survey, it is concluded that the teaching behaviors of different thesis supervisors are potential confounding factors that directly affect the opening and mid-term stages. Specifically, differences in teachers' personalized guidance strategies, interaction frequencies, and varying degrees of continuous attention to students' research progress affect students' direct performance in the opening and mid-term stages, which indirectly affects the final quality of the thesis.

4.2 Causal Diagram with Confounding Variables

(1) Construction of the Causal Diagram. Considering the potential confounding effect of the supervising instructor in the opening and mid-term stages on the final thesis, a causal diagram model with confounding variables was constructed. In this diagram (Fig. 8), the supervising instructor (S) is explicitly identified as a confounding variable.

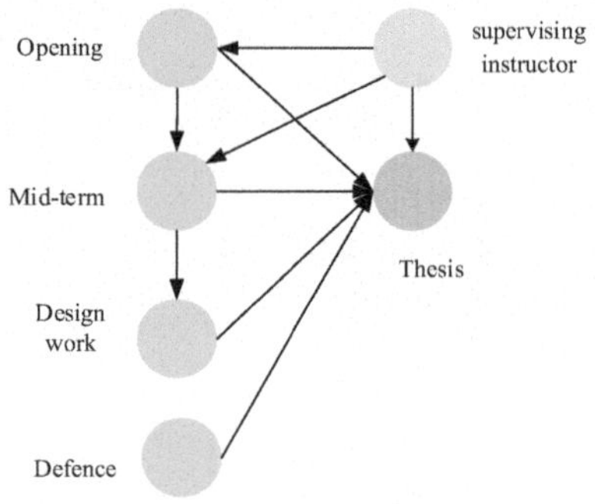

Fig. 8. Instructor's causal diagram for confounding variables.

After constructing the causal diagram with confounding variables, it is crucial to perform confounding adjustments to obtain the true causal effects of each learning stage on the final thesis. For this purpose, the 2019 student performance data were further divided into three categories based on the frequency of supervision: high, medium, and low.

Estimating the causal effect of the intervention from observational data involves using an equation that accounts for potential confounding variables. This equation was derived in detail in the Section II: Methodology (Eq. 7). By incorporating adjustments for Z within the equation, we aimed to isolate the true impact of the intervention on the outcome of interest, thereby obtaining a de-mixed causal effect that more accurately reflects the intervention's efficacy without the influence of confounding factors related to the supervising instructors.

$$P(Y = y|do(X = x)) = \sum_{z} P(Y = y|Z = z, X = x)P(Z = z), \tag{7}$$

we then calculated the conditional probability of the corresponding variable for the values of the variable S:

- *Probability of different guidance: P(Si),*
- *The conditional probability of the corresponding variable: P(Xo |Si),*

a weighted average of the different variables taken according to this conditional probability was made.

$$P(T|X) = \sum_{XOXM} P(Y|X_O = x_O, X_M = x_M, S)P(S) \tag{8}$$

where S (supervising instructor), X_O (opening), X_M(mid-term), and T(thesis).

The resulting causal effects, after accounting for the confounding influences, are presented in Table 6. Upon adjusting for the confounding impact of the supervising instructor, we observed enhancements in the direct effects of both the opening stage (increasing from 0.008 to 0.360) and the mid-term stage (rising from 0.126 to 0.260) on the final thesis. These findings suggest that the genuine associations between the opening and mid-term evaluation stages and the final thesis output were initially obscured by the influence of the supervising instructor acting as a confounding variable.

The process of educational evaluation can encounter a myriad of confounding factors, each potentially influencing the evaluation outcomes. This complexity further complicates the task of extracting authentic feedback from the evaluation data. Traditional evaluation methodologies often struggle to address this challenge effectively. However, by employing structural causal modeling, we can disentangle these confounding variables and uncover the true causal relationships. This enables us to propose more targeted and effective improvement measures (Fig. 9).

Table 6. 2019 Adjusted casual effect

Stages	Total Effect	Direct Effect	Indirect Effect
Opening	0.289	0.080	0.209
Mid-term	0.465	0.126	0.339
Design work	0.751	0.754	0
Defense	0.427	0.427	0

(2) Improvement Measures. By effecting local modifications to the causal graph model, various scenarios can be subjected to tests. Based on the analysis of the aforementioned causal effects, the following intervention measures are proposed to counteract the influence of the supervising instructor:

1) *Explore the new joint guidance mode of graduation design by schools and enterprises to cultivate engineering practice ability.* Graduation design should align with real engineering projects and follow the "Excellence in Engineer Education Training Program" to meet industry ability demands. Traditional graduation design emphasizes theoretical analysis and thesis writing, neglecting the cultivation of engineering thinking such as system design, project management, and the solving of real engineering problems. Emerging industry issues are more complex, demanding higher

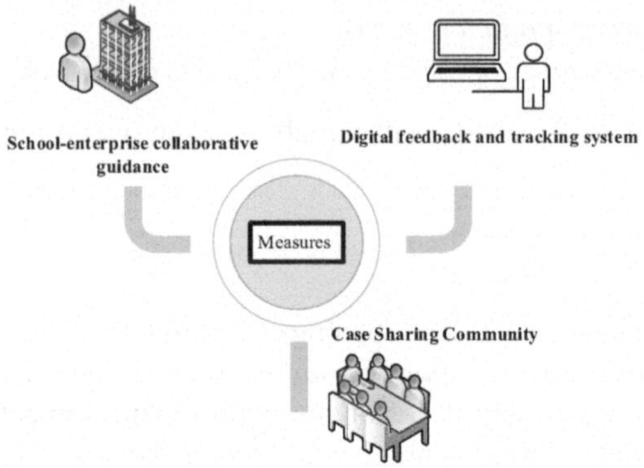

Fig. 9. Improvement measures.

engineering practice skills. Therefore, through the joint guidance of graduation design by the school and enterprise, implemented in specific project practices and in building a teaching team with the synergy of professional and enterprise tutors, this approach effectively resolves the issue of single-instructor guidance differences and significantly enhances students' engineering practice ability to better fulfill graduation requirements.

2) *2)Implement a digital feedback and tracking system for efficient teaching management.* To enhance teaching management efficiency, a digital feedback and tracking system is proposed. It will create a web platform for students and teachers to update real-time opening reports, mid-term progress, and mid-term feedback. Relevant information is automatically recorded, allowing teachers to remind students to submit materials regularly and provide follow-up guidance. The system also has data analysis functions to help teachers understand students' learning status, identify problems, and make timely adjustments, minimizing omissions, ensuring supervision transparency, and improving teaching quality.
3) *3)Establish a case-sharing community for teacher cooperation and development.* A case-sharing community will be set up to organize regular meetings for sharing teachers' experiences and showcasing outstanding cases. It aims to facilitate teachers' collaboration and foster a positive interaction culture. The community will hold regular exchange meetings and excellent case presentations, offering a platform for idea sharing. During these sessions, teachers can share teaching experiences, methods, and results, and discuss problems and solutions. The showcase will display successful cases for others to learn from. Through this community, teacher cooperation is enhanced to promote resource sharing and optimization, improving teaching levels and professionalism.

5 Continuous Improvement

This study applies causal inference methods to engineering education accreditation evaluation, expanding the existing evaluation method system. By analyzing the impact of the learning process on the quality of the graduation thesis using causal inference methods, this research reconstructs the process of teaching improvement and validates the effectiveness of teacher measures from the perspective of data. In addition, by using data analysis to identify the main confounding variable "supervising instructor" and effectively eliminating the influence of this confounding variable by analyzing the observational data set under non-intervention conditions, we controlled for confounding bias. The method of this study is a useful exploration that combines teaching practice and is a multi-perspective and multi-faceted evaluation method that is consistent with the "output-oriented" logic of engineering accreditation.

Future work could be extended to different types of higher education institutions and different teaching and learning scenarios. By fostering cross-institutional research collaborations, more diverse datasets can be collected to improve the universality of causal inference methods.

Acknowledgments. This study was funded by the Teaching Reform Research Project for General Undergraduate Colleges and Universities in Hunan Province (202401000570); Research and Practice Program of New Engineering Science in Hunan Province (202012).

References

1. Zhai, S., Liu, Z.: Artificial intelligence technology innovation and firm productivity: evidence from China. Finance Res. Lett. **58**, 104437 (2023)
2. Imbens, G., Rubin, D., Sacerdote, B., et al.: Unpacking the black box of causality: learning about causal mechanisms from experimental and observational studies. Am. Polit. Sci. Rev. **105**, 765–789 (2011)
3. Heckman, J., Humphries, E., Veramendi, G.: Return to education: the causal effects of education on earnings, health, and smoking. J. Polit. Econ. **126**(S1), 197–246 (2018)
4. Pearl, J., Mackenzie, D.: The Book of Why: The New Science of Cause and Effect. Allen Lane, New York (2018)
5. Vdai, P.: Judea pearl: causality: models, reasoning, and inference. Politische Vierteljahresschrift **42**, 313–315 (2001)
6. Sander, G.: Quantifying biases in causal models: classical confounding vs collider-stratification bias. Epidemiology **14**(3), 6–300 (2003)
7. Angrist, J., Pischke, J.-S.: Mostly Harmless Econometrics: An Empiricist's Companion. Princeton University Press, Princeton (2009)
8. Rui, L.: Causal Inference Based on Graphical Inference. China Machine Press, Beijing (2022)
9. Vispoel, W.P., Guanlan, X., Schneider, W.S.: Interrelationships between latent state-trait theory and generalizability theory within a structural equation modeling framework. Psychol. Methods **27**, 773 (2022)

Assessment of an AI-Assisted Power Supply Experiment Design Based on Causal Inference

Cong He(✉), Bin Duan, Yi Kuang, and SiHai Yang

School of Automation and Electronic Information, Xiangtan University, Xiangtan, China
2903138436@qq.com

Abstract. Significant advancements in artificial intelligence (AI) capabilities have been demonstrated recently by OpenAI, with ChatGPT's educational applications particularly standing out. AI assistance offers individualized learning plans and real-time replies and has been incorporated into power supply design education at every step of the experimental design process, from circuit concept analysis to experimental execution and result analysis. A paradigm shift in classroom evaluation has resulted from this integration, emphasizing students' higher-order thinking abilities and various assessment points. We used the causal inference approach, particularly the front-door criterion, to evaluate a course and determine how instructional tactics precisely affected course effectiveness. The benefit of the causal inference approach over conventional course evaluation methods is that this approach successfully eliminates extraneous variables that might affect students' experimental performance, accurately measuring the actual contribution of AI-assisted instructional strategies to achieve course objectives. An accurate way to evaluate the efficacy of this teaching strategy is to compare the changes in course objective achievements before and after removing confounding variables, as well as the differences in course objective achievements between instructional tactics with and without AI assistance. This validation shows how the use of AI assistance has improved students' ability in all areas of laboratory course study, resulting in a marked improvement in the quality of education.

Keywords: designing assessment points · AI-assisted · front-door adjustment method · course objective achievement

1 Introduction

Artificial intelligence (AI) has been widely used in recent years, significantly changing many facets of our daily lives, including education. According to a recent study, Chat Generative Pre-trained Transformer (ChatGPT) performed only moderately well on the GRE math test at first, placing it in the 25th percentile. However, its performance improved significantly with its second version, GPT-4, reaching an astounding 80th percentile [1], demonstrating the rapid growth of AI technology. According to a recent study, collaborative modeling in academic writing increases efficiency, originality, and output significantly when done by AI and humans. This study shows how huge language models can cooperate with human abilities to enhance academic writing quality

Q. Liu et al. (Eds.): EITT 2024, CCIS 2600, pp. 234–246, 2026.
https://doi.org/10.1007/978-981-95-1525-7_19

[2]. Simultaneously, learning analytics advancements have demonstrated the potential of multimedia language learning tools to supplement educational research. Using these tools, educational data can be thoroughly analyzed to identify patterns and trends [3]. This analytical approach gives educators a solid scientific foundation upon which to build more potent teaching methods.

The requirement for pupils to strengthen their higher-order thinking skills, such as critical thinking and problem-solving abilities, to better meet curriculum objectives served as the driving force for this study. To increase teaching efficacy, this study suggests a novel evaluation system that makes use of causal reasoning and AI to support peer teaching. It seeks to determine the path of future educational technology reform while enhancing students' skills and the overall quality of education through the integration of AI into the experimental design process.

2 Artificial Intelligence-Assisted Appraisal System and Practice Framework

The GPT-3.5 language model from OpenAI was used in this work because of its sophisticated natural language processing powers which facilitate interactive learning and difficult problem-solving. GPT-3.5 is suitable for refining instructional tactics and covers every step of the experimental design process, from strategy planning and conceptualization to outcomes analysis [4], which improves the experimental design process at every stage.

2.1 Structural Equation Modeling of Circuit Parameters

In a circuit experiment design, structural equation modeling (SEM) was used to analyze the intrinsic relationships among circuit parameters (Fig. 1). Represents the load current, represents the load impedance, the output capacitor, the crossover frequency, represents the phase margin and represents the circuit load transient response, specifically the voltage surge amplitude.

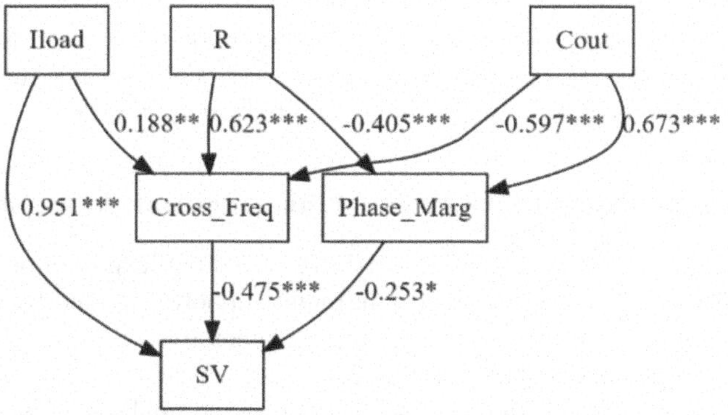

Fig. 1. Structural equation modeling of circuit parameters.

The strength and direction of these relationships could be ascertained by looking at the path coefficients' magnitude and direction (positive or negative).

Students could identify crucial circuit nodes, measure important aspects of transient response, and set up monitoring systems with the aid of AI-assisted power supply experiments. Students tracked their progress, posed queries, anticipated problems, and evaluated their conceptual understanding with the use of AI technology.

2.2 Designing Assessment Points

The essential factors influencing the power load's transient response have been determined based on the previously described parameters. Following the evaluation of students' performance with AI-assisted support, an assessment system was constructed based on their performance on particular knowledge points, as indicated in Table 1. Along with emphasizing single parameter tweaks, this method also highlighted the designer's capacity for creativity, problem-solving talents, and application of AI-assisted technologies.

An in-depth examination of the trial aims, parameters, and methodology indicates how AI assistance could be used when building experimental assessment observation points. AI can optimize experimental design, minimize the number of repetitions, increase experimental efficiency and accuracy, and automatically calculate pertinent experimental parameters, including capacitance, inductance, and feedback resistance [5]. The purpose of these assessment points is to create a set of focused and empirically valuable evaluation benchmarks that will allow participants' knowledge acquisition, operational abilities, and general competence to be thoroughly and multidimensionally assessed.

This evaluation framework carefully outlines the stages involved in experimental design as well as valuable indicators to motivate students to grow and showcase their all-around skills through hands-on activities.

Elements such as "Engineering Knowledge", "Problem Analysis", and "Design/Development of Solutions" are essential for developing students' higher-order thinking abilities within the framework of the Washington Accord's 12 graduate qualities. For instance, students must employ systematic engineering knowledge for thorough analysis while tackling complicated power engineering challenges, which naturally calls for problem-solving and systems-thinking skills. Furthermore, the "Use of Modern Tools" attribute, which includes AI, completely incorporates critical thinking, guaranteeing that these assessment points closely match the entire graduation criteria.

2.3 Optimizing an Experimental Process Framework with AI Assistance

Students studying power supply experiment design can significantly improve their skills in this area with the rich resources and intelligent support provided by ChatGPT and other next-generation AI technologies and highly generative AI tools. The AI-assisted experiment design process is shown in Fig. 2.

Through its assistance in the design, execution, and analysis of experiments, AI improves students' learning and equips them with professional skills. In particular, while

Table 1. Breakdown of observation points for laboratory teaching assessment with the assistance of AI assistance

Assessment Point	The Experimental Design Steps Are Specified	AI Assistance	Higher-order Thinking Skills
Crossover frequency Analysis	Bode plot	Calculate and explain the impact of crossover frequency	Data analysis capacity
Phase margin optimization	Bode plot	Calculate and analyze its relationship with stability	
Load transient response improvement	Curve analyze recovery time	Adjust output voltage response to load changes and evaluate improvement	Problem-solving skills, Optimizing decision-making capacity
System stability evaluation	Identify instability, adjust compensation parameters	Evaluate system stability under different strategies and optimize	Holistic view, dynamic adjustment
Parameter sensitivity analysis	Describe the impact of changes in input voltage and output capacitance on transient response	Predict the effects of parameter variations	Analytical reasoning, systematic thinking
AI-assisted effectiveness evaluation	Reflect on the assistant's role and its enhancement of high-order thinking skills		Critical thinking, technology application

developing power supply trials, it places a strong emphasis on problem-based learning, individualized supervision, and higher-order thinking.

(1) Providing Instant Answers and Knowledge Expansion. AI provides students with prompt, comprehensive responses to inquiries regarding the design of power supply experiments, covering technical details, methods, and essential ideas. To help with comprehension of subjects such as voltage feedback loop compensation and current-mode controlled buck regulators, it provides relevant background information along with concise explanations.

(2) Error Analysis and Personalized Exercises. Utilizing AI technology can precisely identify core knowledge points and key experimental steps in power supply design experiments. It also systematically tracks and records common misconceptions and errors that students encounter during their learning practice. For example, students often inaccurately select feedback impedance and output capacitance, which leads to poor output

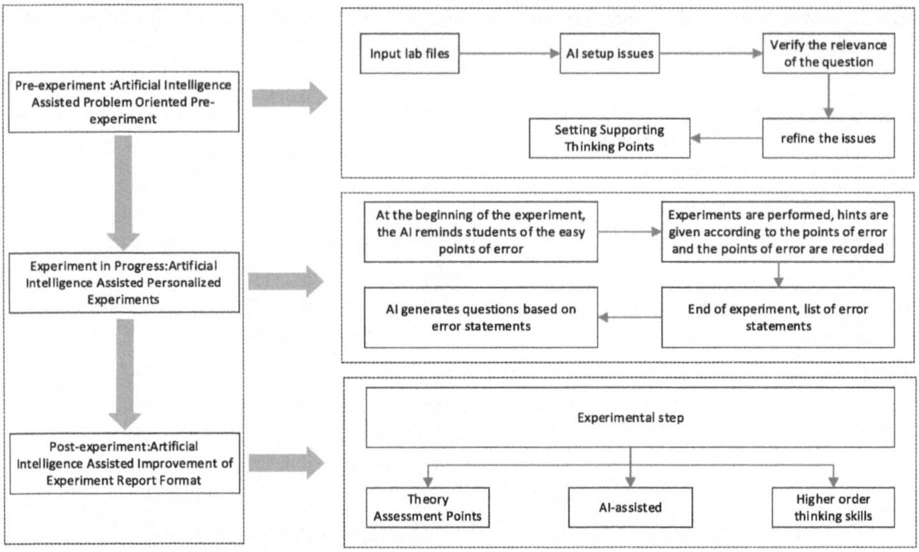

Fig. 2. AI-assisted optimization experiment flowchart.

voltage stability and suboptimal performance. Based on these data, AI can automatically generate targeted error case collections and practice exercises, effectively helping students identify mistakes in practical operations.

(3) Generating Experiment Reports and Technical Documents. Writing experiment reports and technical documents is a crucial part of the power supply experiment design. AI can help students generate report drafts, provide suggestions on themes and structure, and improve writing efficiency and quality. For example, when writing reports on experimental results and analysis, AI can offer framework suggestions, paragraph outlines, and guidance, helping students organize and present their findings and conclusions effectively.

The fundamental goal of education is to promote the comprehensive development of individuals. Achieving this goal in contemporary society heavily depends on effectively utilizing emerging technology platforms, especially AI tools, to cultivate the new intelligence required to meet future challenges. Today's students face an unprecedented era characterized by the deep penetration of AI technology into daily life [6]. This situation necessitates that the education system adaptively guides students to master effective learning and live in an AI environment. AI assistants are potent tools for promoting lifelong learning, adaptability, and innovation in students.

3 Elimination of Confounding Causal Inference Methods

3.1 Causal Theory

The development of causal inference theory originated from a 1998 article in Science magazine titled "Processing Programs for Big Data," which first introduced the concept of "big data." Fig. 3 illustrates a graphical causal model association set, a structural causal model (SCM) concept proposed to address strict causal relationship issues in datasets. In Fig. 3, the actual causal effect is mixed with the spurious correlation between X and Y induced by the fork. Here, W is called a confounding factor. If W acts as an unobservable confounder affecting the causal effect, the causal effect cannot be calculated through conditional expectations. In such cases, it is necessary to use randomized experiments or traditional statistical methods such as regression analysis to estimate the causal effect.

However, in the field of engineering education, using randomized experiments to analyze the achievement of course objectives is impractical due to the comprehensive coverage of engineering education, making random trials unrealistic. Similarly, using regression analysis to calculate causal effects is overly cumbersome and unsuitable for widespread application in education. Causal diagrams provide an alternative language that combines data with causal information, allowing intuitive methods (such as front-door adjustment and instrumental variables approach) to exclude the interference of non-causal correlations and confounding variables to assess the effectiveness of educational interventions accurately.

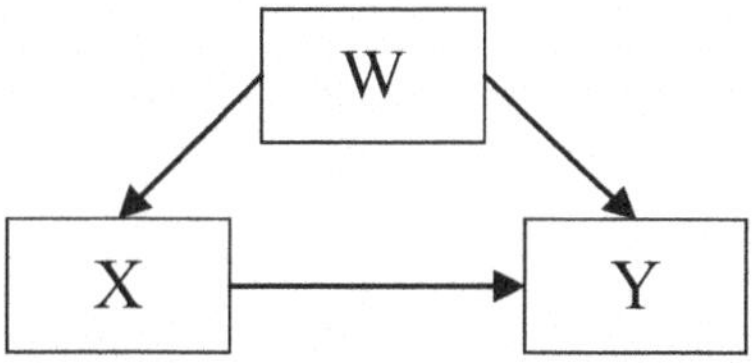

Fig. 3. Causal models with confounders.

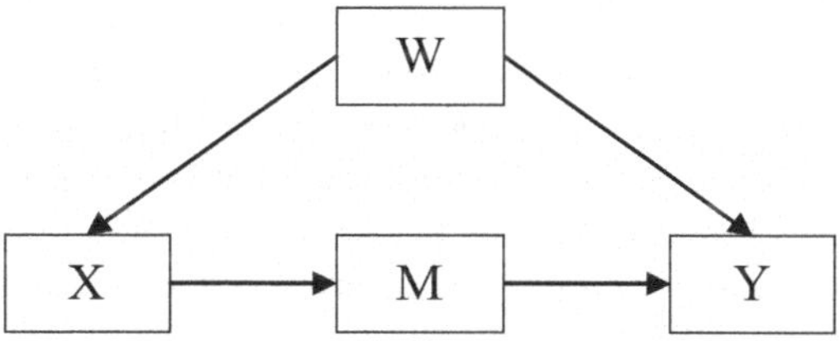

Fig. 4. Causal model with the addition of the mediating variable M.

We examine the causal relationship between AI-assisted teaching tactics and the achievement of course objectives by analyzing experimental teaching data in the context of power supply design studies. Conventional course evaluations, on the other hand, focus only on evaluating student achievement and ignore immeasurable elements such as the experimental setting, the conditions of the group, and outside disruptions to the experiment design. These elements may lead to discrepancies in student performance, which may not fairly represent their actual capacity for learning and hinder the achievement of course goals. They are regarded as confounding variables. Our method entails removing these confounders by incorporating causal inference.

3.2 The Back-Door Criterion

To get the actual causal effect, it is essential to remove the influence of confounding variable W. The relationship between the three is shown in Fig. 3. In studying the causal inference from X to Y, confounding variable W (measurable) is categorized, which is the essence of the do-operator. The do-operator, a significant distinction between causal theory and traditional statistics, expresses predictions by forcibly cutting off all confounding influences of W on X.

We define the back-door path as the road between X and Y, eliminating the direct route, with an emphasis on the relationship between X and Y. The definition of the back-door criterion is as follows: If a set of variables W does not contain any offspring of X and prevents all back-door paths from X to Y, then W meets the back-door requirement. We refer to this set as an adequate set. When we have a large enough sample, we can determine the following:

$$P(Y|do(X)) = \sum_{w} P(Y|do(X), W)P(W|do(X)) = \sum_{w} P(Y|X, W)P(W|do(X)), \tag{1}$$

after that, the back-door adjustment mechanism is derived from this identification process. Concerning regularization and modularity assumptions, we can determine that X causes Y in the following way:

$$P(Y|do(X)) = \sum_{w} P(Y|X, W)P(W). \tag{2}$$

3.3 Front-Door Adjustment Method

When confounding variable W is unmeasurable, back-door paths cannot be blocked; the front-door adjustment method addresses this scenario [7]. It posits that as long as there is a mediator M between X and Y, causal effects can be identified even without satisfying the back-door criterion. The causal model is shown in Fig. 4.

For variable M to satisfy the front-door criterion between X and Y:

- M completely mediates the effect of X on Y.
- There is no unblocked back-door path from X to M.
- X blocks all back-door paths from M to Y.

To identify the causal effect using the front-door criterion:

1) Identify the causal effect from X to M. Since Y is the collider, Y blocks the back-door path. Therefore, according to the back-door adjustment can be obtained:

$$P(M|do(X)) = P(M|X), \tag{3}$$

2) Identifying effects from M to Y, as X block, we can calculate by adjusting X to satisfy the back-door criterion that:

$$P(Y|do(M)) = \sum_{X} P(Y|M, X)P(X), \tag{4}$$

3) The causal effect of X to Y is obtained from the previous two steps:

$$P(Y|do(X)) = \sum_{M} P(M|do(X))P(Y|do(M)), \tag{5}$$

the result is to find all the causal effects from X to Y under the value of M, and all of them are added up to be the natural causal effect from X to Y.

4 Practical Teaching Case Studies

4.1 Constructing the Causal Model

Current educational evaluations typically focus on *P(course objective achievement|teaching strategy)*. However, real-world experimental teaching includes confounding variables, including experimental settings, student motivation and attitudes, and team dynamics, which skew the results. To remove confounding and precisely calculate the true causal impact of teaching strategies on experimental performance, causal inference techniques are required.

The back-door criterion is inappropriate since such confounders are immeasurable. Instead, we used the front-door modification method to uncover the natural causal effect [8] between the teaching strategy and the achievement of course objectives. We employed the students' experimental oral report scores as mediators; Fig. 5 displays the causal model, which reflects the students' comprehension and competency in various assessment points.

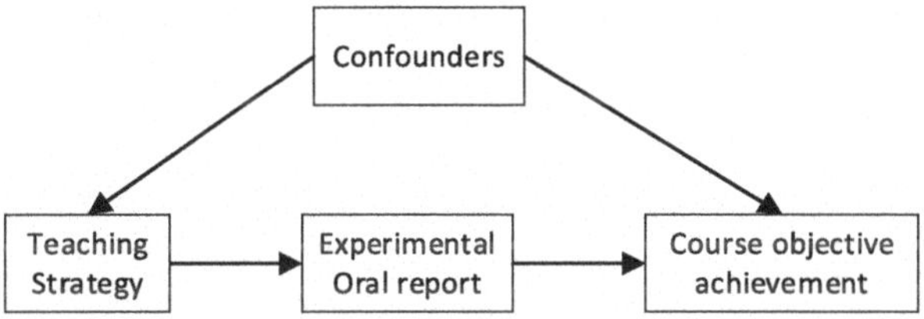

Fig. 5. Experimental causal modeling for power supply design course.

This work helps educators comprehend the influence of confounding factors on experimental teaching by applying the front-door adjustment method in causal inference to AI-assisted course experiment design and quality assessment. By identifying mediator variables, we can lessen confounding impacts and precisely quantify the real impact of the AI-assisted teaching technique on course objective achievement.

4.2 Data Collection and Processing

In this paper, we used data collected from students enrolled at a university specializing in electronics and information technology in the 2021 and 2022 academic years. The students were enrolled in the laboratory course "Power Supply Technology". The 2022 cohort used AI-assisted support in the course.

The goal was to evaluate how well experimental teaching strategies accomplished the course objectives. Y = 0, 1, 2, which stand for unqualified, qualified, and exceptional levels of performance, respectively, indicate the achievement of the course objectives. These values were based on the whole experimental performance, which includes pre-lab reports, experimental operations, final presentations, and lab reports. X = 0, 1, 2, which also denotes unqualified, qualified, and outstanding, is the explanatory variable that represents how well students performed in experimental procedures, as the score in the experimental operation. The mediator variable M = 0, 1, 2 represents the students' performance in the experimental oral report, which was influenced by the experimental operation and affected the final experimental grades.

We used this method to measure and analyze course objective achievement and calculate the aim of course achievement for both cohorts using different teaching strategies. The scores of the 2022 cohort, who used AI-assisted support, are distributed in Table 2.

4.3 Modeling and Analysis

To understand the true impact of AI-assisted teaching strategy on course objective achievement, we needed to eliminate confounding variables using causal inference methods, requiring data such as those in Table 3. The front-door adjustment method was used to obtain the actual value of

$$P(course\ objective\ achievement|do(teaching\ strategy)).$$

Table 2. Number of students by experimental operation score and experimental oral report score

X	M = 0	M = 1	M = 2	Total
0	21	7	7	35
1	9	46	13	68
2	2	16	31	49

According to Table 4, analysis of the causal effect estimates for X = 0, X = 1, X = 2 reveals that when the experimental operation score is unqualified, the probability of the course objective being evaluated as qualified is 86.01%($P(Y \geqslant 1|do(X = 0))$). For qualified experimental operation scores, the probability is 87.20% ($P(Y \geqslant 1|do(X = 1))$), and for excellent scores, it is 90.48% ($P(Y \geqslant 1|do(X = 2))$). The weighted average probability of achieving the course objective across all levels is 87.90%, which is also the actual average course objective achievement after de-confounded using the front-door adjustment [9] (Table 6).

From the above method, these data reveal the primary status of the implementation of the program. Table 5 provides the actual average course objective achievement evaluation for the course with peer support that does not involve AI (82.71%) in the class of 2021. According to these data, the course achievement evaluation of the power supply design course taught with AI assistance is 5.19 percent higher than the course

Table 3. Distribution of course objective achievement by experimental operation and experimental oral report scores

Category	Y = 0	Y = 1	Y = 2	Total
X = 0, M = 0	15	6	0	21
X = 0, M = 1	3	4	0	7
X = 0, M = 2	0	7	0	7
X = 1, M = 0	0	7	2	9
X = 1, M = 1	4	32	10	46
X = 1, M = 2	2	4	7	13
X = 2, M = 0	0	1	1	2
X = 2, M = 1	0	4	12	16
X = 2, M = 2	0	12	19	31

Table 4. Causal effect of AI-assisted teaching on course objective achievement

P(Y|do(X))	X = 0	X = 1	X = 2
Y = 0	14.00%	12.80%	9.52%
Y = 1	54.82%	52.43%	50.61%
Y = 2	31.19%	34.77%	39.87%

Table 5. Causal effect of non-AI-assisted teaching on course objective achievement

P(Y|do(X))	X = 0	X = 1	X = 2
Y = 0	9.12%	7.16%	6.11%
Y = 1	43.91%	45.01%	46.90%
Y = 2	29.21%	43.00%	40.09%

Table 6. Comparison of different teaching strategies on course objective achievement

Course objective achievement		$P(Y \geqslant 1 \mid do(X))$
Teaching strategy	AI-assisted	87.90%
	Non-AI-assisted	82.71%

objective achievement evaluation of the course taught without AI assistance. Therefore, AI-assisted peer teaching strategies have improved the achievement of course objectives.

4.4 Comparison of Causal Effects Under De-Confounded and Non-de-Confounded Conditions

To investigate changes in course objective achievement before and after removing confounding variables, this paper uses the same sample data for control testing. Table 7 presents a comparison table of course objective achievement before and after de-confounding.

Encouraging students to use AI Companion Assistance increased their performance in meeting course objectives. After excluding confounders using the Front-Door Adjustment approach, this improvement became very noticeable.

Table 7. Estimates of causal effects under front-door adjustment versus raw observation

Course objective achievement		Numerical value
The Front-Door Adjustment	De-confounded $P(Y \geqslant 1 \mid do(X))$	87.90%
	Non-de-confounded $P(Y \geqslant 1 \mid X)$	79.92%

The de-confounded course objective achievement was 7.98% greater than the non-de-confounded course objective achievement, indicating that confounding variables were not the only factors at play and that the experimental teaching technique was more beneficial overall.

An underestimation of the intrinsic value of the course may have resulted from confounding variables influencing previously uncorrected assessments of course effectiveness. This research clearly shows that the real educational gains from using AI-assisted teaching techniques considerably outweigh the preliminary findings. The course's actual effectiveness in promoting student learning outcomes exceeds the level initially observed, as confirmed by the possibility that the initial assessments were masked by several confounding variables that were not related to the teaching intervention itself but had an equal impact on the evaluation of learning outcomes.

5 Feasibility and Future Prospects of AI-Assisted Teaching Strategy

Even though AI-assisted teaching has been shown to be very beneficial in lab settings, there are still certain limitations and difficulties in its practical application. Particular analyses of these difficulties follow, and the corresponding solutions are suggested:

5.1 Interdisciplinary Application

1) Limitations: While AI-assisted teaching methodologies have the potential to be applied across various fields in theory, practical implementations face certain challenges. Disciplines differ in their evaluation point settings, which raises the bar for the flexibility and adaptability of AI assistance. To guarantee the accuracy and utility of the AI system calls for a great deal of prior planning in addition to a great deal of time and work from teachers and pupils.

2) Countermeasures: Creating modular AI systems that can be tailored to different topics is one way to potentially alleviate the limits of AI-assisted teaching across disciplines. The AI can more effectively react to the particular needs of each subject by developing discipline-specific modules that enable the modification of evaluation points. These modules could be created with user-friendly interfaces and pre-configured templates that simplify the setup procedure to cut down on the time and effort needed by instructors and students.

5.2 Technical Reliability and Safety

1) Limitations: Preserving data privacy, preventing misuse, and guaranteeing content accuracy all depend on the technology's dependability and security.
2) Countermeasures: To stop inappropriate use of technology, we must improve the security architecture of AI systems, guarantee the privacy of students' and teachers' data, and verify the caliber of material through stringent auditing procedures.

5.3 Limitations on Educational Equity and Accessibility:

1) Limitations: Although AI-assisted instruction can lessen regional differences in access to learning resources, it can potentially widen the digital gap because not all students have equal access to the necessary hardware and internet connections.
2) Countermeasures: Promoting policies to ensure technology advancements to reduce, rather than widen, social inequality, and using both technology and legislation may provide all students with equal access to quality learning resources.

More AI integration in education is anticipated as a result of these projects, which will improve student learning and offer educators significant support, propelling the system toward more efficacy, inclusion, and creativity.

6 Conclusion

In summary, students' mastery of specialized subject matter knowledge was strengthened and deepened by the use of AI-assisted teaching in a power supply design course. By comparing the achievement of the course objectives of an innovative experimental teaching strategy during two years of a school using AI-assisted teaching for one of the years, the course objective was more significantly met with AI assistance than it was without such support. Moreover, we were able to accurately quantify the actual effectiveness of the teaching interventions, remove confounding variables, reveal the precise contribution of the AI-assisted experimentation strategy to the achievement of the course objectives, and find that the actual achievement of the course objectives was better than what was initially observed due to the pedagogical effect being masked by confounding factors. These findings strongly demonstrate the broad prospect and potential of the application of AI technology in the field of educational innovation, and future educational reforms should concentrate on the strategy's refinement and the deep optimization of practices.

Acknowledgments. This study was funded by the Teaching Reform Research Project for General Undergraduate Colleges and Universities in Hunan Province (202401000570) and by the Research and Practice Program of New Engineering Science in Hunan Province (202012).

References

1. OpenAI et al.: GPT-4 Technical Report. arXiv (2024). https://arxiv.org/abs/2303.08774
2. Nguyen, A., Hong, Y., Dang, B., Huang, X.: Human-AI collaboration patterns in AI-assisted academic writing. Stud. High. Educ. **49**(5), 847–864 (2024)
3. Järvelä, S., Nguyen, A.: SHARP – Socially Shared Regulation of Complex Learning Process in Groups (2022)
4. Euchner, J.: Almost human. Res. Technol. Manag. **66**(2), 10–11 (2022)
5. Wollny, S., Schneider, J., Di Mitri, D., Weidlich, J., Rittberger, M., Drachsler, H.: Are we there yet? A systematic literature review on chatbots in education. Front. Artif. Intell. **4**, 654924 (2021). https://doi.org/10.3389/frai.2021.654924
6. Lipman, J., Distler, R.: Schools shouldn't ban access to ChatGPT (2023). https://example.com. Accessed 11 June 2025
7. Pearl, J.: Causality: Models, Reasoning, and Inference. Cambridge University Press, Cambridge (2000)
8. Gillies, D., Pearl, J.: Causality: models, reasoning, and inference. Br. J. Philos. Sci. **52**, 613–616 (2001). https://doi.org/10.1093/bjps/52.3.613
9. Long, X., Duan, B., Ke, Q., Peng, X.: Strategy and practice on continuous improvement of achievement portfolio based on causality. In: 2022 Eleventh International Conference of Educational Innovation through Technology (EITT), New York, pp. 98–103. IEEE (2022)

A Measurement Method of Knowledge Contribution by Learners During a Design Process

Zhaoxue Wang and Fati Wu(✉)

School of Educational Technology, Beijing Normal University, Beijing, China
wft@bnu.edu.cn

Abstract. Assessing the knowledge contributions of learners during the design process is crucial for understanding and improving the quality and effectiveness of design education. However, existing methods for evaluating learners' knowledge contributions in collaborative design settings have limitations. Traditional measure methods primarily rely on the final design product, neglecting the cognitive activities and knowledge construction processes of learners during the design process. Moreover, these studies typically focus on individual learner performance, disregarding the collaborative interactions and knowledge sharing among learners. Consequently, this study innovatively quantifies learners' knowledge contribution in collaborative design processes through the application of objective and quantitative techniques, including keyword frequency and information gain. Additionally, this study explores the integration of design-based problem-solving with teaching practices, providing educators with a novel pedagogical strategy to foster students' deep learning and innovation.

Keywords: knowledge contribution · design education · learning analysis

1 Introduction

Within collaborative design processes, the interplay of designers' perspectives is indispensable for driving innovation and refining design schemes. Measuring knowledge contribution in these processes is imperative for evaluating and enhancing the quality of design education, ensuring that educational strategies effectively support collaborative creativity and innovation.

Assessing knowledge contribution not only evaluates the effectiveness of design activities but also fosters effective communication and collaboration among team members. Existing research has proposed various methods to evaluate knowledge contribution, including self-assessment, peer assessment, automated analysis based on external behaviors, content analysis, and keyword calculations. However, existing studies have predominantly focused on individual-level knowledge contributions, neglecting the significance of team-level knowledge integration and synergy. Despite advancements in technology, the application of emerging technologies in measuring knowledge contribution remains relatively unexplored in current research.

Q. Liu et al. (Eds.): EITT 2024, CCIS 2600, pp. 247–255, 2026.
https://doi.org/10.1007/978-981-95-1525-7_20

In order to overcome the aforementioned limitations, this study presents a quantitative method for evaluating learners' knowledge contributions in collaborative design settings. By concentrating on the dimensions of scientificity, systematicity, and innovativeness, and by extracting keywords from design outcomes, this research aims to provide a more comprehensive and quantifiable analysis of knowledge contribution. Through empirical research, we have identified significant positive correlations between these dimensions and the scores of design schemes. These findings not only highlight the relevance of these dimensions in the design process but also offer valuable insights for the field of design education. By revealing the characteristics of knowledge contribution, this study aims to guide the development of learners' capabilities in design and provide a framework for future research.

2 Related Work

2.1 Construction of Knowledge in the Design

The construction of knowledge in the design domain is a multifaceted endeavor that encompasses the accumulation, organization, and application of expertise within the creative process. Design knowledge is not static but evolves through iterative practices, collaborative engagements, and the integration of diverse disciplinary insights. In design practice, the construction of ideas is not only the result of the thinking processes of the individual but also the result of teamwork. Designers need to express and communicate their viewpoints within the team, enriching and perfecting the design scheme through collective wisdom. This collaborative process of knowledge construction not only facilitates the sharing of knowledge and the complementarity of skills but also enhances the quality and innovation of design solutions. Whether it is project development, product manufacturing, or creative work, designers need to gather professional knowledge from various fields and utilize the concepts and toolkits of each field to analyze, synthesize, and construct new insights and ideas [1]. The quality and efficiency of the designer's knowledge contribution in the collaborative design process can reveal the designer's processing strategy at the cognitive level. According to the Socialization—Externalization—Combination— Internalization model (SECI), the process of externalization and combination of knowledge can be defined as the knowledge contribution viewpoint. The key element in the design process is that learners can mobilize multidisciplinary knowledge to extract and construct innovative ideas. Collaborative knowledge construction promotes individual and collective learning in the field of education and has received widespread attention in recent years.

Despite the advancements in design education and practice, the literature reveals a gap in the comprehensive understanding of how design knowledge is constructed, shared, and leveraged within collaborative settings. This gap is particularly evident in the lack of robust frameworks that can systematically capture the multi-dimensional aspects of knowledge contribution and its impact on design outcomes. Addressing this gap is crucial for enhancing the effectiveness of design teams and the innovation potential of their work. It is against this backdrop that the present study aims to explore and contribute to the understanding of knowledge construction in the design process.

2.2 Measurement of the Knowledge Contribution

It is essential for enhancing knowledge-building endeavors to assess knowledge contribution. Current measurement methods include self-assessment and peer assessment, automatic analysis based on external behaviors, content analysis, and keyword calculation. Cheng and Warren [2] discussed the importance of self-assessment and peer assessment. Almatrafi and Johri [3] analyzed research on assessing behavior. Hong and Scardamalia [4] proposed a keyword measurement method but pointed out that there are limitations to measuring contributions solely based on frequency. Keyword extraction techniques were divided into statistical-based, machine learning-based, and semantic-based methods, each with its strengths and limitations.

Currently, there are three main methods for keyword extraction, each with its own advantages and disadvantages. Statistical-based methods are relatively simple and easy to implement; hence, they have been widely welcomed. However, their drawback lies in their susceptibility to data noise, which may reduce the accuracy of the extraction results.

Machine learning-based methods can provide more accurate keyword extraction, but the downside is the need to construct a large training set, which not only increases the difficulty of model training but also the cost of sample annotation. Semantic-based methods have the advantage of logic that aligns with human thinking, thus offering higher accuracy. However, they largely depend on dictionaries, which may limit their application scope and flexibility.

However, as keyword extraction technology has been continuously improved and refined, with the continuous improvement of dictionaries and the advancement of technology, it has been widely applied in various fields. This technology is also of vital importance to text mining in the design process.

3 Method

3.1 The Research Process

An experiment was conducted involving 103 participants aged between 18 and 22 who were randomly assigned into three separate groups. Prior to the experiment, each participant was briefed on the background and objectives of the experiment to guarantee the successful execution of the collaborative task.

The core of the experiment was a "Sponge Campus Design" project, where participants were tasked with collaboratively devising an eco-friendly and water-conserving campus layout, culminating in the formulation of a sponge campus proposal. Participants were furnished with detailed task sheets outlining the project requirements and evaluation metrics, as well as educational materials on sponge campus features such as green spaces, pavements, and landscaping.

The sequence of the experiment was as follows: Firstly, the researchers set up and tested the lab equipment, briefing the participants on the process and safety measures. Secondly, the groups tackled the design task, with access to learning materials and the internet for research. The entire session was monitored by the researchers to ensure smooth task completion with minimal disruptions. Lastly, the textual interactions gathered during the problem-solving phase were recorded using an Iflytek intelligent recorder and subsequently reviewed for accuracy.

3.2 Data Preprocessing

The study mainly preprocessed the data from the following four aspects. Firstly, irrelevant textual elements, including redundant phrases, spaces, and punctuation marks, were eliminated. Secondly, the technique of oversampling was implemented to address the issue of skewed data distribution in deep learning models. Thirdly, the segmentation tool of choice was the 'jieba' package. Finally, words that had minimal impact on the text's overall meaning, such as pronouns and conjunctions, were pruned to enhance the readability and logical flow of the content.

3.3 Select Evaluation Indicators

Through our review of the literature, we have found that scholars hold various perspectives on the measurement criteria for knowledge contribution. Rauth [5] has identified "perspective-taking" as one of the key competencies in design thinking and posits that the cultivation of this skill can ultimately lead to the development of creative confidence in students. Here, the scientific accuracy of the viewpoints presented is the main criterion for assessment. Individuals share their insights through communication and foster the development of communal knowledge by questioning, clarifying, and negotiating with one another. Additionally, some researchers have divergent views on the evaluation of "perspective proposal" in collaborative design. Among these, a particularly representative perspective is the emphasis on the creativity inherent in the viewpoints presented. For instance, after reviewing 169 papers on "design thinking" McLaughlin concluded that the purpose of cultivating design abilities is to enhance an individual's creative and innovative thinking skills [6]. Wiggins [7] posits that design problem-solving is essentially the creative resolution of complex and intractable issues. Some researchers also advocated for a focus on the "systematic nature of perspectives". For example, Jonassen believes that design requires learners to develop systematic products that meet certain standards [8]. Furthermore, some researchers argue that the systematic orientation and creative orientation of perspectives are equally important [9]. Therefore, this study intends to adopt scientificity, systematicity, and innovativeness as the primary criteria for assessing the construction of learners' perspectives.

Moving forward, we employed the Delphi method for two rounds of iterative revisions to ensure the scientific validity and effectiveness of the measurement criteria. Ultimately, the coefficient of variation for the Delphi method ranged from 0.03 to 0.25. The coordination coefficient was less than 0.001, indicating a consensus among the experts' opinions.

3.4 Method to Calculate Knowledge Construction

Information theory is the scientific study of the quantification, storage, and communication of information. Klüver established a mathematical framework for the analysis of information dissemination, which is still widely applied today [10]. In this study, the concept of self-information taken from information theory is utilized to calculate the knowledge contribution of key terms in learners' speech data, serving as the foundation for assessing knowledge contribution. Wu and colleagues applied information theory to the context of collaborative knowledge construction, proposing the "Information Metric for Knowledge Contribution (IMKC)" method. This approach measures the informational content of a text by summing the self-information of its key terms, thereby exploring knowledge contribution at various levels [11].

Building upon the research findings of Wu, this paper explores how to measure knowledge contribution within the context of collaborative knowledge construction from an information-theoretic perspective, and to assess the scientific, systematic, and innovative nature of learning perspectives. We propose a novel evaluation method aimed at providing a more comprehensive, objective, and accurate analysis of the quality of learning perspectives.

Table 1 presents a summary of the key terms. When these key terms were triggered in the learners' speech, the next step of the calculation was undertaken.

We adapted the calculation method of IMKC from key terms in online posts to key terms in collaborative discussion texts. Consequently, in this study, the information quantity is the sum of the information quantities of all key terms in the text. The specific method for calculating the quantification metrics was as follows:

(1) Scientificity. Scientificity refers to the ability of learners to correctly and effectively acquire information in the process of problem-solving. In this study, the frequency of key term occurrences is used to assess scientificity.

According to Shannon's information theory, the total information content of learner interaction texts refers to the sum of the information content of all key terms within the text, where M refers to the number of learners, n refers to the number of sentences per learner, fre_{ij} is the frequency of key terms appearing in the j-th sentence of the i-th learner, fre_{ij} is the frequency of the k-th learner using key terms, and p_k is the total probability of the k-th learner's speech containing key terms. The calculation formula is shown as follows:

$$p_k = \frac{fre_k}{\sum_{i=1}^{M} \sum_{j=1}^{n} fre_{ij}}. \tag{1}$$

(2) Innovativeness. Innovativeness refers to the proportion of new ideas, new leads, new plans, and new information first proposed by students in the total information content. We organized discussions in groups of three participants each and performed a separate keyword frequency analysis on the discussion texts of each group. First, we arranged the discussion texts of each group on a timeline in chronological order. During the analysis, we allowed for the repetition of keywords. If a sentence contained keywords that have appeared in previous texts without introducing new keywords, we considered the contribution of this sentence to the topic to be less significant. On the contrary, when new keywords appeared in the text, these keywords were included in our analysis to identify information that substantially contributes to the topic. For this calculation, D refers to the number of keywords in the previous sentence, R refers to the number of keywords in the following sentence, card|R-D| refers to the number of new keywords

Table1. Compilation of Textual Keywords in the Design Process (Translated From Chinese).

Keywords	Primary Keywords	Methodological Keywords
Infiltration	Road surface infiltration, road infiltration	Pervious paving, Pervious Asphalt, Pervious Pavement
	Infiltration pond	Infiltration Pond
	Rainwater runoff	Infiltration Pipe, Infiltration Channel, Artificial Soil Infiltration, Grassed Swale
Water storage		Reservoir, Sunken Green Space
		Eco- Roof, Rain Garden
Water purification	Primary Treatment, Physical Treatment Method	Sedimentation, Filtration, Flotation, Centrifugation
	Secondary Treatment, Biological Treatment Method	Activated Sludge Process, Membrane Bioreactor, Oxidation Pond, Anaerobic Biological Treatment, Oxidation Method
	Tertiary Treatment, Chemical Treatment Method	Neutralization, Redox Process, Electrolysis, Chemical Precipitation
Water accumulation location	Exercise Area	Playground, Sports Hall, Track
	Living Area	Dormitory, Canteen, living, Parking Space
	Study Area	Teaching building, office building, library
Type of water accumulation		Rainwater, Domestic Sewage, Domestic Wastewater, Laboratory Wastewater
Plan drawing		Plan, Existing Campus, New Campus

introduced, and ΔI refers to the information gain of newly added keywords. Given that each group consists of three students and there are dialogues in the interactive text, the subscript i represents the i-th learner. Thus, it represents the information gain for the i-th learner.

$$\Delta \mathrm{I} = -\sum_{O=1}^{card|\mathrm{R}-\mathrm{D}|} log_2 p_o, \tag{2}$$

$$\Delta I_i = -\sum_{O=1}^{card|\mathrm{R_i}-\mathrm{D_i}|} log_2 p_{io}, \tag{3}$$

(3) Systematicity. Systematicity can indicate whether the scope of knowledge explored by students is consistent with the knowledge scope that teachers consider to be relevant. This study, from the perspective of information theory, examines the knowledge breadth and coverage of different learners in the "Sponge Campus" activity across three dimensions: knowledge points, thematic content, and related concepts. We opted to use the coverage ratio of key terms to represent the systematicity of students' viewpoints in discussions. When a keyword appears in the learner's speech text, $uni(x)$ is recorded as 1. SYS_i represents the coverage of each learner's speech text to the keyword list. The calculation method is as follows:

$$SYS_i = \frac{\sum_{j=1}^{N} uni(fre_{ij})}{N}, \tag{4}$$

$$uni(x) = \begin{cases} 1, count(x) \geq 1 \\ 0, count(x) = 0 \end{cases}. \tag{5}$$

4 Results

4.1 Knowledge Contribution Score

We conducted descriptive statistical calculations on the scientificity, systematicity, and innovativeness of 103 learners. After standardizing the data within the perspective construction dimension, Table 2 presents the statistical analysis of the level of Knowledge Contribution for the 103 learners in various dimensions.

Table 2. Statistical Analysis of Feature Values in the Perspective Construction Dimension.

	N	MAX	MIN	Mean	SD
Scientificity	103	0.66	0	0.28	0.10
Systematicity	103	1	0	0.20	0.20
Innovativeness	103	0.83	0	0.31	0.18

4.2 Design Scheme Score

The scores for the design scheme were calculated as the weighted sum of the subjective scores given by experts on the students' design plans and the self-assessments and peer assessments within the group. The experts' subjective scoring was conducted using a blind scoring method. When there was a discrepancy of more than 5 points between two scores, a second round of scoring was initiated, where the reasons for the scoring were discussed to arrive at a final score. This process ensured fairness and impartiality in the scoring and prevented the randomness that might arise from a single person's scoring. Student Design Scheme Examples are shown in Fig. 1.

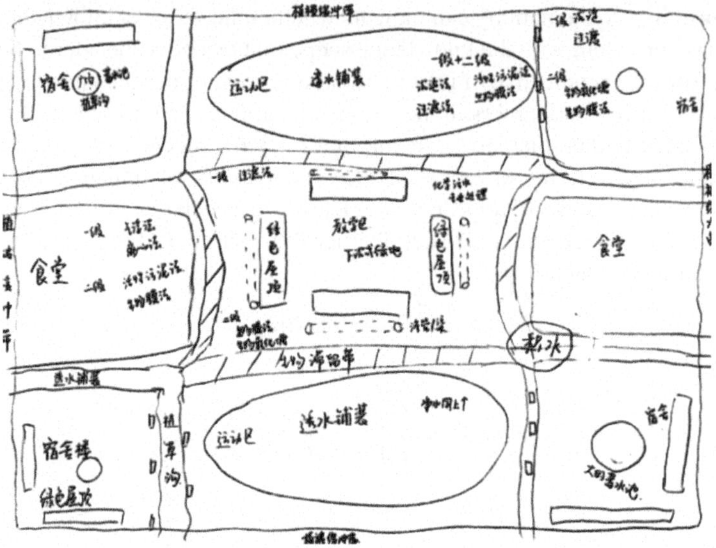

Fig. 1. Example of Sponge Campus Design Scheme Diagram.

Table 3. The relationship between the performance of design schemes and the quantified knowledge input.

	Scientificity	Systematicity	Innovativeness
Grade	0.765**	0.597*	0.682*

* $p < .05$, ** $p < .01$ $N = 103$.

Moving forward, we employed a Pearson Correlation analysis to compare the knowledge contribution of learners with their design scheme scores as shown in Table 3.

The scores for the design schemes show a significant positive correlation with the scientificity, systematicity, and innovativeness of the learners' knowledge contribution during the design process. Scientificity has the strongest correlation with the score at a coefficient of 0.765, indicating the importance of in-depth scientific understanding and application in enhancing the quality of design. Innovativeness follows with a coefficient of 0.682, demonstrating the positive impact of novel perspectives on design scoring. Systematicity, with a coefficient of 0.597, indicates the influence of learners' ability to integrate knowledge and construct coherent design schemes on the score.

5 Conclusion and Future Research

The core goal of this study was to explore and quantify the multidimensional characteristics of learners' knowledge contribution during the design process. By employing a keyword extraction method, we conducted an in-depth analysis and measurement of learners' design outcomes from three critical dimensions: scientificity, systematicity, and innovativeness. Through empirical research, we have demonstrated a significant positive correlation between these three dimensions and the scores of the design schemes, thereby providing important insights for the fields of design education and practice.

This study provides a direction for future research, suggesting further exploration of how these dimensions interact with other aspects of the design process and how they are influenced by different educational backgrounds and design practices. With ongoing research, we can gain a deeper understanding of the complexity of knowledge contribution in the design process and offer more precise and robust support for design education and practice.

References

1. Melles, G., Howard, Z., Thompson-Whiteside, S.: Teaching design thinking: expanding horizons in design education. Procedia Soc. Behav. Sci. **31**, 162–166 (2012)
2. Cheng, W., Warren, M.: Making a difference: using peers to assess individual students' contributions to a group project. Teach. High. Educ. **5**(2), 243–255 (2000)
3. Almatrafi, O., Johri, A.: Systematic review of discussion forums in massive open online courses (MOOCs). IEEE Trans. Learn. Technol. **12**(3), 413–428 (2019)
4. Hong, H.-Y., Scardamalia, M.: Community knowledge assessment in a knowledge building environment. Comput. Educ. **71**, 279–288 (2014)
5. Rauth, I., Köppen, E., Jobst, B., Meinel, C.: Design thinking: an educational model towards creative confidence. In: DS 66-2: Proceedings of the 1st International Conference on Design Creativity (2010)
6. McLaughlin, J.E., Wolcott, M.D., Hubbard, D., Umstead, K., Rider, T.R.: A qualitative review of the design thinking framework in health professions education. BMC Med. Educ. **19**(1), 1–8 (2019)
7. Wiggins, G.P., McTighe, J.: Understanding by Design. ASCD (2005)
8. Jonassen, D.H., Strobel, J., Lee, C.B.: Everyday problem solving in engineering: lessons for engineering educators. J. Eng. Educ. **95**(2), 1–14 (1995)
9. Camacho, M.: An integrative model of design thinking. In: The 21st DMI: Academic Design Management Conference, 'Next Wave', London, United Kingdom, p. 627 (2018)
10. Klüver, J.: A mathematical theory of communication: meaning, information, and topology. Complexity **16**(3), 10–26 (2011)
11. Wu, L., et al.: Information measures of knowledge contribution: a new method to measure knowledge contribution. In: Collaborative Knowledge Building: An Information Theory Perspective. Journal of Educational Computing Research, vol. 59, no. 7, pp. 1319–1342 (2021)

Investigating How Self-recording and Playback Influence Music Students' Self-assessment

Yu Wang[1](✉), Shen Ba[2], and King Yue Li[3]

[1] Department of Early Childhood Education, The Education University of Hong Kong, Hong Kong SAR, China
s1151862@s.eduhk.hk

[2] Department of Cultural and Creative Arts, The Education University of Hong Kong, Hong Kong SAR, China

[3] Department of Curriculum and Instruction, The Education University of Hong Kong, Hong Kong SAR, China

Abstract. The Higher education is increasingly devoted to promoting students' ability to self-assess their work. However, little is known regarding music students' self-assessment practices. The present study investigated the impact of video and audio recording on the self-assessment of tertiary musical instrument students. A preliminary study was conducted with six piano students from three public universities, randomly assigned to video, audio, and non-intervention conditions. This research spanned three weeks and included one baseline task followed by four formal tasks. During the baseline task, all groups self-assessed their performance of a twenty-five-minute piano practice without any assistance. In the formal tasks, the video and audio groups used playback recordings for self-assessment, while the non-intervention group continued as in the baseline task. Self-assessment accuracy was meas-ured by comparing self-ratings to expert ratings. We conducted semi-structured interviews to understand the difference between conditions. Results indicated that both conditions promote accuracy in music students' self-assessment. The difference was that video feedback, which provides both visual and auditory information, significantly improves overall self-assessment accuracy, but is slightly less accurate than audio feedback in assessing auditory elements. Despite the limited sample size, this study pro-vides valuable recommendations for self-assessment strategies utilizing audi-ovisual techniques to enhance effective instrument practice.

Keywords: self-assessment · effective practice strategies · multimedia feedback · music education

1 Introduction

Self-assessment is a crucial component of contemporary education, enhancing students' independent learning skills and self-efficacy [1, 2]. While the effectiveness of self-assessment may be influenced by learning motivation, evaluation frameworks [3], and multimedia feedback techniques [4, 5], sound self-assessment strategies will help music students to regulate their performance process effectively [6].

Q. Liu et al. (Eds.): EITT 2024, CCIS 2600, pp. 256–267, 2026.
https://doi.org/10.1007/978-981-95-1525-7_21

Although existing studies have investigated the effects of different audiovisual feedback modalities on musicians' performance and self-regulation [7], few studies have compared the effects of their use and their role in musicians' assessment of their own practices. The present study fills this gap by simultaneously exploring the role of two modalities, self-audio-recording and self-video-recording, in assisting students' with self-assessment of their practice outcomes. Our findings provide insights for optimizing musical practice strategies and self-assessment tools.

2 Literature Review

2.1 Self-assessment in Education

Self-assessment involves reviewing and evaluating one's own work and learning methods in order to understand and reflect on their effectiveness [8]. It is a vital aspect of metacognition and self-regulation, aiding in better academic performance [9, 10]. In music education, self-assessment is commonly used in self-practice [11], with some students relying on technological feedback methods [12].

However, concerns exist regarding the accuracy of unguided self-assessment, which may lead students toward biased results and reduced validity [13]. Experts suggest that educators develop training courses to enhance the effectiveness of self-assessment [3]. Future research should further explore self-assessment strategies and their effectiveness in different contexts, especially in music education.

2.2 Self-video-Recording and Self-audio-Recording as Self-assessment Tools

Video feedback is effective for self-assessment, allowing learners to view and listen to their performance simultaneously from another perspective, thus enhancing self-reflection [14]. In music education, video feedback aids in improving practice strategies, musicianship, and error detection [7, 12]. It helps musicians develop a "self-teaching" role when rehearsing alone [15]. The combination of video recording and self-assessment has been a suitable teaching strategy for decades [16], with various studies exploring its positive impacts.

Audio feedback also plays a significant role in self-assessment. Studies show that listening to self-audio-recordings helps in more accurate error detection [17, 18]. However, short experimental periods and limited assessment metrics suggest a need for longer studies with comprehensive metrics and more task sessions to fully explore audio feedback's role in self-assessment.

While previous research highlights the positive impact of audiovisual feedback on self-assessment, few studies have compared the effectiveness of different audiovisual modalities and explored the underlying mechanism. For instance, while Hamond, Welch, and Himonides report that visual feedback supports piano learning, though they do not discuss the specific roles or constraints of auditory feedback [7]. Additionally, most studies focus on comparing the impact on learning outcomes rather than on self-assessment itself. Therefore, this study aims to uncover differences in the impact of different audiovisual feedback modalities on the accuracy of tertiary-level music students' self-assessments.

3 Methodology

3.1 Research Design

This study employs a mixed-method research design [19], combining a repeated measures design with semi-structured interviews. A total of six female students (20–28 years old, one undergraduate and five postgraduates) were recruited through purposive sampling from three public universities in mainland China and Hong Kong, China. Participants on average had more than ten years of systematic piano study and were music education majors. We selected these participants as they normally have little access to one-on-one musical instrument lessons and are thus required to practice independently.

Participants were randomly assigned to one of two experimental conditions (video and audio) or a control condition. The experiment spanned three weeks, including one baseline and four formal practices, each practice session separated by about three days (as shown in Fig. 1). In the baseline practice, participants engaged in a five-minute warm-up, a twenty-minute practice, and a ten-minute self-assessment without audio or video feedback. During the formal practice sessions, the video and audio groups recorded their performance during the final ten minutes and used this recording for self-assessment. Specifically, they were required to record their final practice outcomes from that session (e.g., if the task was to practice the first four phrases of the designated piece, they needed record a complete and coherent performance of the four phrases). They then playback the recording three times from beginning to end and filled out a self-assessment scale at the same time. The video group's recordings contained both picture and sound, while the audio group's contained sound only.

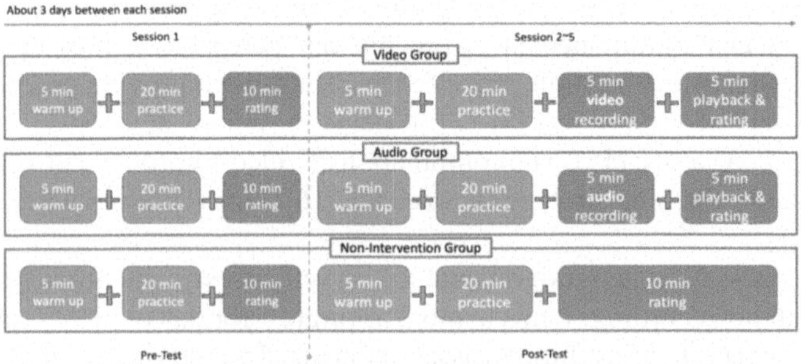

Fig. 1. Practice Sessions and Self-Rating Processes

Training was provided before the experiment, covering the study's purpose, self-assessment procedures, and materials. During the baseline session (Session 1), participants practiced the same piece in familiar practice spaces without external feedback tools. In Sessions 2–5, the audio group recorded audio only, while the video group ensured head, upper body, and hands were recorded visually and audibly. This design aimed to control confounding variables and ensure valid practice and assessment.

Following the experiment, an expert musician from a Chinese public university was invited to rate all the recordings in two rounds, with the second round occurring one

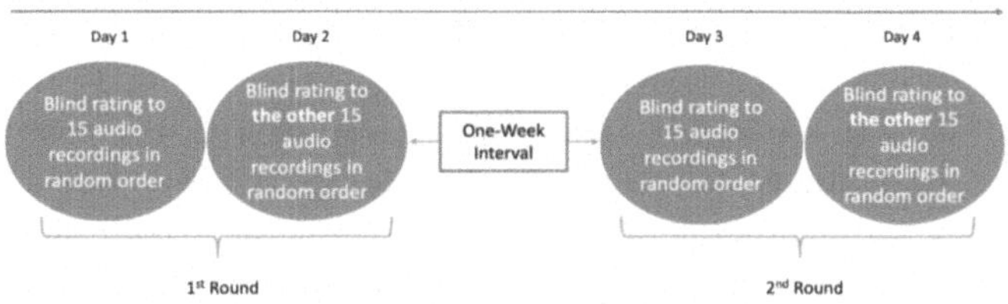

Fig. 2. Expert Rating Processes

week after the first (as shown in Fig. 2). Additionally, all participants agreed to a 45-min semi-structured interview to share their detailed experiences.

3.2 Recording Locations and Equipment Selection

Data from the self-assessment scales showed that all participants took part in the experiment at the place where they practiced on a daily basis. Four of them (P01, P02, P03, P06) practiced at home and two (P04, P05) practiced in well-soundproofed piano rooms at school. All participants used their daily recording devices for recording and playback (as shown in Table 1). The recording equipment used remained consistent throughout the experiment.

Table1. Equipment Selection

Group	Participants	Gear
Video	P01	iPhone 13
	P05	iPhone 14
Audio	P03	iPhone 12
	P06	Shure MV88+iPhone 13 pro max

All of the submitted recordings, as well as the self-assessment scales and interview data, indicate that the participants strictly adhered to the requirements outlined in the training. The practice sessions, recording and playback, and self-assessment strategies were conducted consistently across groups.

3.3 Practice and Self-rating Materials

The self-assessment scale was adapted from Saunders and Holahan's Woodwind Brass Solo Evaluation Form (WBSEF) and contains seven sub-elements: tone, melody, rhythm, tempo, technique, pedaling, and interpretation [20]. The scale has a maximum score of 70 points, with each element accounting for 10 points. Although originally designed for wind-brass players, the WBSEF has been adopted for self-assessment by other instrumentalists, including pianists [6].

Mendelssohn's Songs Without Words Op. 30 No. 3 in E major ("Lied Ohne Worte") was chosen as the practice piece. This piece includes elements that correspond to all sub-elements of the self-assessment scale, such as clear indications of tempo, dynamics, and pedaling. Also, participants had not practiced this piece before, and the difficulty level of this piece was appropriate for both the target group's skills and the provided practice time.

3.4 Data Analysis

Quantitative data were processed using descriptive statistical methods, including the calculation of means, variances, frequencies, and correlation coefficients. These statistical measures were employed to evaluate differences and trends in both the overall self-assessments and the individual sub-element self-assessments across the different audiovisual feedback modalities.

The qualitative data, obtained from 45-min semi-structured interviews with each participant, were thematically analyzed using the In Vivo approach [21]. By systematically coding these themes, we gain insight into participants' perceptions of the value and drawbacks of self-assessment regarding the different sub-elements, thereby providing a richer context to complement the quantitative findings.

3.5 Ethical Considerations

Written consent was obtained from each participant. Ethical approval was granted by the Institutional Review Board (IRB). Participants were referred to using aliases during the experiment to ensure data anonymity and confidentiality.

4 Results

4.1 Comparing Self-assessments and Expert Ratings

To understand the dynamics and patterns of self-assessment accuracy, we analyzed the difference (Diff) between the mean total self-ratings (MTSR) and the mean total expert ratings (MTER) of the participants in each group for each session. The MTSR is the total ratings from all seven sub-elements of the self-assessment scale for a single practice session; the same is the case for the MTER, but for the expert's ratings. "Difference" indicates the deviation between these means: "Diff = + n" means the mean of self-ratings are higher than the mean of expert ratings by n points, "diff = 0" means they are equal.

The video group's MTSR began at 32 and steadily rose in each session, peaking at 58 in the final session. The MTER for this group started at 30 and gradually increased to 53, though at a more conservative rate compared to the self-ratings. For the audio group, the MTSR began at 28 and significantly increased to 60 by the experiment's end. The MTER also rose from 28 to 48 but dipped slightly to 44 in the final session, diverging from the trend of self-ratings. For the non-intervention group, the MTSR started at 43 and increased slightly each session, reaching 58 by session 5. The MTER for this group began at 27 and rose to 41, showing the smallest increase among the groups.

As for the specific differences between MTSR and MTER, the video group initially slightly overestimated their performance, with a difference of + 2, while the audio group's self-ratings were consistent with the expert ratings, showing no difference (as shown in Fig. 3). As the experiment progressed to the second session, both groups exhibited a decrease in their tendency to overestimate, with difference of -2 and -4, respectively. Then, both groups returned to a slight overestimation, ranging from + 4 to + 6. Significant deviation occurred once in the fourth session (audio group, diff = + 16) and the fifth session (video group, diff = + 10). Conversely, the non-intervention group exhibited persistent overestimation with minimal convergence between self and expert assessments (diff = + 11 ~ + 17).

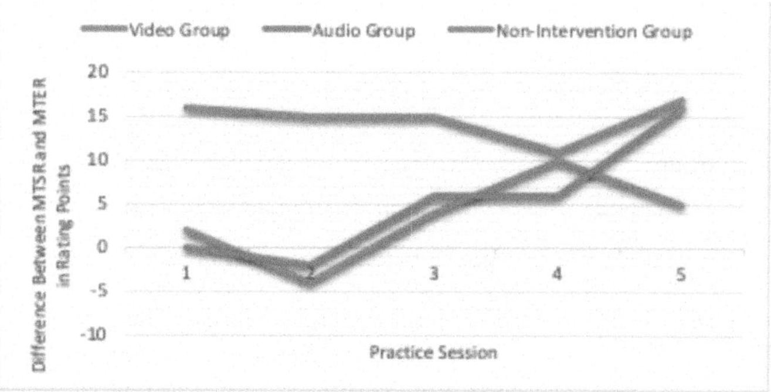

Fig. 3. Difference Between MTSR and MTER Across Five Practice Sessions for Video, Audio, and Non-Intervention Groups

We also analyzed the Spearman's correlation coefficients in order to assess the relationship between MTSR and MTER across the four formal practice tasks for each group (as shown in Tables 2, 3 and 4). Significant positive correlations were found in the audio and non-intervention groups, with coefficients of 0.764 and 0.778, respectively, while the video group had a lower coefficient of 0.484. This suggests a more consistent relationship between self-assessment and expert assessment in the first two groups, and differences in the improvement of perceived accuracy in the latter group.

Table 2. Correlation Coefficients Between MTSR and MTER in the Video Group

Video Group			MTSR	MTER
Spearman's rho	MTSR	Correlation Coefficient	1.000	.484
		Sig. (2-tailed)		.224
		N	8	8
	MTER	Correlation Coefficient	.484	1.000
		Sig. (2-tailed)	.224	
		N	8	8

Table 3. Correlation Coefficients Between MTSR and MTER in the Audio Group

Audio Group			MTSR	MTER
Spearman's rho	MTSR	Correlation Coefficient	1.000	.764*
		Sig. (2-tailed)		.027
		N	8	8
	MTER	Correlation Coefficient	.764*	1.000
		Sig. (2-tailed)	.027	
		N	8	8

*. Correlation is significant at the 0.05 level (2-tailed).

Table 4. Correlation Coefficients Between MTSR and MTER in the Non-Intervention Group

Non-Intervention Group			MTSR	MTER
Spearman's rho	MTSR	Correlation Coefficient	1.000	.778*
		Sig. (2-tailed)		.023
		N	8	8
	MTER	Correlation Coefficient	.778*	1.000
		Sig. (2-tailed)	.023	
		N	8	8

*. Correlation is significant at the 0.05 level (2-tailed).

4.2 Self-assessment and Expert Ratings by Sub-element

In addition to comparing the differences between MTSR and MTER, we compared the differences between mean self-ratings (MSR) and mean expert ratings (MER) for each element of the scale. Subsequently, the frequency (freq) of zero differences (i.e., the number of times the self-ratings and expert ratings were in perfect agreement) was also tracked and compared for each group. We also noted which sub-elements were rated with the most extreme cumulative frequency. These analyses were done to further explore differences in the effects of different recording and playback modalities on self-assessment of different musical elements.

With regard to *tone*, the video and audio groups showed zero difference initially, with minor fluctuations later, while the non-intervention group consistently overestimated (diff = + 1 ~ + 2). For *melody*, video and audio groups were mostly accurate (diff = 0 ~ + 1), except for peaks in later sessions (diff = + 2), whereas the non-intervention group's overestimation grew over time (diff = + 1 ~ + 3). The sub-element of *rhythm* saw high accuracy in the video group (diff = 0), with slight deviations in the pretest. Here, results show complete agreement in the audio group initially, with minor positive differences later; the non-intervention group was mostly consistent but had occasional large overestimations (diff = + 2). In *tempo*, all groups showed significant positive differences (diff = + 4), with the video group fluctuating but aligning closely with expert ratings in later sessions, while the audio and non-intervention groups consistently overestimated (diff = + 4 ~ + 5). On the *technique* side, the video and non-intervention groups consistently overestimated (diff = + 2) except before and after the intervention, while the audio group hovered between slight overestimation and underestimation. Regarding *pedaling*, the audio group showed consistent post-intervention agreement (diff = 0), the video group slightly overestimated and underestimated, and the non-intervention group showed significant overestimation (diff = + 2 ~ + 3). In *interpretation*, the audio and non-intervention groups consistently showed a high degree of accuracy (diff = 0) except for slight inaccuracies at the beginning and end; only the video group showed consistent slight underestimation for this sub-element.

Comparing the frequency of zero differences between the groups after the intervention revealed that the audio group had the highest (freq = 14), the video group had the second highest (freq = 10), and the non-intervention group had the lowest (freq = 4) (as shown in Fig. 4). The frequency of each element in the video group was more extreme, with rhythm in particular appearing four times, while technique and tempo were the lowest (freq = 0). In contrast, the elements of the audio group were more often rated (none of them less than one time), and were especially positively affected in terms of pedalling (freq = 4), interpretation (freq = 3), and rhythm (freq = 3). Among the sub-elements, rhythm ratings were positively affected in both the video and audio groups (freq = 3 ~ 4). However, interpretation and pedalling showed a greater difference across the two groups: both sub-elements were insignificant in the video group (freq = 1) but very significant in the audio group (freq = 3 ~ 4). In contrast, the non-intervention group showed no significant effects overall, with no more than one frequency in any sub-element except interpretation (freq = 2).

4.3 Nuanced Impacts of Audiovisual Feedback Modalities on Self-assessments of Each Sub-elements

In addition to the above quantitative data, we used thematic analyses to analyze the intervention groups' unique insights regarding the interview question "How do you think video/audio recording and playback differentially affected self-assessment for the seven sub-elements?" With this question we hoped to more deeply explore the specific causes and effects of video versus audio interventions on self-assessment in each sub-element.

(1) **Tone:** Participants indicated that both video and audio playback helped them evaluate and adjust specific details of their tone. However, these technical devices have certain limitations, such as the inability to accurately reproduce the tone, which may affect their judgment of the tone during performance. The video group's P01 stated, "While I play, I feel the sound is softer, but the recorded video sounds dry, which is not as pleasant." P06 also commented, "It [the recording] will diminish the variations in your tone." The audio group's P03 mentioned, "[After playing back the video,] I was wondering if I should involve some wrist movement."
(2) **Melody:** Participants using both audio and visual feedback modalities reported that the recordings helped them evaluate melody in more detail, including better assessments regarding the expression and fluency of the melody. For example, participant

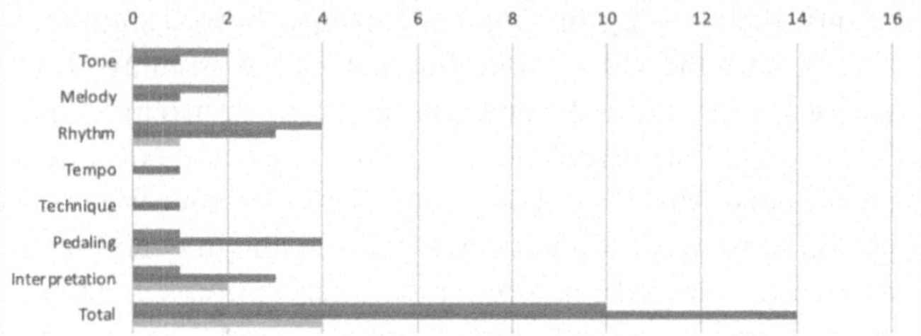

	Total	Interpretation	Pedaling	Technique	Tempo	Rhythm	Melody	Tone
Video	10	1	1	0	0	4	2	2
Audio	14	3	4	1	1	3	1	1
Non	4	2	1	0	0	1	0	0

Fig. 4. Frequencies of 0 Difference in Each Sub-element by Groups in the Post-test

P01 from the video group said, "Without listening [to the recording], you wouldn't know where the phrase ends better." P06 from the audio group noted, "During this imagination, I'm not sure if the effect I played out really amplifies the melodic notes. At this point, I use the recording to verify."

(3) **Rhythm:** The video and non-intervention groups did not express any insights into this sub-element. From the audio group, P06 indicated that "[the audio feedback did not have any impact,] as being a university student itself has the ability to ensure its accuracy."

(4) **Tempo:** Participants using both audio and visual feedback modalities indicated that their recordings did positively impact their self-assessment in this sub-element. However, in terms of self-awareness, the impact of video feedback was greater than that of audio feedback. Participant P01 from the video group stated, "The most important aspect for me is the tempo." Meanwhile, participant P03 from the audio group mentioned, "I might still pay a bit of attention to the tempo."

(5) **Technique:** The audio group was able to notice technical errors and deficiencies in greater detail by listening to their recordings; for instance, P03 identified and corrected a missed break in her performance. Neither participant from the video group reported insights related to this sub-element.

(6) **Pedaling:** The insights received for this sub-element are basically consistent with those in the melody sub-element, namely that participants using either audio or visual feedback modalities reported that their recordings could positively contribute to the evaluation of pedaling techniques. P01 from the video group shared, "When I listened [to the video], the pedal was particularly muddy." Similarly, P06 from the audio group stated, "It [the recording] will amplify [the details], making the timing between pedal changes clearer, or the muddy and discordant parts more apparent."

(7) **Interpretation:** Participants indicated that their audio or visual feedback had a relatively positive impact on their self-assessment in this sub-element. Notably, participants in the video group mentioned that they became aware that both visually and aurally, feedback might affect their judgment of dynamic contrasts, but overall, video feedback played a more facilitative role.

5 Discussion and Suggestions

5.1 Differences in the Impact of Audiovisual Feedback on Overall Self-assessments

These study results show that the two intervention groups, overestimated their self-assessment ratings much less than the non-intervention group. This indicates that both audio and video feedback do help students to more accurately assess their music practice. This aligns with Silveira and Gavin's assertion that playback improves self-assessment accuracy [17]. However, unlike their study, the present study observed fluctuations of accuracy across posttests for both the video and audio groups. These fluctuations may be attributed to the longer, X-week experimental period and more comprehensive sub-elements assessed in our study. This suggests that the duration and scope of audiovisual feedback interventions may have dynamic and unpredictable effects on the accuracy of self-assessment. This is a novel finding compared to existing studies, but needs to be validated by subsequent studies with larger sample sizes.

The persistent overestimation in the non-intervention group aligns with Ehrlinger, Mitchum, and Dweck, who suggest that a lack of external aids can reduce self-evaluation effectiveness [13]. This indicates that external supports like audiovisual feedback can optimize students' self-assessment skills. The video group's differences remained relatively small, while the audio group's differences increased over time. This suggests that the additional visual element provided by the video feedback compared to the audio feedback helped students maintain overall score accuracy. The lower correlation results in the video group might suggest that video group may need time to adapt to and process the extra visual information component over the audio feedback. Additionally, this study's results align with Boucher, who found that visual feedback helps identify technical errors [15]. Combining the insights provided by the video group reveals that their initially larger differences were also due to severe self-criticism induced by video feedback.

5.2 Differences in the Impact of Audiovisual Feedback on Self-assessments for Each Sub-element

By analysing the differences in self-assessment across various sub-elements, it can be inferred that the intervention of two technological feedback modalities had a positive impact on most sub-elements, despite the limitations of participants' devices in capturing sound. In contrast, the non-intervention group, which lacked any external feedback, showed significant overestimation in most sub-elements, with some elements worsening over time.

However, there are some special cases. Firstly, all groups exhibited significant deviations in tempo. Insights from participants indicated that this was due to their inability to use external resources (e.g., metronomes, professional recordings) to help them learn or maintain the intended tempo. Nevertheless, the improvement in accuracy over time in the video group suggested that video feedback can stimulate ongoing self-reflection until the tempo is eventually corrected. This underscores the dual necessity of visual and auditory information when self-assessing tempo. Secondly, all groups demonstrated superior accuracy in rhythm. According to the audio group's interview responses, audio feedback did not significantly aid university-level students' rhythm assessments, as their sense of rhythm had already been internalized through long-term learning experience.

Additionally, the audio group showed more noticeable advantages in accuracy regarding technique, pedaling, and interpretation compared to the video group. The audio group reported that listening to recordings made them more sensitive to the quality of auditory elements such as pedaling, whereas this was not emphasized by the video group. This suggests that audio feedback plays a crucial role in assessing sub-elements primarily involving auditory elements and can enhance students' auditory sensitivity to a higher level. Moreover, only the video group exhibited a persistent underestimation in interpretation, whereas the audio group did not. This indicates that video images may hinder accurate assessment of interpretation and highlights the added value of audio-only feedback for evaluating such non-visual elements.

5.3 Suggestions for Future Research

While this study provides valuable insights and recommendations, its generalizability is limited by the small sample size and the restriction to female participants. Future research should expand the sample size and include a more diverse participant pool to validate and extend these findings. Future studies might also consider updating and refining the self-assessment scale to suit various skill levels of music learners.

6 Conclusion

This study concluded that both video and audio feedback contribute to the effectiveness of self-assessment for higher education musical instrument students. The extent to which each type of feedback impacted the assessment of each musical sub-element varied. In terms of improving the overall self-assessment accuracy of students, video feedback, which was chosen to include both visual and auditory information, was more effective than audio feedback, which included auditory elements alone. However, accurate assessment of auditory information is more dependent on the assistance of audio feedback. Music educators should encourage students to combine both types of audiovisual feedback in self-directed practice situations in order to optimize the effectiveness of self-assessment to a greater extent and help them achieve learning outcomes at a higher level. These findings highlight the importance of incorporating diverse audiovisual feedback modalities in music education, offer empirical support for the value of music practice and assessment, and advance technology-supported music education. Additionally, these insights may be valuable for self-assessment in other self-directed learning fields.

References

1. Boud, D., Lawson, R., Thompson, D.G.: Does student engagement in self-assessment calibrate their judgement over time?. Assess. Eval. High. Educ. **38**(8), 941–956 (2013)
2. Panadero, E., Jonsson, A., Botella, J.: Effects of self-assessment on self-regulated learning and self-efficacy: four meta-analyses.Educ. Res. Rev. **22**, 74–98 (2017)

3. Brown, G., Harris, L.: The future of self-assessment in classroom practice: Reframing self-assessment as a core competency. Frontline Learn. Res. (2014)
4. Powell, S.R.: The influence of video reflection on preservice music teachers' concerns in peer- and field-teaching settings. J. Res. Music Educ. **63**(4), 487–507 (2016)
5. Suzuki, A., Mitchell, H.F.: What makes practice perfect? How tertiary piano students self-regulate play and non-play strategies for performance success. Psychol. Music **50**(2), 611–630 (2022)
6. Hewitt, M.P.: Self-efficacy, self-evaluation, and music performance of secondary-level band students. J. Res. Music Educ. **63**(3), 298–313 (2015)
7. Hamond, L.F., Welch, G., Himonides, E.: The pedagogical use of visual feedback for enhancing dynamics in higher education piano learning and performance. Opus **25**(3), 581–601 (2019)
8. McMillan, J.H., Hearn, J.: Student self-assessment: the key to stronger student motivation and higher achievement. Educ. Horizons **87**(1), 40–49 (2008)
9. Andrade, H.L.: A critical review of research on student self-assessment. Front. Educ. (Lausanne) **4**, 87 (2019)
10. Yan, Z.: Self-assessment in the process of self-regulated learning and its relationship with academic achievement. Assess. Eval. High. Educ. **45**(2), 224–238 (2020)
11. Austin, J.R., Berg, M.H.: Exploring music practice among sixth-grade band and orchestra students. Psychol. Music **34**(4), 535–558 (2006)
12. Waddell, G., Williamon, A.: Technology use and attitudes in music learning. Front. ICT **6**, 11 (2019)
13. Ehrlinger, J., Mitchum, A.L., Dweck, C.S.: Understanding overconfidence: theories of intelligence, preferential attention, and distorted self-assessment. J. Exp. Soc. Psychol. **63**, 94–100 (2016)
14. Tripp, T., Rich, P.: Using video to analyze one's own teaching. Br. J. Educ. Technol. **43**(4), 678–704 (2012)
15. Boucher, M., Creech, A., Dubé, F.: Video feedback and the self-evaluation of college-level guitarists during individual practice. Psychol. Music **49**(2), 159–176 (2021)
16. Johnston, H.: The use of video self-assessment, peer-assessment, and instructor feedback in evaluating conducting skills in music student teachers. Br. J. Music Educ. **10**(1), 57–63 (1993)
17. Silveira, J.M., Gavin, R.: The effect of audio recording and playback on self-assessment among middle school instrumental music students. Psychol. Music **44**(4), 880–892 (2016)https://doi.org/10.1007/978-3-319-21602-7_8
18. Waggoner, D.T.: Effects of listening conditions, error types, and ensemble textures on error detection skills. J. Res. Music Educ. **59**(1), 56–71 (2011)
19. Creswell, J.W., Creswell, J.D.: Research Design: Qualitative, Quantitative, and Mixed Methods Approaches, 6th edn. SAGE Publications Inc. (2023)
20. Saunders, T.C., Holahan, J.M.: Criteria-specific rating scales in the evaluation of high school instrumental performance. J. Res. Music Educ. **45**(2), 259–272 (1997)
21. Creswell, J.W., C. N. Poth.: Qualitative Inquiry & Research Design: Choosing Among Five Approaches, 4th edn. Sage Publications (2018)

Exploring the Influencing Factors of Pupil Gratification Towards Elementary School Library Services: A TAM-Based Study

Qingtang Liu, Jingyi Wu(✉), Ruiqiu Ye, Qing Zheng, and Si Zhang

Faculty of Artificial Intelligence in Education, Central China Normal University, Wuhan, China
1694746822@qq.com

Abstract. Primary school library is an important place for primary school pupils to acquire knowledge, and its service quality directly affects pupils' reading behavior and habits. Understanding the gratification of primary school pupils with the library and its influencing factors is crucial for improving library services. Based on an integrated model of technology acceptance and the model of library user satisfaction and loyalty, we construct a hypothesis model of gratification influencing factors and verify its effectiveness through a questionnaire survey and structural equation model. The results showed that pupils' perceived ease of use and usefulness of the library affects their willingness to use it, and their willingness to use and their own motives further affect their gratification by influencing their behavior in using the library. It is suggested to strengthen library management, optimize resource allocation, teach effective learning strategies, and increase the frequency of library use and reading promotion activities to better meet pupils' reading needs and improve gratification.

Keywords: primary school library · user gratification · influencing factors · path analysis

1 Introduction

In the International Federation of Library Associations and Institutions (IFLA) School Library Guidelines, 2nd Revised Edition [1], it is mentioned that in general, school libraries are changing from a resource-centered model to a learner-centered model. One of its important functions is to serve as an effective interface for the information society, so it is necessary to provide access to digital information resources that reflect both school curriculum and users' interests and culture. The school library should consider creating a "maker space" equipped with computers and other productive equipment needed for practical learning activities. In order to scientifically guide the construction of school libraries, it is necessary to understand students' satisfaction with the library and study the factors that affect students' satisfaction with the library.

Teravainen-Goff and Clark [2] summarized the staffing of British libraries, the provision of library books and technology and the use of libraries through literature review.

Q. Liu et al. (Eds.): EITT 2024, CCIS 2600, pp. 268–280, 2026.
https://doi.org/10.1007/978-981-95-1525-7_22

It only considers students' gender, age, socioeconomic background, and ethnic background, and does not consider the influence of the library itself on students' reading. An anonymous questionnaire survey was conducted among primary school principals, education office directors, and primary school education administrators in Aetolias, Akarnanias, and Attiki [3]. It investigated the infrastructure construction of public and private primary schools in Greece, the existence of printed and non-printed materials in school libraries, and the borrowing situation of libraries, but did not investigate students' evaluation of them. The research of primary school library focuses on the library collection and basic reading environment, and focuses on the views of librarians, ignoring the views of students as the main users. However, in the research results of middle school and university libraries, students' data have been collected by questionnaires according to user satisfaction and user needs, and development suggestions have been put forward accordingly. For instance, Lulu-Pokubo [4] utilized a revised four-point Likert scale to devise their own questionnaire, gathering insights that enabled a profound analysis of the existing issues in school libraries and a discussion on the degree to which pupils recognize the measures taken to enhance reading habits.

However, there is not much research conducted on primary school libraries, resulting in a comparatively limited pool of research results when compared to those of middle school libraries and university libraries. Moreover, the scientificity and innovation are still insufficient, and most of them are qualitative research [5]. A search in CNKI using the keywords "primary school library" and "empirical research" yielded a mere handful of relevant studies, devoid of robust empirical quantitative data. Furthermore, the advent of smart campuses and libraries has integrated various critical elements of information systems—such as resources, technology, information, and services—into primary school libraries [6]. This integration has significantly transformed pupils' reading environments and methodologies. Therefore, it is necessary to understand the pupils' views of and gratification with the library and carry out scientific research on the primary school libraries using the construction of pupil satisfaction.

Based on the integrated model of technology acceptance, we developed a satisfaction model tailored for primary school libraries and then designed a comprehensive questionnaire to collect relevant data. Finally, drawing upon the research outcomes and synthesizing previous suggestions, we aim to offer guidance for the scientific development and construction of primary school libraries.

2 Research Theories and Hypothesis

2.1 Research Theories

(1) Technology Acceptance Model (TAM) and Uses and Gratifications (U&G) Theory

In 1989, Davis [7] made an in-depth analysis of how users accept information systems and established a Technology Acceptance Model (TAM). This model has since served as a pivotal theoretical framework for numerous related studies. TAM is composed of two core components—user motivation and outcome variables. User motivation covers many aspects, including perceived ease of use and perceived usefulness as well as users' overall evaluation and emotional response to technology. Outcome variables include behavioral intention and user behaviors.

While TAM primarily focuses on perceived ease of use and perceived usefulness as key factors influencing technology acceptance, it is acknowledged that the adoption and use of information technology is driven by a multitude of motivations, including not only the desire to complete tasks efficiently but also to satisfy emotional needs and personal goals. To address this gap, Luo, Remus, and Chea [8] skillfully integrated TAM with the Uses & Gratification (U&G) theory. The U&G theory, a paradigm for media use, posits that social influence and personal perspectives drive individuals to engage with specific media to satisfy certain personal needs. Drawing from this combined theoretical foundation, Luo, Remus, and Chea [8] constructed an integrated model of technology acceptance that delved into the influences of both internal and external motivations on users' acceptance of Internet information services. Their research was validated through a questionnaire survey, demonstrating the model's excellent adaptability and stability. Building upon this integrated model, the present paper aims to explore and validate the various factors and mechanisms that influence gratification with primary school libraries.

(2) Model of Library User Satisfaction and Loyalty

In 2003, Martensen and Grønholdt [9] put forward a model of library user satisfaction and loyalty, which aims to evaluate library users' perception, satisfaction, and loyalty to library service quality through quantitative methods. Guan [10] integrated and improved the variables of the model of library user satisfaction and loyalty and explored the relationship between the six variables of library collection resources, librarian service, environmental atmosphere, hardware facilities, daily management and education and training, and user gratification. In this study, the relevant influencing factors that can reflect the characteristics of primary school libraries were selected and integrated into the perceived usefulness and perceived ease of use in a questionnaire. Moreover, motivation is used to measure the influence of primary school students' internal needs and subjective conditions on their use behavior and satisfaction.

2.2 Research Hypothesis

(1) Pupils' Behavioral Intentions

Davis [7] defines perceived usefulness as the degree of perception that an individual can improve his work efficiency when using a specific system. He defines perceived ease of use as the degree to which individuals think that a particular system is easy to operate and does not need too much effort when evaluating its use. Chen et al. [11] argues that if the system is easy for users to understand and operate, the probability of users adopting this technology or system will be higher. And if users can easily understand and operate the system, their tendency to adopt related technologies or systems will increase significantly. Tian's [12] research underscores that college pupils' perception of ease of use in digital academic reading positively correlates with their perceived usefulness. Based on these perspectives, we propose the following hypotheses:

H1: Pupils' perceived ease of use positively predicts perceived usefulness.

H2: Pupils' perceived usefulness positively predicts their behavioral intention.

H3: Pupils' perceived ease of use positively predicts their behavioral intention.

(2) Pupils' Behavioral Usage

In the realm of library science, numerous studies underscore the crucial role of users' intention to use, which significantly shapes their subsequent behavior. Fishbein, reported in Ji, Li, and Zhu [13] emphasized that among various factors influencing use behavior, behavioral intention stands out as the most significant predictor. Similarly, Zhang et al. [14] revealed that users' willingness to utilize digital reading promotion services in libraries positively enhances their actual usage. Papacharissi and Rubin [15] further underscored motivation as an inherent drive to fulfill personal needs and aspirations.

In the context of this study, pupils' intention to use the library is mainly influenced by the library's objective conditions, while motivation stems from either social influence or internal desires. Both intention and motivation are key drivers that influence pupils' behavior towards library usage. This leads to the following hypotheses:

H4: Pupils' behavioral intention positively predicts their behavioral usage.

H5: Pupils' motivations positively predict their behavioral usage.

(3) Pupils' Gratifications

Chen [16] posits that gratification is an internal emotional state stemming from readers' comparison of their psychological expectations with their actual experiences during the utilization of library services. Similarly, Hecht [17] argues that "gratification" originates from the emotional response triggered by the positive reinforcement of one's actions.

In the context of this study, gratification for primary school pupils regarding the library reflects their emotional response after interacting with and experiencing the library's offerings. This response serves as a subjective evaluation of the library's infrastructure and service quality. The frequency of visits may indicate that students are using resources effectively and are satisfied with the library [18]. Therefore, we formulate the following hypothesis:

H6: Pupils' behavioral usage positively influences their gratification.

3 Methods

Based on the integrated model of technology acceptance and the model of library user satisfaction and loyalty, this study fully considers the characteristics of primary school libraries and establishes a model of primary school library gratification with six variables: perceived usefulness, perceived ease of use, behavioral intention, behavioral usage, motivation, and user gratification (Fig. 1). Furthermore, this study tests six hypotheses.

3.1 Research Design

(1) Instruments

In this study, we utilized a questionnaire survey methodology. Prior to distribution, the questionnaire underwent rigorous review and refinement by experts and teachers. Additionally, a small-scale pilot test was conducted to assess its effectiveness. Based on the feedback received and taking into account the unique characteristics of primary

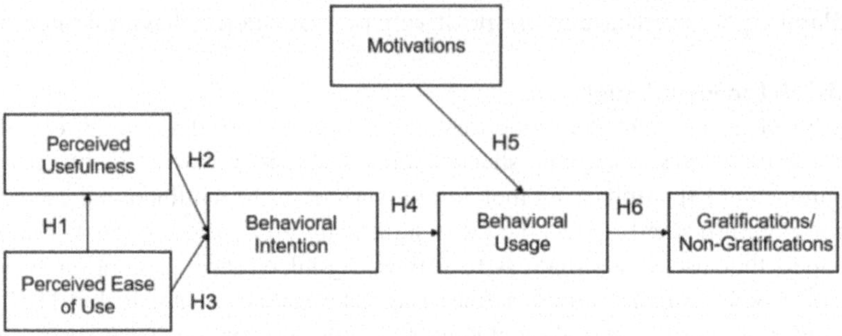

Fig. 1. Hypothesized research model.

school libraries, we adjusted the number, order, and phrasing of the questions in the questionnaire, ultimately culminating in a refined test questionnaire. The questionnaire consists of two parts: the first part investigates the demographic characteristics of the sample participants; the second part consists of items related to observation variables, using a five-point Likert scale format.

1) *Perceived Ease of Use.* Perceived Ease of Use is the degree to which individuals think that a particular system is easy to operate and does not need too much effort when evaluating its use [7]. In this study, we use environmental facilities, quick usage, and daily management to make pupils understand the perceived ease of use.
2) *Perceived Usefulness.* Perceptual usefulness is how well a user uses a technology or a system to improve their work ability or performance [10]. In this study, we use knowledge acquisition, reading ability, and information literacy to make pupils understand the perceived usefulness.
3) *Behavioral Intention.* It is the tendency of users to continue to use the library for additional tasks after using the library for an original task [10]. In this study, we consider it from three angles: willingness to use, continuation of usage, and recommendation for usage.
4) *Motivations.* Reading motivation refers to the internal mental activity and internal dynamic processes of individual reading activities that are guided, stimulated, and maintained by the goals related to reading [19]. In this study, we let pupils evaluate their motivations from the perspective of autonomous activity and information acquisition.
5) *Behavioral Usage.* Behavioral usage involves amount of use, duration of use, and type of use [15]. We use frequency, current status, and borrowing status as the three variables to observe behavioral usage.
6) *Gratifications / Non-Gratifications.* If the user feels the same as he expected in the process of using the library, that is, if the actual performance is in line with expectations and if what is gained in the process is in line with the input, then this will produce a happy mood and a sense of satisfaction [20]. If the performance is not in line with expectations, then the user will not feel a sense of satisfaction [20]. In this study, we designed the library collections, environmental facilities, personnel services, and system items to evaluate pupils' gratification.

Table 1. Index system affecting the gratification of primary school library Use

Latent Variable	Observed Indicators
Perceived Ease of Use	Environmental Facilities
	Quick Usage
	Daily Management
Perceived Usefulness	Knowledge Acquisition
	Reading Ability
	Information Literacy
Behavioral Intention	Willingness to Use
	Continuation of Usage
	Recommendation for Usage
Behavioral Usage	Frequency
	Current Status
	Borrowing Status
Motivations	Autonomous Activity
	Information Acquisition
Gratifications/Non-Gratifications	Library Collections
	Environmental Facilities
	Personnel Services
	System

The latent variables of experimental interest and the corresponding observed indicators are shown in Table 1. In terms of the dimensions encompassing behavioral usage, motivations, perceived ease of use, and behavioral intention, we primarily drew upon the meticulously crafted questionnaire by Luo, Remus, and Chea [8]. As for the dimensions of perceived usefulness and gratifications, we have adopted and modified the respective questionnaires designed by Davis [7] and by Teng [20]. During the design of the questionnaire, for concepts that are difficult for primary school students to understand, we used examples to facilitate their comprehension. For instance, when measuring information literacy in the dimension of perceived usefulness, we used "Using the school library has improved my information literacy. For example, I can now find the information I want more easily and analyze and judge the information I get better." as the question stem. For variables with broader meanings and contents, we summarized them using aspects that primary school students are familiar with in their daily lives. For example, when it comes to the information acquisition aspect of the motivations dimension, we used "I go to the school library to satisfy my desire for books and knowledge."

3.2 Data Analysis

In this research, five primary school libraries in a specific region of Wuhan were targeted as the focal points for investigation, with a particular focus on pupils in grades four to six. All selected schools have established a certain level of smart library infrastructure. The survey yielded a complete recovery of all on-site distributed pupil questionnaires, totaling 441. After rigorous scrutiny, 102 unqualified responses were eliminated, ultimately yielding a final count of 339 valid questionnaires, which constituted 76.9% of the total amount distributed. For statistical analysis and validation of the conceptual model, SPSS 21.0 and AMOS 24.0 were primarily employed in this study.

4 Results

4.1 Description of Sample Characteristic Distribution

Regarding the gender distribution, males accounted for 53.7% of the respondents, while females comprised 45.4%, indicating a slight male dominance. In terms of grade distribution, fourth-grade pupils made up 27.4% of the respondents, fifth-grade pupils 33%, and sixth-grade pupils 39.5%. The data suggest a relatively balanced gender ratio, with the majority being from the sixth grade, but an overall even distribution across grades.

4.2 Reliability Test

During this research, the key factors influencing pupil gratification with primary school libraries were measured using scales. To ensure the accuracy and reliability of the measurement results, we conducted a Cronbach's α reliability test to conduct an in-depth analysis of the internal consistency of each dimension. The Cronbach's alpha coefficient of the primary school library application scale was greater than 0.7, indicating that the scale used in this study exhibits good internal consistency [21].

4.3 Validity Test

In this study, a preliminary analysis of the sample data was conducted using SPSS 21.0 software, specifically involving the calculation of the KMO statistic and the implementation of Barlett's test of sphericity, to verify the applicability and reliability of the data. KMO greater than 0.8 is considered to have an acceptable measure [22]. The calculated KMO value was 0.934, which is greater than 0.8, and the p-value was less than 0.001, indicating good validity.

Furthermore, the convergent validity and discriminant validity of the model were considered. To ensure excellent convergent validity, i.e., the ability of the scale to accurately capture the measured constructs, we required that the factor loadings and composite reliability (CR values) must reach a high level of 0.7 or above, and the average variance extracted (AVE) should also remain above 0.5 [23]. To ensure outstanding discriminant validity, it was necessary to ensure that the square root of the AVE value of each influencing factor significantly exceeds the correlation coefficients between all other influencing factors [24]. Table 2 shows the results of distinctive validity tests for dimensions of the impact factor scale. According to the results stated in the table, the test results indicate that the scale in this study exhibits good convergent and discriminant validity.

4.4 Structural Model and Hypothesis Testing

(1) Goodness of Fit Test

Table 3 presents the model fit test results. The CMIN/DF (chi-square to degrees of freedom ratio) is 4.14, falling within the range of 3–5. The RMSEA (root mean square error of approximation) is 0.08, approximately within the good range. The test results for IFI, TLI, and CFI have all reached satisfactory levels. It can be concluded that the constructed model for influencing factors of gratification in primary school libraries in this study demonstrates a good fit.

(2) Path Analysis

If the CR value (Critical Ratio, the ratio of the path coefficient to its standard error) is greater than 1.96, and the p-value is less than 0.05 [25], then these values indicate that the relationship path is significant. In the path hypothesis relationship test of this study, all the hypotheses were confirmed. Based on the analysis results, a Structural Equation Modeling (SEM) path relationship diagram of the influencing factors of primary school library gratification was drawn as shown in Fig. 2.

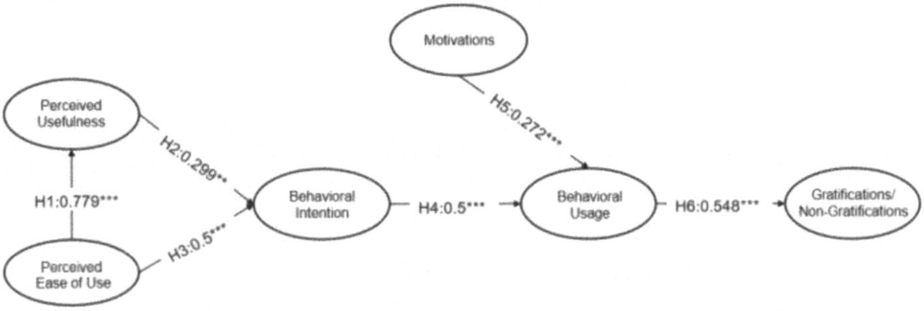

Fig. 2. SEM path relationship of influencing factors of primary school library gratification.

Figure 2 is the SEM path relationship of influencing factors of primary school library gratification. According to the results shown in Fig. 2, pupils' perceived ease of use positively predicts perceived usefulness ($\beta = 0.779$, $p < 0.001$). Therefore, hypothesis H1 is valid. Furthermore, pupils' perceived usefulness positively predicts their behavioral intention ($\beta = 0.299$, $p < 0.01$), confirming the validity of hypothesis H2. Pupils' perceived ease of use positively predicts their behavioral intention ($\beta = 0.5$, $p < 0.001$). Therefore, hypothesis H3 holds. Moreover pupils' behavioral intention positively predicts their behavioral usage ($\beta = 0.5$, $p < 0.001$). Therefore, hypothesis H4 holds. Pupils' motivations positively predict their behavioral usage ($\beta = 0.272$, $p < 0.001$), validating hypothesis H5. Finally, pupils' behavioral usage positively influences their gratification ($\beta = 0.548$, $p < 0.001$). Therefore, hypothesis H6 holds.

Table 2. Distinctive validity test results for IF scale dimensions.

Variant	Behavioral usage	Motivations	Perceived Usefulness	Perceived Ease of Use	Gratifications	Behavioral Intention
Behavioral usage	0.847					
Motivations	0.492	0.800				
Perceived Usefulness	0.451	0.767	0.829			
Perceived Ease of Use	0.562	0.688	0.781	0.774		
Gratifications	0.507	0.763	0.764	0.758	0.771	
Behavioral Intention	0.528	0.626	0.687	0.716	0.762	0.791

Table 3. SEM Model fitness test.

Indicator	Reference Standard	Measured Result
CMIN/DF	1–3 is excellent, 3–5 is good	4.14
RMSEA	<0.05 is excellent, <0.10 is good	0.08
IFI	>0.9 is excellent, >0.8 is good	0.836
TLI	>0.9 is excellent, >0.8 is good	0.805
CFI	>0.9 is excellent, >0.8 is good	0.835

5 Discussion

5.1 Influencing Factors and Relationship of Behavioral Intention

Perceived ease of use can positively affect perceived usefulness, which has been demonstrated in many structural equation model studies, and it is not difficult to understand it in the field of primary school libraries. If the books in the library are clearly catalogued and neatly placed, students can quickly obtain the books they want and find the characteristics of importance and usefulness in the library. Therefore, when formulating, developing, and updating library rules, it is important to ensure their ease of use.

The important discovery of this study is that pupils' perceived usefulness and perceived ease of use positively predicts their behavioral intention. Library entrance education for freshmen is a good opportunity to shape students' awareness of the usefulness and ease of use of libraries. In library entrance education, we should focus on letting all primary school pupils know about the use of the library and help them to realize that they can easily learn to use it, and then gradually realize that using the library to collect information and absorb knowledge can improve the reliability of reading ability.

Libraries provide long-term services for pupils. When pupils report that they have difficulty in finding books, recommending books to them can promote their reading [26]. If the library can implement standardized management, deeply analyze the background borrowing data, and accurately recommend books to primary school pupils so that they can easily obtain books, it will promote their intention to use the library.

5.2 Influencing Factors and Relationship of Behavioral Usage

This study proves that behavioral intention is the antecedent variable of behavioral usage, which is positively influenced by pupils' perception of the usefulness and ease of use of the library. Synthesizing the previous discussion points, it can be concluded that standardizing the management of the library, scientifically improving the service level of the library, and then enhancing the primary school pupils' intention to use the library can improve their library use behavior.

Motivation is another important influencing factor of behavioral usage, and our research demonstrates that pupils' motivations positively predict their behavioral usage. This finding is also consistent with previous studies. With the support and encouragement of parents and society, primary school pupils take the initiative to enter the library and keep reading, which behavior is expected show great satisfaction with book knowledge and the library [27]. This does not mean that the library has no idea about pupils' motivations. The primary library can influence parents' ideas and family reading environment by holding parent-child activities, thus playing a positive role in pupils' motivation.

5.3 Influencing Factors and Relationship of Gratification

Pupils' behavioral usage positively influences their gratification. The increase of use behavior can improve satisfaction to a certain extent, so it is very necessary to lead primary school pupils to use the library and extend their borrowing time. Library visits may be one of the most important variables to seek intervention measures to strengthen pupils' reading frequency and attitude, and it can be promoted by influential social contacts such as teachers and parents [28].

In a word, pupils' perception of the ease of use and usefulness of the library affects their willingness to use the library, and their willingness to use and their own motives further affect their satisfaction by influencing their behavior of using the library. Based on this finding, we put forward the following suggestions for perceived ease of use, perceived usefulness, and motivation. Library cataloging should be carried out reasonably, and the collection structure in the library should be clear, so that primary school pupils can easily know the location of the books they need. Daily standardized management of the library will make primary school pupils realize that the library is a place where they can easily acquire knowledge and find the information they want, which will further affect their use intention. During library entrance education, we should expound the definition, collection resources, functional orientation, and significance of the library, and we should teach primary school pupils how to effectively use library resources to assist self-learning. We should let primary school pupils really feel that the book resources in the library are what they need and are of high quality. We should also help them understand the importance of the library and the great help that the library can provide for themselves. Finally, the school library should play a leading role in planning and guiding reading promotion activities, linking families, schools, and society to create a strong reading culture, stimulating primary school pupils' reading enthusiasm and helping them form good reading habits.

6 Limitations and Conclusions

From a theoretical perspective, this paper applies the integrated model of technology acceptance to the field of primary school libraries and verifies the influencing factors and path structures that affect gratification with primary school libraries. This research on gratification with primary school libraries can fill a gap in empirical research. From a practical perspective, this paper combines previous summary opinions on the improvement of service levels and development opportunities for primary school libraries, providing insights to scientifically guide the construction and updating of primary school libraries.

This study finds that pupils' perception of the ease of use and usefulness of the library affects their willingness to use the library, and their willingness to use and their own motives further affect their satisfaction by influencing their behavior of using the library. In the formulation and development of library rules, ensuring ease of use should be an important prerequisite. Standardization of the management of library book resources and precise allocation is critical. The functions of the library and how to use it efficiently for knowledge learning and ability enhancement should be communicated during orientation training for primary school pupils. It is necessary to frequently guide primary school pupils to use the library and extend their borrowing duration, as increased usage behavior can lead to a certain degree of gratification improvement. School libraries should play a leading role in planning and guiding reading promotion activities, collaborating with families, schools, and society to jointly create a strong reading culture, inspire primary school pupils' enthusiasm for reading, and help them form good reading habits.

There are also certain limitations in this study. For instance, among the primary schools surveyed during the field research, some were still in the initial stages of having just completed the digitization of their libraries. Consequently, there were limitations in terms of the opening hours and the target audience served by these libraries. Recognizing these constraints, the schools invited us to conduct a follow-up visit and conduct further research in the future. Additionally, during the actual survey, we discovered that some items highly related to library intelligence had to be excluded due to issues related to their validity.

Acknowledgments. This work was supported by National Natural Science Foundation of China 'Research on Instructional Interaction Behavior Recognition and Pattern Mining in Informationization Classroom Integrating Multimodal Data' (grant number: 62277021) and the Humanities and Social Sciences Research Project of the Ministry of Education of China (grant number: 22YJAZH067).

References

1. Oberg, D., Schultz-Jones, B.: IFLA school library guidelines, 2nd revised edition (2015). https://repository.ifla.org/handle/20.500.14598/58
2. Teravainen-Goff, A., Clark, C.: School libraries: a literature review on current provision and evidence of impact. National Literacy Trust (2017)
3. Saitis, C., Saiti, A.: School libraries in Greece: a comparative study of public primary schools in rural and urban areas. Libr. Inf. Sci. Res. **26**(2), 201–220 (2004)
4. Lulu-Pokubo, E.: Challenges of school libraries. Int. J. Libr. Sci. **17**, 62–71 (2022)
5. Paton-Ash, M., Wilmot, D.: Issues and challenges facing school libraries: a case study of selected primary schools in Gauteng Province, South Africa. South Afr. J. Educ. 35(1) (2015)
6. Boell, S., Cecez-Kecmanovic, D.: What is an information system? In: 2015 48th Hawaii International Conference on System Sciences (2015)
7. Davis, F.: Perceived usefulness, perceived ease of use, and user acceptance of information technology. MIS Q. 13 (1989). https://doi.org/10.2307/249008
8. Luo, M.M.-L., Remus, W., Chea, S.: Technology acceptance of internet-based information services: an integrated model of TAM and U&G theory. In: Technology Acceptance of Internet-based Information Services (2020)
9. Martensen, A., Grønholdt, L.: Improving library users' perceived quality, satisfaction, and loyalty: an integrated measurement and management system. J. Acad. Librariansh. **29**(3), 140–147 (2003)
10. Guan, L.: Research on continuous use of university libraries —— Perspective of library user satisfaction model and TAM model. Library 96–104 (2020)
11. Chen, R., et al.: Study on the structural relationship of influencing factors of users' willingness to continue using mobile libraries based on ISM. Res. Libr. Sci. 93–101 (2021)
12. Tian, T.: Study on the influencing factors of college students' willingness to use digital academic reading based on TAM. Educ. Inf. Technol. 68–72 (2024)
13. Ji, Z., Li, B., Zhu, J.: Study on the influencing factors of user behavior of smart libraries. Library 17–21 (2018)
14. Zhang, N., et al.: Study on the influencing factors of users' willingness to use digital reading promotion services in libraries-based on TAM/AISAS model. Libr. Work Study 18–28 (2023)
15. Papacharissi, Z., Rubin, A.M.: Predictors of internet use. J. Broadcast. Electron. Media **44**(2), 175–196 (2000)
16. Chen, L.: Research on influencing factors of service gratification of primary and secondary school students in public libraries under the background of 'double reduction'. M.S. thesis, Nanchang Univ, Nanchang, China (2023)
17. Hecht, M.L.: Toward a conceptualization of communication gratification. Q. J. Speech **64**, 47–62 (1978)
18. Fagyan, C.K., et al.: The impact of library use frequency on student satisfaction: an evaluation of resources, services, and facilities. Iconic Res. Eng. J. **7**(1)
19. Ge, M., Zhao, Y.: Analysis of differences in teenagers' reading motivation and training strategies. Res. Libr. Sci. (2011)
20. Teng, H.: College students' use of library resources satisfaction survey research; Take H normal university as an example. M.S. thesis, Harbin Normal Univ, Harbin, China (2016)
21. Taber, K.S.: The use of Cronbach's alpha when developing and reporting research instruments in science education. Res. Sci. Educ. **48**, 1273–1296 (2018)
22. Wu, R.M.X., et al.: A comparative analysis of the principal component analysis and entropy weight methods to establish the indexing measurement. PLoS One **17**(1) (2022)
23. Hair, J.F., Black, W.C., Babin, B.J., Anderson, R.E.: Multivariate data analysis: a global perspective. In: Multivariate Data Analysis, 7th Edition (2010)

24. Shiau, W.-L., Yuan, Y., Pu, X., Ray, S., Chen, C.: Understanding fintech continuance: perspectives from self-efficacy and ECT-IS theories. Ind. Manag. Data Syst. **120**(9), 1659–1689 (2020)
25. Ding, Z.: Empirical study on the influencing factor model of customer satisfaction of the red special collection of libraries. M.S. thesis, Nanchang Univ, Nanchang, China (2023)
26. Loh, C.E., Gan, S., Mounsey, S.: What do children want to read? A case study of how one primary school library supported reading for pleasure. J. Libr. Adm. **62**(7), 931–945 (2022)
27. Bracken, S.S., Fischel, J.E.: Family reading behavior and early literacy skills in preschool children from low-income backgrounds. Early Educ. Dev. **19**(1), 45–67 (2008)
28. Mat-Roni, S., Merga, M.K.: The influence of extrinsic and intrinsic variables on children's reading frequency and attitudes: an exploration using an artificial neural network. Aust. J. Educ. **63**(3), 270–291 (2019)

What Characteristics of Social Media Make Teachers' Learning and Training Effective

Ying Zhang, Yu Zhao, Zhongling Liu, Xiuhan Li(✉), and Xiangyang Cui

Faculty of Artificial Intelligence in Education, Central China Normal University, Wuhan, China
2021214448cxy@mails.ccnu.edu.cn

Abstract. Social media, which enables instant access to resources and assistance while promoting collaboration among educators through the formation of various communities of practice (CoPs), is increasingly being used to facilitate lifelong learning and professional development for teachers. However, not all social media–based CoPs yield satisfactory outcomes for teachers, and there is a lack of comparison of the disparities between effective and ineffective communities. This study conducted a mixed survey (questionnaire and interview) among primary and secondary school teachers in mainland China to explore community characteristics that contribute to teachers' effective learning and training. The findings indicate that significant distinctions were perceived in the characteristics of domain, community, and practice between the highly satisfying CoPs and those that elicited lower satisfaction levels. Variables such as social connection, thematic focus, interaction patterns, offline support, and leadership were found to play a crucial role in determining the effectiveness of social media–based CoPs. Factors like region, gender, or member quantity were less important.

Keywords: Teachers' Learning · Social Media · Communities of Practice

1 Introduction

In the information age, teachers' in-service autonomous learning has become increasingly important. China has a significant number of primary and secondary school teachers, but there is a lack of unified official training opportunities for them. The widespread use of social media provides a pathway and resources for teachers to engage in autonomous learning. Social media provides a platform of communication, resource sharing, and practical collaboration for teachers' professional development and lifelong learning, which has given rise to an instant, demand-driven, peer collaboration-based, spontaneous mode of teacher education. More and more teachers are sharing their experiences, confusions, emotions, and resources on social media and have formed social media–based communities of practice (CoPs). However, due to the informal nature, dynamism, and diversity of such CoPs, their role and effectiveness in teachers' learning are still subject to significant questioning and challenges. A key issue in current research has thus been how to leverage the advantages of social media effectively to promote teachers' professional development and enhance the effectiveness of social

Q. Liu et al. (Eds.): EITT 2024, CCIS 2600, pp. 281–288, 2026.
https://doi.org/10.1007/978-981-95-1525-7_23

media–based CoPs. Many high-quality, social media–based CoPs provide great support in solving instructional problems and enhancing teaching literacy, thus receiving wide acclaim and satisfaction from teacher participants. This study explored teacher participants' actual experiences with social media–based CoPs to investigate highly recognized communities with high satisfaction levels and to determine what characteristics those communities possess. The results can provide a reference for constructing high-quality, social media–based CoPs and promoting teachers' professional development.

2 Literature Review

Social media–based CoPs, despite being widely acknowledged to offer tangible benefits for teachers' professional development, also create challenges and potential barriers. As such, teachers have mixed opinions about the actual effectiveness of social media–based CoPs. Communities experience a lifecycle that is influenced by various factors. Xing and Gao, for example, identified three dimensions—social, interactive, and cognitive—as discourse features that influence teachers' sustained engagement in professional learning through Twitter [1]. Ghamrawi proposed a different set of three factors—personal, interpersonal, and background—as elements influencing the effects of teacher CoPs [2]. First, members' career motivation and practice needs affect their engagement and contribution to social media–based CoPs. For instance, several studies found a lack of internal motivation and external incentives among members to participate in community activities [3–5]. Second, interpersonal factors encompass continual communication opportunities, social connection networking possibilities, and the presence of a visible and capable leader [6]. Third, the effectiveness of a CoP may be influenced by various background factors, such as external support, well-structured tasks, a positive e-culture, permitted differentiation, and easily accessible materials and activities that avoid being overly theoretical. Additionally, knowledge construction and management within social media–based CoPs is a crucial issue that has a significant impact on their effectiveness. Teachers' participation and engagement in social media–based CoPs are typically not controlled by the creators or moderators of the group [4]; thus, there tends to be limited control, assessment, or routine evaluation of engagement in professional development by those delivering the resources [7].

Research on teachers' online learning has indicated that social media, as a potent tool for instant communication, is increasingly being used to support teachers' professional development activities. However, teachers have diverse assessments of and levels of satisfaction with different social media–based CoPs. The primary factor contributing to this disparity lies in the spontaneity and informality inherent in CoPs, which leads to variations in their actual efficacy. There has only been limited research describing the specific characteristics and construction of high-quality social media–based CoPs for teachers. Current research on teachers' CoPs has also primarily focused on higher education instructors. It is thus a crucial concern to determine how to establish efficient and sustainable teacher CoPs for primary and secondary education. Accordingly, this study concentrated on primary and secondary school teachers as research subjects, employing a case study approach to solicit their evaluation of social media–based CoPs and report the distinguishing features of two CoP cases that elicited their highest and lowest levels of satisfaction.

In light of the above, it is possible to draw a few conclusions about the construction of high-quality CoPs for teachers. Following Wenger's definition of CoPs, we have defined teachers' CoP as an online community in which members interact to enhance their knowledge and expertise, as well as to improve teaching practices [8].

3 Methodology

3.1 Survey

This study used a survey with mixed instruments to explore the factors influencing teachers' learning experience from social media communities. The survey included questionnaires and interviews. The questionnaire survey aims to investigate teachers' participation in social media–based CoPs and their evaluation of the usefulness of such CoPs. The questionnaire was developed by two researchers in teacher education and one consultant with expertise in survey research. Several items in the questionnaire were adapted from previous frequently cited studies. For instance, the items concerning the impacts of social media–based CoPs were derived from Patton and Parker's comprehensive synthesis of the four key roles played by CoPs in facilitating teachers' professional development: knowledge acquisition, skill enhancement and competence building, promotion of collaborative practice, and provision of affective support [9].

The interviews focused on teachers' perceptions of influential factors for learning effectiveness of social media-based CoPs. The interviews were conducted in a semi-structured manner, employing focus groups. The interview question outlines are listed as follows:

(1) Which do you think is more effective: social media-based CoPs or other teacher professional development approaches? Why?
(2) Under what circumstances would you join a social media-based CoP voluntarily?
(3) How can the effectiveness of a teacher's social media-based CoP be evaluated in your opinion?
(4) What significant characteristics do you believe an effective teacher's social media-based CoP should have? You can explain it from community structure, members, activities, and other features.
(5) What do you think enhances the effectiveness of teachers' media-based CoPs?

3.2 Sampling

This study employed a case sampling method to distribute the questionnaires, with sampling procedures designed to ensure an equitable chance of selecting K–12 teachers from different areas. Ensuring equal selection opportunities is crucial for obtaining a probabilistic sample while minimizing response bias to allow the drawing of inferences about the teacher population based on survey results. This study selected a total of eight representative elementary and secondary schools from an administrative region (X) of a first-tier city in central China. A questionnaire survey link was distributed to the teachers' networks within these eight schools, which resulted in responses from a total of 255 K–12 teachers. The inclusion criteria for the sample required that participating teachers possess an understanding of teachers' CoPs and have prior experience engaging them. Of the 255 respondents, 206 were female, with 63% being elementary school teachers and 37% being secondary school teachers. The average age among participants was approximately 30 years old; only a small proportion (11%) exceeded the age of 40.

We asked the respondents at the end of the questionnaire if they are willing to participate in a further interview. A total of thirty-two respondents (including twelve males) agreed to attend the focus-group interviews. We implemented seven focus-group interviews to collect qualitative data.

3.3 Data Analysis

The quantitative data from the questionnaire were analyzed using SPSS 24.0. Simple descriptive statistics were applied to the forced-choice questionnaire items and Likert scales. Mean scores and standard deviations were used to present teachers' perceptions of the various subscale items. Frequencies and percentages were calculated for all respondents.

Interview data were recorded by voice recorders and worksheets. All qualitative data were coded by two independent coders through NVivo 10. The coding schemes for interviews were developed based on Wenger's CoP theory: domain, community, and practice [10]. Domain-related strategies refer to making the theme, goals, scopes, and content of the community more specific. Community-related strategies focus on facilitating members coming together around ideas or topics of interest (the domain) and interacting with each other to learn and develop. Practice-based strategies emphasize the resource production and knowledge innovation that members share and develop together. The inter-rater reliability of coding was measured by Cohen's Kappa Coefficient (the percentage of agreement between coders) in NVivo 10. The technique of iterative discussion between two coders was used until the Kappa Coefficient of each node was above 0.75, which was considered excellent agreement [11]. Throughout the coding process, constant communication between the coders led to adjustments resulting in a higher Kappa value of 0.827.

4 Results

4.1 Questionnaire Results

Teachers' Participation in Social Media–based CoPs. Because WeChat and QQ are the most widely known social media software in China, we investigated teachers' use of these two platforms for online CoPs. The quantity of teachers' CoPs established through social media and their different characteristics. The average number of CoPs in which the respondents were involved through these two social media platforms is 5.13, with an SD of 3.89. Female respondents have an average of 1.91 more CoPs than male respondents. Respondents in primary schools reported 5.79 social media–based CoPs, while those in middle schools reported 3.89. No significant correlation was found between respondents' age and their social media–based CoPs use.

Teachers' Evaluation of the Usefulness of Social Media–based CoPs. Respondents were asked to rate the usefulness of social media–based CoPs activities compared with other forms of professional development activities. The results are shown in Table 1. Respondents considered social media–based CoPs to be the second most useful professional development activity after school-based training activities. National or provincial training activities were a close third, at 3.65. MOOCs related to teacher education came last, which were less constrained.

Table 1. Perceived usefulness of teachers' professional development activities.

Learning Forms	Usefulness Ranking
School-based training activities	2.90[a]
Social media–based CoP activities	3.56
National/provincial training courses	3.65
MOOCs	4.34

a. The smaller the ranking, the more the usefulness; CoP, community of practice; MOOCs, massive open online courses.

Influencing Factors of Teachers' Engagement in Social Media–based Communities of Practice. There are many factors that can influence a teacher's participation in social media–based CoPs. Respondents' perceived ratings for the importance of these factors are shown in Table 2. According to the respondents, group members' social connection and leadership play the most important role in facilitating engagement in social media–based CoPs; these both had a rating of 2.69, followed by the factor of "themes" (3.15), which indicates that respondents consider the value of the practice topics and content in the CoPs. The following influential factors were all related to the CoP member characteristics, such as members' geographical location and discipline background, which ranked third and fourth respectively. Members' age was at the bottom of the list, and its importance ranking coefficient was 4.89.

Table 2. Perceived ranking of factors influencing their engagement in social media–based communities of practice.

Influential Factors	Importance Ranking
Social connection	2.69[b]
Leadership	2.69
Themes	3.15
Members' geographical location	3.52
Members' discipline background	4.05
Members' age	4.89

b. The smaller the number of "ratings", the greater the importance.

4.2 Interview Results

Participants' interview responses were coded based on Wenger's CoP theory. A total of 51 valid responses were collected and further coded from three dimensions: domain, community, and practice. As shown in Table 3, teachers mentioned the strategies of effective communities (N = 28) most often, followed by effective practice (N = 16), and finally effective domains (N = 7). They also mentioned 16 other strategies.

First, and most importantly, the respondents think that the key to the success of a social media–based CoP is enhancing its community-based features. The main purpose of community-based strategies is to increase members' interactions, which results in enhanced membership and practice outcomes.

Table 3. Coding of strategies for enhancing the effectiveness of teachers' online learning communities.

Dimension	Frequency	Themes
Domain	7	Topic, content, demand, field, target
Community	28	membership, management, regulation, leadership, activities, face-to-face/offline support, communication
Practice	16	resources, cases, evaluation, knowledge management

Four respondents unanimously put forward the need to carry out regular activities (including both online and offline) that are helpful for teachers' development. These activities, they emphasized, should be in various forms, such as expert lectures, training sessions, teaching, and research. Sharing and discussing the activities is essentially beneficial for teachers' professional development. Several respondents highlighted the importance of offline activities that could promote their engagement, enhance their social connections, and facilitate communication. One teacher proposed the integration of gamification elements into the activities to increase members' engagement, and shared a CoP case that included various gamified activities. Several management rules of social media–based CoPs to maintain their initiatives, such as clock-in, on-duty, or red-packet, were described by 35 teachers, who also thought that having a responsible and knowledgeable group leader bringing the members together to learn and practice with each other is vital for social media–based CoPs.

Second, the practice-based strategies aim at improving knowledge production and innovative creation. Members of a CoP are practitioners, who build a repository that includes experiences, stories, techniques, experiences, and solutions to persistent issues [10]. It is crucial to provide effective and timely resources, such as teaching cases, tools, and documents, but members could also create new knowledge through discussion and collaboration. For example, one teacher reported her experience in a social media–based CoP on using geometer sketchpad software in elementary school mathematics courses. She and other members took turns presenting their self-created works, which were also saved on a user-friendly and easily accessible digital platform. Considering the informality and fragmentation of social media–based learning activities, it is necessary to manage the knowledge within social media–based CoPs. Knowledge management activities reported by teachers in the High = satisfaction CoPs (H-CoPs) included hosting showcase events, creating weekly reports, or posting chat tweets.

Third, the respondents thought that the domain features of a social media–based CoP should be focused and clear, as well as be closely related to teachers' most concerned issues. A social media–based CoP should focus on one specific field or project. Members of a community should have similar characteristics, such as teaching in the same discipline, in the same area, or on the same project.

5 Discussion and Limitations

Social media–based CoPs have become an important approach for teachers' autonomous learning and professional development, which meet their just-in-time needs for exchanging knowledge, negotiating meaning, and accessing resources. This study implemented a questionnaire survey among Chinese primary and secondary school teachers and found that they widely embrace the concept of lifelong learning and actively participate in social media–based CoPs for their professional development. They believe that compared to other forms of online learning, such as MOOCs, social media–based CoPs have greater practical effectiveness. Through CoPs, teachers can communicate and seek help promptly, share useful information and resources, thus enhancing their teaching practices. However, not all social media–based CoPs can achieve such effects; community leaders are needed to plan and regulate the community effectively to maintain effectiveness and sustainability.

This study identified three key structural elements (i.e., domain, community, practice) that characterized H-CoPs. Teachers tended to be more satisfied with communities focused on specific subjects or teaching issues that provide effective knowledge, tools, and resources. H-CoPs also require necessary offline support, including support through activities, institutions, and policies. Principal teachers, as well as instructional researchers from education bureaus, play a key role in the construction of these communities.

This study had some limitations. First, teachers' social media–based CoPs focused on WeChat and QQ, while overlooking other specialized platforms designed for teacher CoPs. Additionally, it is important to consider that cultural factors may influence teachers' perceptions of CoPs because Chinese teachers were the subjects of this study. Caution should therefore be exercised when generalizing the research findings to informal online learning contexts based on social media in other countries.

Second, the sample size of this study needs to be further expanded, with a more detailed characterization of teacher groups. The relationship between topics and community structural elements could also be further explored.

Third, the data primarily relied on self-reported questionnaire surveys, and qualitative interview data and content data from social media–based CoPs were not considered. As such, analyzing chat content data from social media–based CoPs through cognitive network analysis, for example, could deepen our understanding of the structure and characteristics of teachers' large-scale, informal online professional learning networks. Further research is also needed regarding other professional activities within teachers' social media–based CoPs, such as network tweets.

Acknowledgments. This research was supported by the Youth Program of Humanities and Social Sciences of the Ministry of Education, "Research on the Group Dynamics Characteristics of E-Learning Communities of Elementary and Middle School Teachers in the Perspective of CoP" under Grant No. 21YJC880041.

References

1. Xing, W., Gao, F.: Exploring the relationship between online discourse and commitment in Twitter professional learning communities. Comput. Educ. **126**, 388–398 (2018)
2. Ghamrawi, N.: Teachers' virtual communities of practice: a strong response in times of crisis or just another fad? Educ. Inf. Technol. **27**(5), 5889–5915 (2022)
3. Gutiérrez-Páez, N.F., Santos, P., Hernández-Leo, D., Carrió, M.: Designing a pre-service teacher community platform: a focus on participants' motivations. In: De Laet, T., Klemke, R., Alario-Hoyos, C., Hilliger, I., Ortega-Arranz, A. (eds.) Technology-Enhanced Learning for a Free, Safe, and Sustainable World, pp. 352–357. Springer International Publishing, Cham (2021)
4. Luo, T., Freeman, C., Stefaniak, J.: 'Like, comment, and share': professional development through social media in higher education: a systematic review. Educ. Technol. Res. Dev. **68**(4), 1659–1683 (2020)
5. Zhang, S., Liu, Q.: Investigating the relationships among teachers' motivational beliefs, motivational regulation, and their learning engagement in online professional learning communities. Comput. Educ. **134**, 145–155 (2019)
6. Macià, M., García, I.: Informal online communities and networks as a source of teacher professional development: a review. Teach. Teach. Educ. **55**, 291–307 (2016)
7. Donelan, H.: Social media for professional development and networking opportunities in academia. J. Further High. Educ. **40**(5), 706–729 (2016)
8. Wenger, E.: Communities of Practice: Learning, Meaning, and Identity. Cambridge University Press, Cambridge (1999)
9. Patton, K., Parker, M.: Teacher education communities of practice: more than a culture of collaboration. Teach. Teach. Educ. **67**, 351–360 (2017)
10. Wenger, E.: Communities of practice: a brief introduction(2011)
11. Neale, J.: Iterative categorization (IC): a systematic technique for analyzing qualitative data. Addiction **111**(6), 1096–1106 (2016)

Undersea Adventure (UA): A Stealth Approach to Collaborative Skills Assessment in Schools

Fei Wang(✉)

Department of Film and Television Arts, Shanghai Publishing and Printing College, Shanghai, China
56334810@qq.com

Abstract. This paper describes the design of Undersea Adventure (UA), a novel computer game to assess students' collaborative skills development. UA is an innovative product that integrates theory-driven content with customized game mechanics. Game assessment approach facilitates covert assessment, enabling real-time collection of dynamic evidence of social skills, while players' choices during gameplay yield essential data. To develop an engaging and valid game, we used an iterative data-driven validation process. This included creating, testing, and revising UA based on student performance and feedback, and retesting UA until gameplay aligned with independent ratings of collaborative skills. We first investigated the potential for utilizing detailed student behavior logs to enhance assessment accuracy. Our findings demonstrate that integrating external metrics assessing players' collaboration skills with comprehensive behavioral logs pertaining to collaboration can effectively enhance the precision of built-in assessments. Next, we investigated the effectiveness of UA as a collaborative skills assessment tool. The assessment of the game demonstrated high levels of student engagement and favorability, as well as ensuring the reliability of data collection and analysis. These findings support the use of the interactive and engaging computer game format for the stealth assessment of students' collaborative skills. The innovative design methodologies developed are expected to be advantageous in the improvement and refinement of educational computer games.

Keywords: collaborative skills · assessment · computer game · game testing

1 Introduction

Collaborative skills have been considered a core competency for 21st century learners, which is crucial for their success in the knowledge economy and globalized world [1, 2]. At present, numerous countries are revising and restructuring their curricula to incorporate collaborative skills as a fundamental area of study [3]. For instance, the Singapore Ministry of Education has implemented "Project Work" as a curriculum, in which effective communication and collaboration are emphasized as crucial skills and are delineated as educational objectives. The National Education Association (NEA) explicitly emphasizes the necessity for students to acquire collaborative, team-based learning in order to effectively prepare for the social competition of the 21st century [4]. Through

Q. Liu et al. (Eds.): EITT 2024, CCIS 2600, pp. 289–299, 2026.
https://doi.org/10.1007/978-981-95-1525-7_24

collaborative learning, students are provided with the opportunity to engage in critical thinking and exchange their interests, experiences, and expertise. The National Research Council of the United States has emphasized that the standards for English language arts, encompassing "collaborative listening to comprehend and assess others' discourse," and the standards for mathematics, involving collaborative teamwork, play a pivotal role in school education [5]. In the global academic community, the International Commission on Education (ICE) has integrated collaborative problem-solving (CPS) learning and development into school curricula to equip students for their prospective careers [6]. Furthermore, the International Education Commission and academic researchers have developed and refined effective pedagogical approaches to facilitate collaborative problem-solving, with a specific emphasis on enhancing educational technology integration in the classroom [7].

The most common approach for measuring teamwork and social skills is for individuals to rate themselves (or others) on a five-point agreement scale [8]. Much of the current research on collaboration and collaborative problem-solving relies on the traditional Likert-scale measure of collaboration. These psychometric scales focus on measuring one construct at a time. However, collaboration is a complex multi-person performance involving both social and interpersonal skills. A key issue with assessing collaborative skills is that there are invariably multiple component skills being assessed together. Moreover, collaborative interactions are dynamic and not amenable to assessments through static methods. Being able to assess multiple constructs simultaneously in a dynamic way is challenging.

A desire for alternative tools that provide data on collaborative processes to support the assessment of students' collaborative skills has led to interest in computer games for assessment and learning [9]. Computer games provide a promising avenue for the assessment of collaborative skills and have several advantages over traditional methods. Situations that are challenging to observe in natural settings can be incorporated into the assessment. Once an assessment system is established, it can effectively gather collaborative process data from a large cohort of students. The issue of observer reactivity can be overcome with stealth assessment techniques, in which assessments are "woven directly and invisibly into the fabric" of the game and players' choices during gameplay to provide the data needed for assessment [10, 11]. This assessment approach occurs alongside the collaborative process, greatly reducing the likelihood that students will change their behavior to please observers.

In comparison to traditional assessment methods, the utilization of computer games as an assessment tool offers a more authentic approach. Through the creation of collaborative learning scenarios, it facilitates interactive environments for students' cooperative communication, thereby eliciting a comprehensive demonstration of learners' cognitive and social skills within the collaborative process. This technology for creating realistic virtual collaborative environments is becoming more common [12, 13].

In the present study, we used customized game mechanics to design and evaluate a stealth assessment computer game to examine students' collaborative skills. We used game log data, user feedback, and expert feedback to revise the game during an iterative testing procedure. After the game was developed, we used extensive game log data and user feedback to create performance indices to establish the validity of our computer game for collaborative skills assessment.

2 Development

2.1 Development of UA

The primary design goal of Undersea Adventure (UA) was to create virtual situations in order to assess students' collaborative skills. To accomplish this, we created UA as a multiplayer collaborative game situated in an underwater-like setting where students engage in role-playing and collaboration to complete a series of challenges.

(1) Game Design. Within the virtual gaming environment, the National Ocean Exploration Department dispatches three marine biology specialists to form an investigative team tasked with exploring the surroundings of a sunken ship. As depicted in Fig. 1, each exploration team comprises one specialist in marine herbivores and two specialists in marine carnivores, with these roles being assumed by members of the collaborative group. In this virtual gaming scenario, each team member collaborates to accomplish various tasks. Upon successful completion of a collaborative task, the team can progress to the next level.

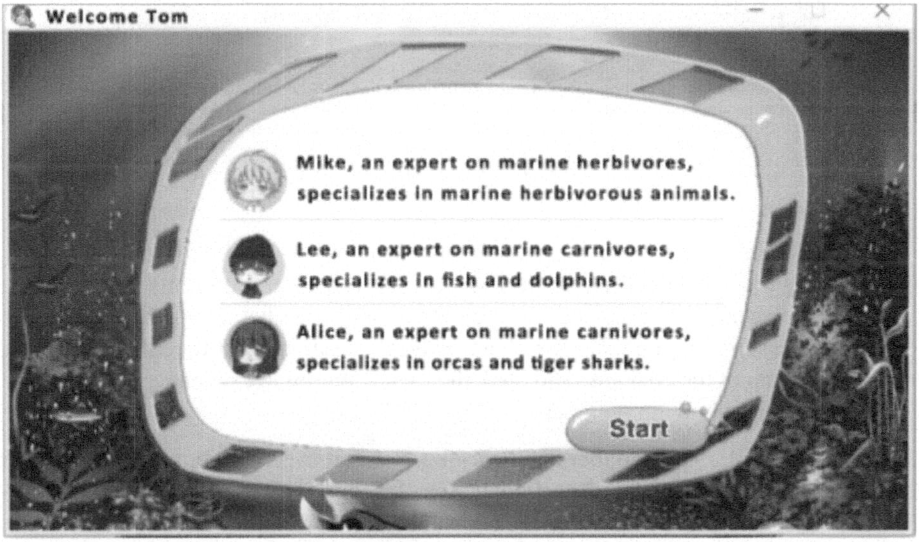

Fig. 1. Description of the game role.

(2) Game Mechanics. Each scene employs the same set of point-and-click mechanics, enabling players to access the corresponding functional panel by clicking on the navigation controls located at the bottom of the game user interface (e.g., "Skill," "Clue," "Chat," and "Next Level"). Figure 2 shows different functional panels. With the skill panel, students access information regarding the characteristics of marine animals, encompassing their physical attributes, reproductive strategies, behavioral patterns, and dietary preferences (see Fig. 2). The clue panel provides insights into the interconnectedness of species within the food chain (see Fig. 3). Within the chat panel, team members engage in collaborative discussions to exchange ideas, share pertinent information and resources, and ultimately arrive at a consensus (see Fig. 4).

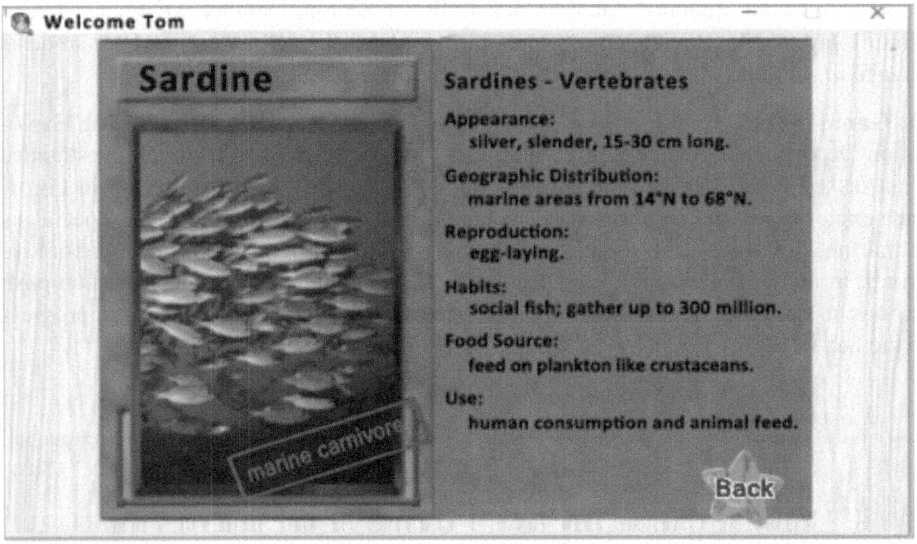

Fig. 2. The skill panel.

In UA, players can communicate with other team members by clicking the chat panel. The chat panel is positioned on the left side of the screen and presents collaborative tasks for each scene. The right side of the panel features a chat area that displays the conversation history among team members. Players can input their thoughts in the text box at the bottom of the chat area and click the "send" button to exchange messages and collaborate with other team members.

(3) Scene Content. The collaborative learning materials and collaborative tasks were developed by two seasoned middle school biology teachers with extensive teaching experience. The theme of this collaborative task was selected from the "food chain" content in middle school science curriculum, and students had no prior acquaintance with this topic, thereby guaranteeing the homogeneity of knowledge levels among learners. The learning materials encompassed information concerning the characteristics of six marine organisms, three auxiliary cues related to the food chain, and the first two stages of test questions, vigorously supporting student collaboration. Four progressive collaborative scenes were designed based on the collaborative materials and tasks to assess students' collaborative abilities.

In each scene, players are presented with a collaborative problem that needs to be solved. Content was developed to elicit measurable player behaviors known to be integral to the collaborative skills competency being assessed in each scene.

For example, in the second collaborative scene, the players use the provided food web information to determine the short-term impact of plant death on the number of six different organisms near the sunken ship. Players can interact with two teammates to share information and ideas and collaborate to complete the task. To successfully complete the task, players can find feature information related to marine animals through the skill panel and share it with teammates through the chat panel. The chat data is used to evaluate the social dimension of collaborative skills.

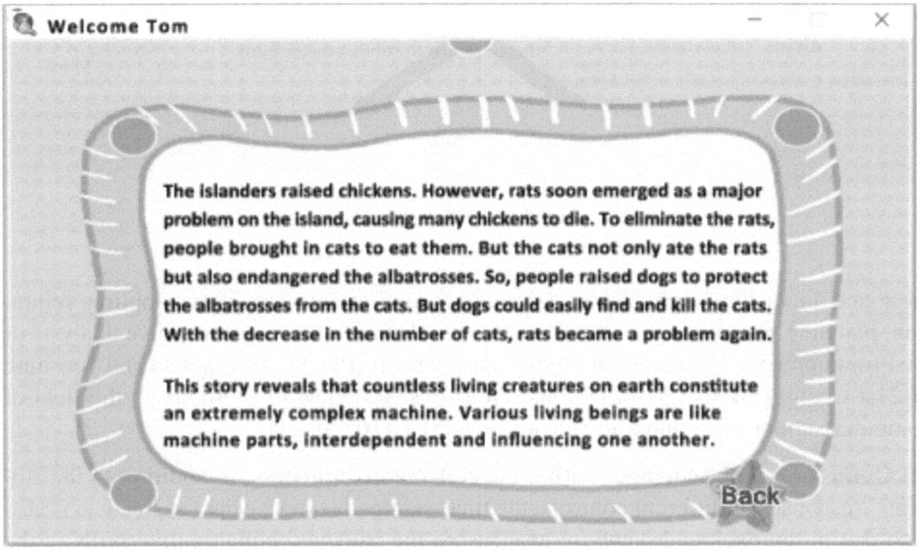

Fig. 3. The clue panel.

The first two task scenes have tests, and the test scores and other information of players are automatically captured by the computer game for detailed analysis of the cognitive dimension of collaborative skills. Additionally, UA offers immediate feedback to students through the game interface, displaying their scores in real time so that they can adjust their cognitive understanding of knowledge to make informed judgments in subsequent scenes.

2.2 UA as a Valid Measure of Collaborative Skills

(1) Assessment Metrics. The assessment of collaborative skills encompasses two dimensions, namely the cognitive dimension and the social dimension [14, 15]. Refinement of the cognitive dimension involves three key metrics: cognitive level, metacognitive level, and cognitive efficiency, drawing upon Sternberg's [16] theoretical component model of triadic intelligence and O'Neil's [17] research [18–20]. In assessing

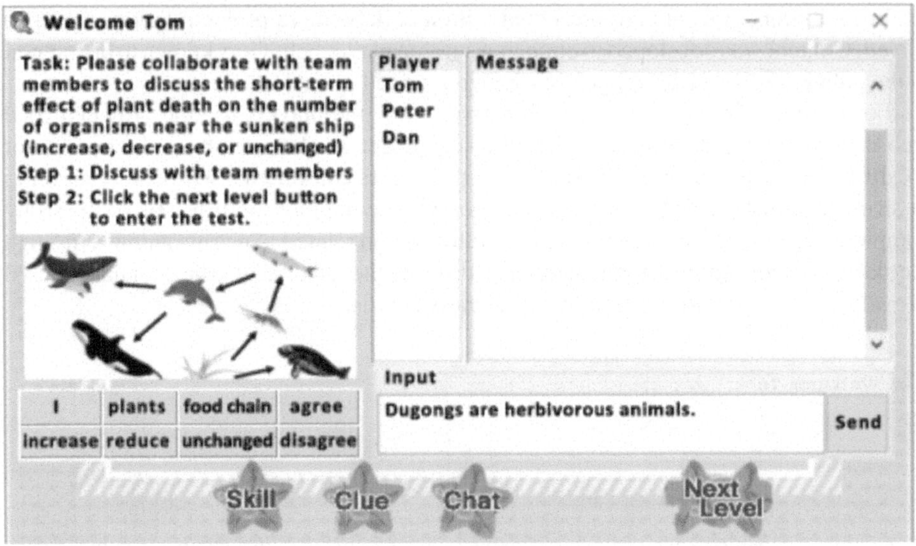

Fig. 4. The chat panel.

the social dimension, eight metrics (sharing, execution, interpretation, adoption, refutation, planning, monitoring, and evaluation) are derived from a comparative analysis of the Programme for International Student Assessment (PISA), Assessment and Teaching of 21st Century Skills (ATC21S), and an extensive review of literature in the field of computer-supported collaborative learning (CSCL) [6, 20, 21].

(2) Game Log and Scoring. During the collaborative process, in addition to the students' responses to different controls, the time-stamped response data as log data is also recorded in the back-end database to measure the students' collaborative skills in the social dimension. The cognitive level index is the students' test score. When the students answer a test question correctly, the test score data item is automatically increased by 10 points; otherwise, it is set to 0. The meta-cognitive level index is calculated as the ratio of task planning time to total test time. These two indicators are automatically captured from time-stamped response data and correspond to both total read time of students during collaborative tasks and overall game testing duration for measuring their meta-cognitive levels regarding time allocation. The cognitive efficiency index represents how frequently students operate controls; each click on any control such as the "Skill," "Clue," or "Chat" buttons increases this indicator value by 1.

In UA, there is a chat panel that facilitates team members' communication with each other via text chat. The collaborative conversations in the chat panel are logged in the database file and used to measure the students' collaborative skills in the cognitive dimension. Upon clicking the "send" button, a student's collaborative discourse data is automatically captured and stored in the database. These collaborative discourse data are coded to access students' social level. Phrases such as, "I'll take the first task on," indicate proactive behavior and problem-solving initiative, aligning with the "execution" indicator. Phrases like, "I think," and, "I agree," demonstrate the expression of personal opinions, corresponding to the "adoption" indicator.

3 Testing

3.1 Student Testing

In order to optimize both the gameplay and assessment quality of UA, two independent iterative evaluations were conducted with 7th grade students in two schools in Shanghai. The first iterative test included students from one 7th grade class of students ($n = 36$), who were divided into 12 groups. Data collected from this initial sample were used to revise the content and gameplay, as well as verify the difficulty of in-game challenges within collaborative scenes. The revised scenes were then tested by a group of 33 students from one 7th grade class, who were divided into 11 groups. Data collected from this second sample were used to test the technical acceptance and gaming experience of UA as a collaborative skill assessment tool.

In both iterative test groups, testing took place during a regularly scheduled one-hour computer lab class at the school. Each student was assigned a computer on which to play UA. Following the project introduction and computer orientation, students were given 40 min to complete the four UA scenes. Teachers were asked to observe without providing assistance. Game logs tracked the location and time of each mouse click, as well as the option chosen by the student, in order to map students' interactivity. The resulting data from the first iterative test group were used to adjust the difficulty of cooperative tasks, enrich the gaming experience, strengthen the interconnectedness of collaborative tasks, and provide suggestions for improving specific game mechanics. The log data from the second iterative test were used to verify the refinements from the first iterative test as well as to test the technical acceptance and gaming experience of the UA game. These log data were also used to calculate the students' collaborative skills performance to verify the effectiveness of UA as a collaborative skills assessment tool.

3.2 Student Ratings of UA

After playing UA, a semi-structured individual interview was conducted and involved all the students. We asked students to rate how fun/interesting the game was, how easy it was to use and understand, how easy they finished the collaborative task, and how much they liked it overall. Research staff then utilized comments from students to propose suggestions for how to make the navigation or user experience better. In creating the second iteration of the game, we paid special attention to the problems that existed and refined the game accordingly.

After the second iteration of the test, the students participated in a survey on the technical acceptance of computer games to measure the usefulness, usability, and gaming experience of the evaluation tool. The technology acceptance questionnaire was modified from the classic information technology acceptance questionnaire. It consisted of 12 items on a five-point Likert scale, where "5" represented "strongly agree" and "1" represented "strongly disagree." Six items were dedicated to "usefulness" and another six to "ease of use." Cronbach's alpha values of 0.967 and 0.854 were attained for the questionnaire for its dimensions of "usefulness" and "ease of use," respectively, showing good reliability in internal consistency.

3.3 Experts Ratings of UA

After the first iteration of the test, research staff invited two junior high school biology subject experts to assess the appropriateness of the materials used in the collaborative process, examining whether the text materials were compatible with the collaborative task. Meanwhile, the two experts also verified the collaborative task content and the fun of the game. In creating the second iteration of the game, we revised the four collaborative tasks and used them for the second round of iterative testing.

4 Results

4.1 Student Ratings of UA

Overall, students rated UA very positively. In terms of students' technical acceptance of UA, Table 1 shows the survey results of students, emphasizing the "usefulness" dimension ($M = 4.29$, $SD = 0.674$) and the "ease of use" dimension ($M = 4.10$, $SD = 0.617$); this represents students' acceptance of computer games as a method for assessing collaborative skills. In addition, the interview data showed that, overall, students sensed the helpfulness of the game in improving their collaborative communication. There were high levels of student engagement and numerous positive comments about the game (e.g., "This is awesome!"; "The game is very fun!"). Researchers observed very few misunderstandings or requests for technical help. Students easily understood how to navigate UA, and analysis of usage data indicated a low rate of errors (e.g., misclicks).

Table 1. Technical Acceptance of UA

Technical Acceptance	Mean	*SD*
Usefulness	4.29	0.674
Ease of Use	4.10	0.617

4.2 Experts Ratings of UA

Both experts unanimously agreed that the learning materials used in the game were scientifically sound and aligned with the cognitive level and existing cognitive experiences of the target student group. Meanwhile, the modified collaborative tasks passed the rationality test conducted by the experts and could be used for the second round of iterative testing.

4.3 UA as a Valid Measure of Social Skills

After the second iteration of the test, we collected data, including clickstream data, collaborative discourse data, and test data. These data correspond to the cognitive and social dimensions of collaborative skills, ensuring the independence of the two dimensions of evaluation. In the social dimension, UA gathered 887 instances of collaborative discourse data, indicating an average generation of 27 instances per individual. In the cognitive dimension, UA collected 7,278 pieces of clickstream data and 33 test data, with an average generation of 221 pieces of clickstream data per person. The results show that the game developed in this study can collect the expected data, and the overall results of the data are quite satisfactory.

To ensure that UA assessed collaborative skills, we conducted descriptive analyses of the data produced during the collaborative process. Table 2 displays these results. As expected, the data collected from the game in UA can assess students' collaborative skills from both cognitive and social dimensions. However, based on the data from UA, we found that students exhibited differences in social dimension indicators. Among them, sharing discourse occurred most frequently, with an average of 8 sharing discourse per student. Following sharing discourse were discourse related to interpretation, planning, and monitoring, which appeared relatively frequently. Discourse related to evaluation, refutation, and execution appeared the least. This indicates that students performed better in knowledge sharing, presenting opinions during collaborative activities. These findings contribute to ensuring the reliability of data collection and analysis to a certain extent, thereby demonstrating the effectiveness of the UA game in assessing students' collaborative skills.

Table 2. Descriptive Analysis of assessment indicators for collaborative skills in UA

Dimension	Assessment Indicators	Mean	SD
Cognitive Dimension	Cognitive Level	65.76	21.21
	Meta-cognitive Level	0.02	0.01
	Cognitive Efficiency	221.52	83.69
Social Dimension	Sharing	8.06	4.10
	Execution	0.97	1.40
	Interpretation	6.24	3.62
	Adoption	1.76	2.03
	Refutation	0.82	1.10
	Planning	3.61	3.54
	Monitoring	3.03	3.14
	Evaluation	0.73	0.84

5 Conclusion

Computer games provide a promising avenue for the assessment of student collaborative skills. It is likely that computer-based assessments will be utilized heavily in the future, as the student assessment paradigm shifts from single-indicators to multiple skills, and from summative evaluation to process evaluation. In the present study, we created and tested an innovative stealth assessment of students' collaborative skills. This study contributes to the improvement of computer games in education by describing a novel design methodology for developing these kinds of assessments.

The results of this assessment study are linked to the promise of educational games' potential to offer opportunities for instruction and intervention to build students' collaborative skills. Game logs can record students' conversations, behavior choices, and task and non-task time. Using these data, UA can assess students' collaborative skills from multiple dimensions. This process-oriented multidimensional assessment approach facilitates a more comprehensive and realistic portrayal of learners' collaborative skills. Consequently, UA can be employed for assessment to discern students' potential in collaborative skills. Utilizing UA's assessment function, we are developing tutorials for students with the intent of improving students' collaborative skills strategies.

In this study, we established the validity of UA with the data from the Technology Acceptance Questionnaire and feedback from the game. We believe that, because UA was developed specifically to provide game-based collaborative scenarios analogous to authentic situations students encounter, it can best reflect students' authentic collaborative learning experiences and developmental levels.

UA leverages innovative technologies to provide an engaging and powerful collaborative skills assessment tool with real-time reporting functions for educators to easily track students' progress toward collaborative goals. UA uses a stealth assessment approach, which embeds evaluation directly into the game task, capturing various types of data during the learner's interaction with the game, and utilizing this evidence to evaluate the learner's various skill states and traits. Therefore, it becomes an attractive collaborative skills assessment option. Compared to standard measures, using computer games for skill assessment not only provides rich data but also allows the assessment to be conducted without disrupting the game and affecting the student's gaming experience. This approach partially mitigates the influence of external environmental factors and individual subjectivity, thereby effectively ensuring genuine student engagement and performance.

Acknowledgments. This study is supported by the Shanghai Educational Science Research Project under contract number C2023188.

References

1. Trilling, B., Fadel, C.: 21st Century Skills.: Learning for Life in Our Times. John Wiley & Sons, Hoboken, NJ (2009)
2. Ananiadou, K., Clarol, M.: 21st Century Skills and Competences for New Millennium Learners in OECD Countries. OECD Publishing, Paris, France (2009)

3. Care, E., Anderson, K., Kim, H.: Visualizing the Breadth of Skills Movement Across Education Systems. The Brookings Institution, Washington, DC (2016)
4. National Education Association: Preparing twenty-first century students for a global society: an educator's guide to the 'four Cs'. ETS Res. Rep. Ser. (2010)
5. National Research Council, Education for Life and Work: Developing Transferable Knowledge and Skills in the 21st Century. The National Academies Press, Washington, DC (2012)
6. OECD: Draft PISA Collaborative Problem Solving Framework. OECD Publishing, Paris, France (2013)
7. von Davier, A.A., Halpin, P.F.: Collaborative problem solving and the assessment of cognitive skills: psychometric considerations. ETS Res. Rep. Ser. (2013)
8. John, O.P., Naumann, L.P., Soto, C.J.: Paradigm shift to the integrative big-five trait taxonomy: History, measurement, and conceptual issues. In: John, O.P, Robins, R.W., Pervin, L.A. (eds.). Handbook of Personality: Theory and Research, pp. 114–158. Guilford Press, New York, NY (2008)
9. Shute, V.J. Ke, F.: Games, learning, and assessment. In: Ifenthaler, D., Eseryel, D., Ge, X. (eds.) Assessment in Game-Based Learning: Foundations, Innovations, and Perspectives. Springer, New York, NY (2012)
10. Heinzen, T.E., Landrum, R.E., Gurung, R.A.R., Dunn, D.S.: Game-based assessment: the mash-up we've been waiting for. In: Reiners, T., Wood, L.C. (eds.) Gamification in Education and Business, pp. 201–217. Springer International Publishing, Switzerland (2015)
11. Chung, G.K.W.K., Baker, E.L.: An exploratory study to examine the feasibility of measuring problem-solving processes using a click-through interface. J. Technol., Learn. Assess. **2**(2) (2003)
12. Steinkuehler, C.: Massively multiplayer online gaming as a constellation of literacy practices. E-Learn. Digit. Media **4**, 297–318 (2007)
13. Young, M., Schrader, P.G., Zheng, D.: MMOGs as learning environments: an ecological journey into 'Quest Atlantis' and 'The Sims Online'. Innov. J. Online Educ. **2**(4) (2006)
14. Strijbos, J.W.: Assessment of (computer-supported) collaborative learning. IEEE Trans. on Learn. Technol. **4**, 59–73 (2011)
15. Howley, I., Mayfield, E., Rosé, C.P.: Linguistic analysis methods for studying small groups. In: O'Donnel, A., Chan, C., Chin, C. (eds.) International Handbook of Collaborative Learning. Taylor and Francis, London (2013)
16. Sternberg, R.J.: Implicit theories of intelligence, creativity, and wisdom. J. Personality Soc. Psychol. **49**(3), 607–627 (1985)
17. O'Neil, H.F., Chuang, S.H., Chung, G.K.W.K.: Issues in the computer-based assessment of collaborative problem solving. Assess. Educ. Princ. Policy Pract. **10**, 361–373 (2003)
18. Howley, I., Adamson, D., Dyke, G., Mayfield, E., Beuth, J., Rosé, C.P.: Group composition and intelligent dialogue tutors for impacting students' academic self-efficacy. Intell. Tutoring Syst. **7315**, 551–556 (2012)
19. Kwon, K., Liu, Y.H., Johnson, L.P.: Group regulation and social-emotional interactions observed in computer supported collaborative learning: comparison between good vs. poor collaborators. Comput. Educ. **78**, 185–200 (2014)
20. Meier, A., Spada, H., Rummel, N.: A rating scheme for assessing the quality of computer-supported collaboration processes. Int. J. Comput.-Support. Collaborative Learn. **2**, 63–86 (2007)
21. Griffin, P., Care, E., McGaw, B.: The changing role of education and schools. In: Griffin, P., McGaw, B., Care, E. (eds.) Assessment and Teaching of 21st Century Skills, pp. 1–15. Springer, Heidelberg, Germany (2012)

Author Index

Q. Liu et al. (Eds.): EITT 2024, CCIS 2600, pp. 301–302, 2026.
https://doi.org/10.1007/978-981-95-1525-7

The manufacturer's authorised representative in the EU is Springer Nature Customer Service Centre GmbH, Europaplatz 3, 69115 Heidelberg, Germany. If you have any concerns regarding our products, please contact ProductSafety@springernature.com

Printed and bound by CPI Group (UK) Ltd, Croydon, CR0 4YY
01/10/2026
02225908-0009